P-graphs for Process Systems Engineering

Ferenc Friedler • Ákos Orosz
Jean Pimentel Losada

P-graphs for Process Systems Engineering

Mathematical Models and Algorithms

Springer

Ferenc Friedler
Széchenyi István University
Győr, Hungary

Ákos Orosz
University of Pannonia
Veszprém, Hungary

Jean Pimentel Losada
Budapest University of Technology and Economics
Budapest, Hungary

ISBN 978-3-030-92215-3 ISBN 978-3-030-92216-0 (eBook)
https://doi.org/10.1007/978-3-030-92216-0

This Springer imprint is published by the registered company Springer Nature Switzerland AG
The registered company address is: Gewerbestrasse 11, 6330 Cham, Switzerland

In memory of Dr. L.T. Fan

Preface

As a consequence of our ever-increasing need for higher efficiency and better sustainability, the world is going through rapid and significant changes. Naturally, the goals concerning efficiency and sustainability define proper ambition and target for both short and long term. Going in this direction is progressive; however, it requires a higher level of exploitation and better use of the potential resources available locally or globally. This can, however, be realized with additional interconnections among the already complex network of subsystems, *i.e.*, with a higher-level integration. Because of that, the difficulties of our time very much differ from those of decades ago. The gradually better quality of life afforded by recent developments comes together with a major influence on our everyday routine. In the old days, a local catastrophe occurring to an independent system had essentially no direct influence on its surroundings. In contrast, a local catastrophe nowadays may become regional, or in some cases global, due to the high integration of systems. For instance, an accidental cut-off in an electric transmission network may induce a continental-level catastrophe. It is, therefore, crucial that further improvement of the already highly interconnected systems be made by considering their resilience, controllability, and operability in addition to the general requirements of efficiency and sustainability. To do so, however, the appropriate knowledge of complex systems is needed.

Currently, main research efforts are aimed at understanding the behavior of two types of complex networks; the difference between them relates to the types of mathematical models for describing the individual nodes of the networks. In one type of complex networks, simple mathematical models describe the nodes, *e.g.*, the Internet. In the other type, complex mathematical models represent the network's nodes. The scope of our work belongs to the second type of complex networks which is closely related to a variety of engineering systems, including the chemical industry, production systems, and supply chains. The difficulty in examining this type of systems lies in both the size of the network, in terms of the number of nodes, and the complex mathematical models of the nodes. Because of the fundamental difference between the two types of complex networks, the research achievements of one of them can hardly be beneficial to the other type. The examination of the second type of complex networks belongs to the area of process systems engineering (PSE) where the complex systems are called complex processing systems, or simply, processing systems. The latter terminology will be used here.

The need for the systematic design and operation of processing systems originated in the petroleum industry back in the 1970s. That time saw the beginning of the creation of simulation programs for analyzing the expected behavior of a processing system prior to its construction. These simulation programs have become highly sophisticated, and by now, they are capable of optimizing large processing systems of a given network structure; nevertheless, they are incapable of systematically optimizing the network connections. Because of that, the question of how to connect the systems elements to each other, *i.e.*, what the best network of the individual elements is, cannot be answered by such tools. In other words, they are incapable of the systematic synthesis of complex engineering networks.

Other types of approaches for the synthesis of processing systems are based on the optimization of a mathematical model. The key point in developing the mathematical model of processing systems' synthesis, however, is to establish an exact relationship between the real problem and its mathematical model. This is extremely crucial, and also difficult, because the mathematical model of a synthesis problem is not the description of a real object, it is an artificial entity. Therefore, this model's behavior cannot be compared with any real object for verification. The philosophical and practical consequences of this issue will be one of the main focuses of the book. It has to be emphasized that process design based on simulation programs and process synthesis based on mathematical optimization have major contributions to the development of industry, but still, fundamental questions must be answered to satisfy future needs.

With this work, our purpose is to provide a theory of systems by focusing on processing systems, *i.e.*, networks, whose nodes are represented by complex mathematical models. This theory has been constructed on a unique reformulation of network synthesis that can simply be referred to as inside-out modeling of systems. A set of axioms of process structures is the basis of the theory, where the axioms are the expressions of the fundamental structural knowledge embedded into feasible networks. This axiomatic setup serves both the rigor and the long-term development of the theory. A framework formulated on this theory, called the P-graph framework, provides new methods for synthesizing complex systems by taking into account recently identified requirements on resilience, reliability, and controllability. The framework was originally developed for processing systems in the chemical industry; however, its application area now spans a wide range of engineering systems from reaction engineering to supply chain management.

The three main parts of this book cover the theoretical foundation and the basic algorithms of the framework (Part I), the framework's applications (Part II), and the proofs of the algorithms (Part III). The book gives rigorous mathematical formulations of the methods together with illustrations of the theory for satisfying the needs of researchers, systems engineers, and research students. Demonstration software is also available for readers for better understanding of the theory and to collect experience in synthesizing network structures with various properties.

Finally, a paradigm shift is defined as "an important change that happens when the usual way of thinking about or doing something is replaced by a new and different way" (Merriam-Webster). In this regard, a natural question can be asked: Does the proposed framework represent a paradigm shift in process systems engineering, especially, in process synthesis? We will let the reader answer this question.

Győr, Hungary Ferenc Friedler
Veszprém, Hungary Ákos Orosz
Budapest, Hungary Jean Pimentel Losada
November 2021

Acknowledgments

The authors express their utmost gratitude to the late Prof. L. T. Fan, the co-founder of the P-graph framework.

The authors are grateful to late Balázs Imreh, Botond Bertók, late Csanád Imreh, Zoltán Kovács, István Heckl, Csaba Holló, Zoltán Blázsik, and Andres Argoti for their contribution to the development of the framework. All P-graph figures in the book have been drawn by the freely available software P-Graph Studio, which is also acknowledged.

Contents

Notation

Symbols and Functions

$O(f(x))$	Order of magnitude of function $f(x)$ (see Appendix A)
$\varnothing$	Empty set
$\wp(m)$	Power set of set m
$\wp'(m)$	Power set of set m excluding empty set
$\times$	Cartesian product of two sets
$\mathbb{R}$	Set of real numbers
$\in$	Element of a set
$\notin$	Not an element of
$\neq$	Different
:	Such that
$\cup$	Union of sets
$\cap$	Intersection of sets
$\setminus$	Set difference
$\subseteq$	Subset
$\subset$	Strict subset
$\not\subseteq$	Not a subset
&	And (as a logical operation)
$\forall$	For all
$\Leftrightarrow$	If and only if
$\exists$	There exists
$\mathbb{S}$	Structural model of PNS
$\sim$	Equivalence relation between structural models
P	Set of desired products
M	Set of materials
R	Set of raw materials
O	Set of operating units
$S(\mathbb{S})$	Set of solution-structures of structural model $\mathbb{S}$

$\widehat{S}(\mathbb{S})$	Set of feasible solution-structures of structural model $\mathbb{S}$
$\mu(\mathbb{S})$	Maximal structure of structural model $\mathbb{S}$
$\delta_1[m_1] \leq \delta_2[m_2]$	Decision-mapping $\delta_2[m_2]$ is an extension of $\delta_1[m_1]$
$\delta_1[m_1] < \delta_2[m_2]$	Decision-mapping $\delta_2[m_2]$ is a proper extension of $\delta_1[m_1]$
$\delta_1[m_1] \prec \delta_2[m_2]$	Decision-mapping $\delta_2[m_2]$ is a regular extension of $\delta_1[m_1]$
T_r	Set of materials to be removed in reduction phase of algorithm MSG
W_s	Set of materials to be included in composition phase of algorithm MSG
$\Delta[m]$	Set of units capable of producing the materials of set m
$\Delta(x)$	Set of units capable of producing material x
$\delta[m]$	Decision-mapping defined on set m
$\delta(x)$	Operating units assigned to material x by decision-mapping δ
$\widehat{\delta}[\widehat{m}]$	Maximal neutral extension of decision-mapping $\delta[m]$
$\overline{A}$	Complement of set A
$\lvert A \rvert$	Cardinality of set A

Abbreviations

AA	Adipic Acid
ABB	Accelerated Branch-and-Bound algorithm
ABB-RB	Accelerated Branch-and-Bound Algorithm with Refined Bounding function
B&B	Branch-and-Bound
CE	Circular Economy
CSTR	Continuous Stirred Tank Reactor
LABBA	Look Ahead Branch-and-Bound Algorithm
LP	Linear Programming
MILP	Mixed-Integer Linear Programming
MINLP	Mixed-Integer Non-linear Programming
MSG	Maximal Structure Generation algorithm
M-type	Node representing materials
NLP	Non-linear Programming
O-type	Node representing operating units
PFR	Plug Flow Reactor
PNS	Process Network Synthesis
PNS_w	PNS for specific cost function
RNS	Reaction-Network Synthesis
SSG	Solution-Structure Generation algorithm

Part I
Foundation of the P-graph Framework

The limits experienced in the design of processing systems reveal that a new theoretical foundation is needed; this is especially valid for process synthesis where the structure of the best process is to be determined. The main source of difficulty arises from the structural complexity of the current engineering systems. Part I introduces the concept of a new framework that exploits the inherent structural properties of feasible engineering systems. This framework is the basis of the new inside-out modelling methodology.

Chapter 1
Basic Concepts of Automatic Process Design

Abstract In this chapter, the theoretical hindrances to effect automated process synthesis are summarized. It is shown that the conventional mathematical optimization approach is inappropriate for process synthesis because the optimal solution of the mathematical model cannot be verified as the best result for the original problem. The main components of the P-graph framework are introduced for preparing the systematic generation of verified mathematical optimization models for process synthesis.

Keywords Mathematical modelling · Automatic process synthesis · Optimization · MINLP · P-graph

Regardless of the field, any engineering design is to be based on decisions that result in the best solution among the potential options; this is usually termed the optimal solution. This is valid, *e.g.*, for the design of a house, an engine, or a process; where the meaning of "optimal" may vary substantially in the different applications.

In the current work, we are focusing on the design of processing systems that may be called simply *process design*, which consists of determining the best configurations of manufacturing processes. These processes must yield the required products from the available raw materials via plausible transformation operations. Thus, the solutions of the problem should specify the best set of operations and their connections, that satisfies the necessities of the designer. The first stage of process design is identified as *process synthesis, or process network synthesis (PNS)*, referring to the "creation" of a process from a set of building blocks. These building blocks are called *operating units* or simply *units*, thus, PNS determines the network of the connections of the operating units or simply the structure of the process. In practice, usually a large number of different networks of operating units satisfy the original design goal.

Since the selection of the process structure from the set of options has major influence on the quality of the process, that is usually expressed by indicators such as cost, profitability, sustainability, and controllability; it is a natural goal to make the process synthesis procedure automatic as much as can be. Figure 1.1 illustrates automatic synthesis in determining the specification of the best process utilizing the

F. Friedler et al., *P-graphs for Process Systems Engineering*,
https://doi.org/10.1007/978-3-030-92216-0_1

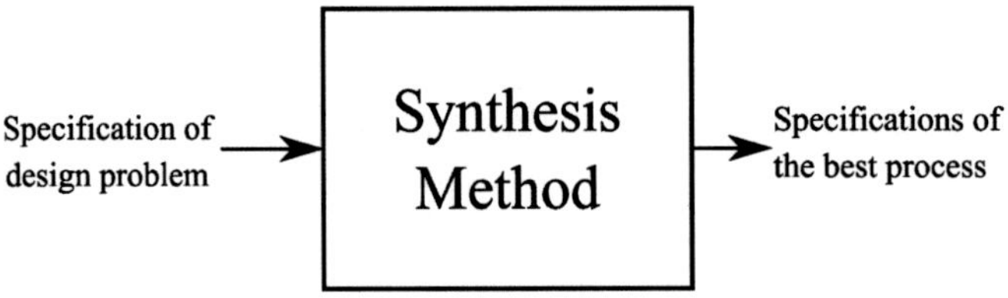

Fig. 1.1 Concept of automatic process synthesis

Fig. 1.2 Typical formulation of a MINLP mathematical optimization problem

$$\min f(x,y)$$
$$\text{s.t.}\quad g(x,y) \le 0$$
$$x \in \mathbb{R}^k\ ,\ y \in \{0,1\}^m$$

originally available information of the design problem as input. Currently, there is no method for process synthesis that fulfills these requirements for a simple reason, the available mathematical modelling tools fail to provide a proper base for that. In the following paragraphs, these theoretical limitations are studied and then a novel modeling tool, called inside-out modelling, will be presented for the development of partially or fully automatic process synthesis.

Because of the competitive environment of the globalized world, it is essential that products and services exhibit high performance features. In addition, when designing a processing system, the solution must also satisfy specific conditions, *e.g.*, sustainability, reliability, controllability or profitability. Nowadays is frequent that strategic decisions, relevant for the solution's performance, are made on the basis of mathematical models and optimization, as it encompasses a set of tools for verifying the superiority of a solution.

Mathematical optimization relies on a set of constraints that attempt at representing the expected behavior of the system under design, whereas the designer's expectations are expressed by the *objective function*. The constraints of the mathematical model may appear in the form of equations or inequalities. A typical formulation of a mathematical optimization problem is presented in Fig. 1.2, where the goal is to minimize the objective function, f, satisfying the conditions expressed by the constraint on function g. Both the constraints and the objective function depend on a set x of k continuous variables, and a set y of m integer variables; therefore, the model is referred to as a Mixed-Integer Programming (MIP) model. More specifically, the model is classified either as a Mixed-Integer Linear Programming model (MILP), or as a Mixed-Integer Non-Linear Programming model (MINLP) depending on the nature of the objective function and the constraints.

Highly sophisticated and effective optimization procedures have been developed to minimize the objective function, thus guaranteeing the optimality of the result in most cases. The usual practice in solving a mathematical optimization problem is

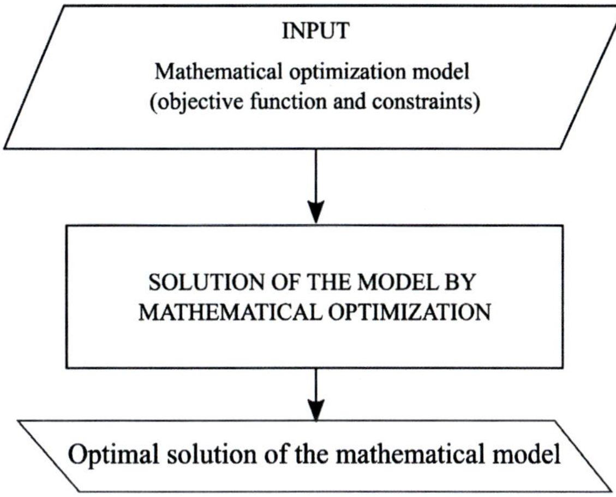

Fig. 1.3 Conventional approach to mathematical optimization

illustrated in Fig. 1.3 where this particular approach demands the mathematical optimization model as an input in the form of Fig. 1.2. Since a process synthesis problem never appears in this form, the key point here relies on determining an appropriate mathematical optimization model for process synthesis. This is a challenging task both in theory and practice. The first question to be answered is, what mathematical model is considered appropriate?

The natural requirement for qualifying a mathematical model as appropriate is that its optimal solution must be exactly the same as the optimal solution of the original, *i.e.*, industrial, problem. In principle, different mathematical models may be appropriate for the same synthesis problem. Our natural goal is to develop a systematic procedure for generating an appropriate mathematical model for a process synthesis problem. Since the currently available tools are incapable for that, because of fundamental theoretical difficulties, a new modelling concept is to be introduced.

There are two components of the mathematical optimization model of a processing system. The first comprises the models of the operating units and the second represents the relationships, *i.e.*, connections, among them. The mathematical models of the individual operating units can be verified by comparing their outcome to the behavior of operations existing in reality. Naturally, the behavior of the mathematical model is not absolutely identical to that of the original operation, there is no theoretical limit in determining the mathematical model that fits to the original operation at any level of accuracy. In what follows, it will be assumed that the mathematical model of an operating unit represents the original operation with a required accuracy.

In the case of optimizing or simulating an existing process, *i.e.*, in *process analysis*, the connections among the operating units are explicitly available. Thus, the relationships among the units are specified exactly the same way as they are given in the original process. Therefore, the mathematical model appropriately describes the processing system, *i.e.*, the optimal solution of the mathematical model may correspond to the optimal solution of the real system at any level of accuracy. Similarly, if the process analysis is performed for evaluating design options, the connections among the units absolutely correspond to the connections in the process to be realized. Consequently, there is no theoretical difficulty in developing a mathematical model that describes a process in its analysis on any level of accuracy.

This is, however, not the case for process synthesis, where the set of operating units to be built in the process and their connections are to be determined. Even if the mathematical models of the plausible operating units can be verified (as far as it is supposed), the connections among them are the matter of optimization. Though, several connection options can be embedded in the widely used superstructure, the superstructure-based mathematical model cannot be verified by comparing to any real object since this model is artificial, *i.e.*, it is not the model of any real object. Numerous errors have been published in the literature because of this theoretical issue: the optimal solution of the mathematical model of these synthesis procedures was not the optimal solution of the original problem, *i.e.*, the mathematical models were not appropriate. The in-depth study of the inside properties of the imagined, appropriate mathematical models of process synthesis will be the basis of developing the model generation procedure. This will be done in the current and following chapters.

As a consequence of the above assessments, an obvious question is to be answered prior to further development: What is our purpose in process synthesis? Do we want to optimize certain mathematical model, or do we want to provide the best solution for the original (industrial) problem? Naturally, only the second option is acceptable even if the first option is usually practiced.

In most cases, the optimization model of process synthesis is constructed on the basis of a so-called superstructure that is intended to represent all the plausible relationships among the system's components. However, this structure is usually drawn by intuition or heuristics, which poses an issue as the behavior of a complex system is difficult to predict (even if it is possible at all). This is a consequence of the interactions among subsystems that frequently yield unexpected results (see, *e.g.*, [1]). Moreover, there are several design rules and heuristics that are not valid in the search for the optimal solution, as they may affect inappropriately the structure of the process to be designed, therefore, an appropriate model is necessary for automation in process synthesis.

To define an appropriate model, let us suppose that there is an evaluation method *e.g.*, a simulation program, capable of exactly determining the objective function's value for a set of process specifications when implemented in real life (*i.e.*, in its industrial realization). Therefore, if we have specifications for two different process design options, this evaluation method is capable of determining which one will be better in its industrial realization. Naturally, two different solutions may have the same practical value. Also, let us suppose that a certain mathematical optimization model identifies a set of process specifications as its optimal solution. Then, such a model for a process synthesis problem is defined to be *appropriate* if no other mathematical optimization model renders a better solution when assessed by the evaluation method.

In addition to the concern of whether the model is appropriate, there is another problem about conventional mathematical optimization. Such a problem is related to the selection of the most preferred solution by the decision makers. The majority of the known works merely yield a single solution considered to be the best, *i.e.*, the nominal optimum for the selected objective function. However, since various design aspects, such as complexity of piping configuration or discrete sizes of commercial equipment, have not been taken into account in the model, the obtained single solution is not necessarily suitable for implementation. Therefore, it is vital to determine the list of n-best solutions (where n is a natural number not less than two) from where the decision makers can perform further selection.

Therefore, the conventional approach of mathematical optimization presented in Fig. 1.3 is not appropriate for the concept of automatic process synthesis for two reasons. First, the model is difficult to formulate and cannot be verified which could derive in wrong solutions; and second, the solvers usually generate a single solution that could be not adequate for the original engineering problem. Thus, a different framework is required.

Figure 1.4 depicts an extended approach that overcomes the aforementioned hindrances. First, this approach relies on a more suitable input for automatic synthesis, *i.e.*, the input of the problem is specified in its original form from the industry and such original data are employed for the automatic generation of a synthesis model. Second, it generates the set of n-best alternatives of design that fulfill the requirements of the problem, thereby providing the designers with a comprehensive amount of information about the system. In addition, this extended

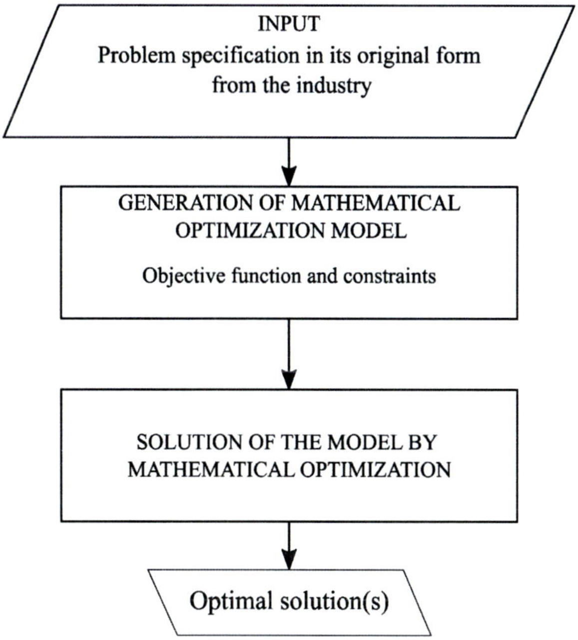

Fig. 1.4 Extended mathematical optimization procedure for process synthesis

approach must also account for the efficiency of the synthesis algorithms that solve the design problems whose sizes are ever increasing and pose higher structural complexities.

The Elements of the P-graph Framework

The P-graph framework, a joint effort initiated by L.T. Fan and F. Friedler in 1990, aims at overcoming these issues and providing strong theoretical basis for synthesis. The framework has been further advanced by numerous researchers and design engineers from around the world in the last decades.

The P-graph framework is based on an in-depth study of structures of feasible processes. It is supported by three pillars: (i) an unambiguous structure representation based on process graphs (P-graphs), that eliminates uncertain situations in the synthesis procedure (see Chap. 2); (ii) a set of combinatorial axioms of structures of feasible processes [2] (see Chap. 3); and (iii) a set of effective algorithms for generating the best or n-best solutions of the original design problem [3] (see Chaps. 5 and 6).

The P-graph framework is presented herein as a step towards achieving the goal of automatic synthesis. The framework directly concerns the structural relationships of the system's elements and exploits their properties for solving the synthesis problem. It employs the basic structural relations, the mathematical models of the individual operating units, and original industrial data as the input for synthesis. Such data may comprise specifications of the available raw materials, the goals of the design (*e.g.*, minimum cost, maximum safety, *etc.*), the products specification, and any additional constraints. This input is employed to automatically generate a rigorous superstructure of the problem, from which mathematical optimization models for synthesis can be derived. In addition, the combinatorial algorithms of the framework are formulated to effectively determine the set of n-best solutions, and to reduce computational effort by exploiting the knowledge embedded in the problem's structure. The advantages of the algorithmic acceleration based on the problem's knowledge can be demonstrated by the following simple example. Suppose that a set S of n elements is given. Then, the task is to determine a set T, whose elements are the subsets of S comprising two elements. This task is performed by two distinct approaches.

The first approach relies on a simple but inefficient procedure. It evaluates all possible subsets of S, and then tests whether they belong to set T. In this approach, the number of steps is $(2^n - 1)$. The second approach relies on an efficient realization involving the direct determination of all subsets of two elements. The number of steps required for this is $\left(\frac{n^2+n}{2}\right)$. Table 1.1 lists the number of steps that each approach requires for different sizes of set S, and the reduction ratio of the search space. The advantage of the second approach becomes more evident as the size of the problem increases.

Table 1.1 Number of steps required for different sizes of set S

n	Number of steps		Reduction ratio of search space
	Approach 1	Approach 2	
	$2^n - 1$	$\frac{(n^2+n)}{2}$	
10	1 thousand	55	$\approx$ 18
20	1 million	210	$\approx$ 4500
30	1 billion	465	$\approx$ two million

The second approach exploits inherent knowledge about the problem to reduce the search space and to eliminate unnecessary steps of the evaluation from the outset. The P-graph framework follows a similar philosophy, where the knowledge of the inherent combinatorial properties of process' structures is exploited for acceleration.

The subsequent chapters of this book are designed to provide a comprehensive understanding of the theoretical basis of the P-graph framework, as well as to describe the advantages of its application in various fields.

References

1. Kovács Z, Friedler F, Fan LT (1993) Recycling in a separation process structure. AICHE J 39: 1087–1089
2. Friedler F, Tarján K, Huang YW, Fan LT (1992a) Graph-theoretic approach to process synthesis: Axioms and theorems. Chem Eng Sci 47:1973–1988
3. Friedler F, Tarjan K, Huang YW, Fan LT (1992b) Combinatorial algorithms for process synthesis. In: 42nd Canadian Chemical Engineering Conference. Toronto, pp 207–209

Chapter 2
Representation of Process Structures in Process Network Synthesis: P-graphs

Abstract This chapter proves that the conventionally used directed graph representation is not appropriate for process synthesis; therefore, P-graphs are introduced to overcome this difficulty. The mathematical definition of P-graph is given together with the set operations associated to the graphs.

Keywords P-graph · Complex networks · Graph operations · Bipartite graph · Process synthesis

The most fundamental question to be answered in preparing the appropriate mathematical model for synthesizing processing systems is how to represent process structures. This question was originally studied in [1]; the present chapter is devoted to provide a firm answer.

It is a natural requirement that the structure representation must be unambiguous, *i.e.*, two structurally different synthesis problems must have different structures expressing the differences between them. Also, it is supposed that the structure representation provides all information about the process' structure that is necessary for algorithmic synthesis. In this regard, let us first examine if the most common structure representation, directed graphs, satisfies such requirements.

In the *directed graph representation* (or *digraph representation* in short) of a processing system, the operating units correspond to the nodes of the graph and the connections to its arcs. Now, it will be shown by two simple examples involving three operating units that the most common digraph representation is not appropriate for synthesizing processing systems.

Example 2.1 There is an operating unit, operating unit 01, that needs two different feed streams for the generation of the single product. These feed streams are produced separately by operating units 02 and 03. Clearly, for the generation of the product, all three operating units are necessary, and thus, this synthesis problem has a single solution involving all three operating units. The digraph representing this example is shown in Fig. 2.1.

F. Friedler et al., *P-graphs for Process Systems Engineering*,
https://doi.org/10.1007/978-3-030-92216-0_2

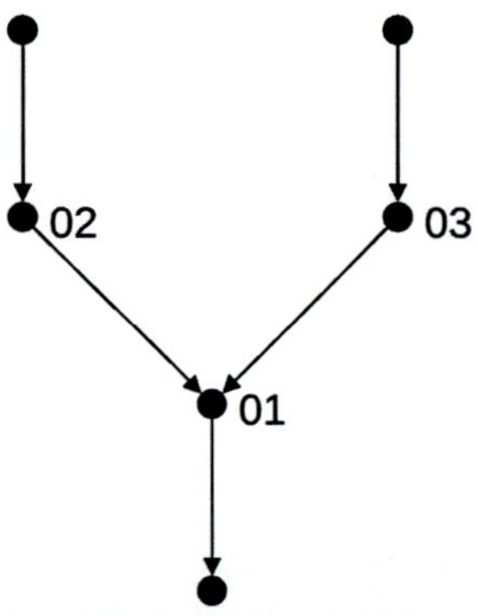

Fig. 2.1 Digraph representation of Examples 2.1 and 2.2

Example 2.2 Operating unit 01 produces the single product from a single feed stream. This feed stream can be produced by both operating units 02 and 03; either of them is sufficient for operating unit 01. As a synthesis problem for the manufacture of the product, there are three structurally different solutions. These solutions are operating units 01 and 02, operating units 01 and 03, and all three operating units. Naturally, this synthesis problem is structurally different from Example 2.1, still it has the same digraph representation as shown in Fig. 2.1.

While the digraph representation can be suitable for analyzing processing systems, these simple examples show that it is incapable of appropriately characterizing the structures of processing systems in their synthesis. Therefore, a more sophisticated graph than the conventional digraph is required to uniquely characterize a synthesis problem. For this reason, a particular graph, the P-graph, is introduced to capture not only the syntactic but also the semantic contents of the process structures. For the formal description of this new type of graph, some background is needed, which is detailed below.

By definition, synthesis is an action to determine the best "combination of elements to form a connected whole system" from the available set of elements for achieving a certain goal. In other words, synthesis is the action to set up a system from available building blocks to complete prespecified technical, economical, or social targets. In the parlance of process systems engineering (PSE), these building blocks, or elements, are called the operating units. Each operating unit has inputs that must satisfy certain conditions. If all appropriate inputs are available, the operating unit generates all of its well-defined outputs. Two operating units can be connected if the specification of one of the outputs of an operating unit fits to the specification of one of the inputs of the other operating unit. A crucial point in preparing the mathematical model for synthesis is to suppose that the set of specifications of the inputs and outputs of all available operating units is finite. Each element of the set of all specifications will be called *material* in the synthesis framework. Term *material* will be used to denote any type of input or output of an operating unit. Therefore, a *material* signifies a more general object than its conventional meaning; for example, a material can be energy or manpower.

Stated formally, let M be a given nonempty finite set of all materials that may appear in the synthesis problem to be solved. Set M can be specified by the set of names or a set of vectors of characteristics of these materials. The exact description, *i.e.*, the contents of set M, may vary depending on the level of precision or detail desired. While a coarse description of the materials may suffice for approximately representing the system, a fine description may be required for detailed representation. For example, both

$$M_1 = \{x_1, x_2, x_3, x_4, x_5, x_6\}$$

and

$$M_2 = \{(A, \varnothing, \varnothing), (\varnothing, B, \varnothing), (\varnothing, \varnothing, C), (A, B, \varnothing), (\varnothing, B, C), (A, B, C)\}$$

may represent the materials involved in a sharp separation scheme of a mixture of components A, B, and C. Clearly, the latter is more descriptive than the former although both may contain identical information.

Now, let us set up a graph for representing the structural part of a synthesis problem. First, a node is assigned to each element of M and also to each available operating unit. Let us suppose that the nodes representing a material are distingushed from the ones representing operating units. To do so graphically, a material-type node, *i.e.*, an M-type node, will be denoted by a small circle (or point) whereas the node of an operating unit, *i.e.*, an O-type node, will be shown as a small horizontal bar. Now, let us establish the arcs of the graph. If a material can be an input to an operating unit, a directed arc will go from the node representing the material to the node representing the operating unit. Similarly, if a material is one of the outputs of an operating unit, a directed arc will lead from the node representing the operating unit to the node representing this material. Let us suppose that all directed arcs related to the available operating units are drawn. A graph specified in this way shows all possible connections among the materials and operating units. Our goal in synthesis is to find the subgraph or subgraphs representing the best process or processes.

The graph that has just been identified for the synthesis of processing systems will be called process graph, or P-graph. An example of a P-graph is shown in Fig. 2.2 where $M = \{A, B, C, D, E\}$ and there are two operating units, the first transforms material C to materials A and E, the second transforms materials D and E to materials B and C.

P-graphs belong to a special class of directed graphs the so-called *bipartite* graphs, since its nodes can be partitioned into two disjoint sets and no two nodes in the same set are adjancent. Let us observe that in this representation, an operating unit is given by two sets of materials, namely, by the set of its input materials and the set of its output materials. For example, operating unit $O1$ in Fig. 2.2 can be given as

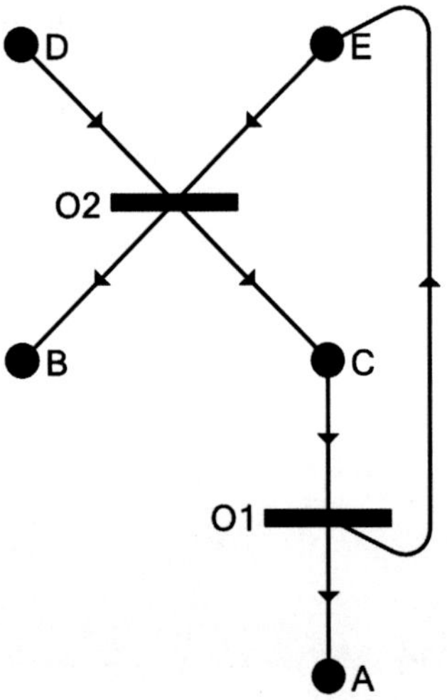

Fig. 2.2 Illustration of a P-graph

pair ({C}, {A,E}). It is supposed that neither the set of inputs nor the set of outputs are empty sets. If m denotes the set of materials, the set of input materials of any operating unit must be an element of the set of all nonempty subsets of m, which is denoted by $\wp'(m)$; similarly, the set of output materials is also an element of $\wp'(m)$. Therefore, any operating unit is an element of the set $\wp'(m) \times \wp'(m)$, *i.e.*, the set of all pairs of nonempty subsets of the set of materials. From each input material, an arc leads to the operating unit and from the operating unit an arc leads to each of its output materials. Using this description of the operating units, the notion of process graph, or P-graph, can formally be defined.

Definition 2.1 Let set M be given as a set of materials. By *process graph* (or *P-graph* in short), we mean a bipartite graph given by pair (m, o) where $m \neq \emptyset$, $o \neq \emptyset$, $m \subseteq M$, $o \subseteq \wp'(m) \times \wp'(m)$; the nodes of this graph are the elements of $m \cup o$ and its arcs are the elements of $A_1 \cup A_2$ where

$$A_1 = \{(X, y) : y = (\alpha, \beta) \in o \,\&\, X \in \alpha\}$$

and

$$A_2 = \{(y, X) : y = (\alpha, \beta) \in o \,\&\, X \in \beta\}.$$

In these expressions for A_1 and A_2, X designates an M-type node; y, an O-type node; α, the set of the input materials of y from each of which an arc is directed to y; and β, the set of the output materials of y to each of which an arc is directed from y.

Clearly, one can assign a P-graph to any processing system by representing the operating units in the given form. It is worth noting that no arbitrary P-graph can represent the structure of a feasible process. The suitable P-graphs must satisfy some combinatorial properties which will be elaborated in Chap. 3.

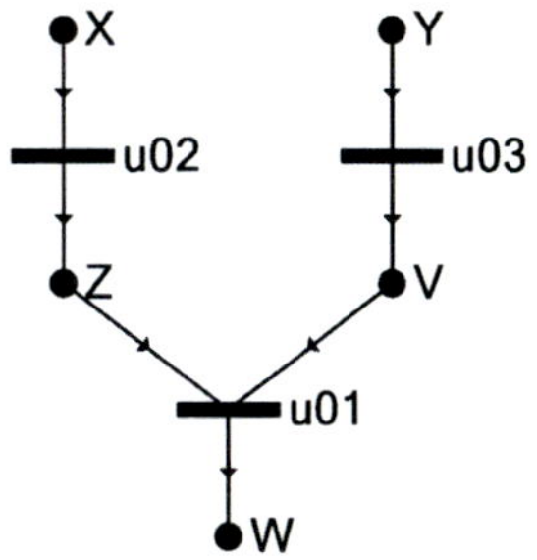

Fig. 2.3 P-graph representation of Example 2.1

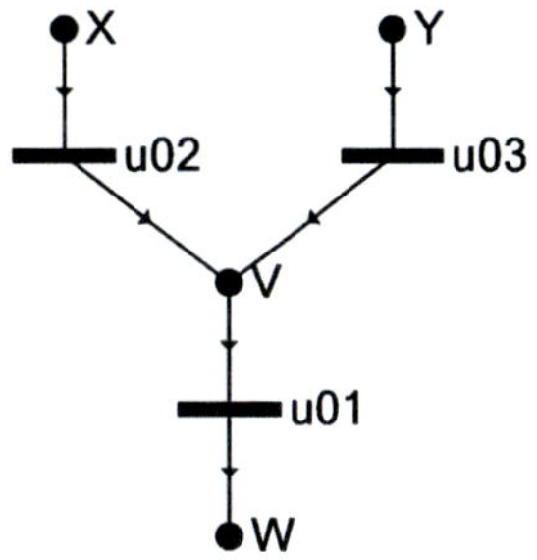

Fig. 2.4 P-graph representation of Example 2.2

Now, let us revisit the simple examples presented at the outset of this chapter that have identical digraph representation. For the process system given in Example 2.1, let $m = \{X, Y, Z, V, W\}$ where it is assumed that operating unit u_{02} produces material Z from material X, and operating unit u_{03} produces V from Y; furthermore, operating unit u_{01} produces W from materials Z and V. Then, the operating units appearing in this process system can be given as follows: $u_{02} = (\{X\}, \{Z\})$, $u_{03} = (\{Y\}, \{V\})$, and $u_{01} = (\{Z, V\}, \{W\})$. The corresponding P-graph describing the structure of this process system is given in Fig. 2.3.

For the process given in Example 2.2, let $m = \{X, Y, V, W\}$ where it is supposed that operating unit u_{02} produces material V from material X, operating unit u_{03} produces V from Y; furthermore, operating unit u_{01} produces W from V. In this case, the operating units occuring in this process are $u_{02} = (\{X\}, \{V\})$, $u_{03} = (\{Y\}, \{V\})$, and $u_{01} = (\{V\}, \{W\})$. The corresponding P-graph representing the structure of this process system is given in Fig. 2.4. Thus, the P-graph representations of Examples 2.1 and 2.2 express the structural differences of these examples, which is not the case with directed graphs.

The notion embodied by P-graph is more general than its application restricted to processing systems. An example for a different science field is given in Fig. 2.5 [2].

To complete the formal description of P-graphs, further definitions are needed, which are provided below.

Definition 2.2 Let two P-graphs (m_1, o_1) and (m_2, o_2) be given with $(m_1 \cup m_2) \cap (o_1 \cup o_2) = \emptyset$. Then, by the *union* of (m_1, o_1) and (m_2, o_2), we mean P-graph given by $(m_1 \cup m_2, o_1 \cup o_2)$.

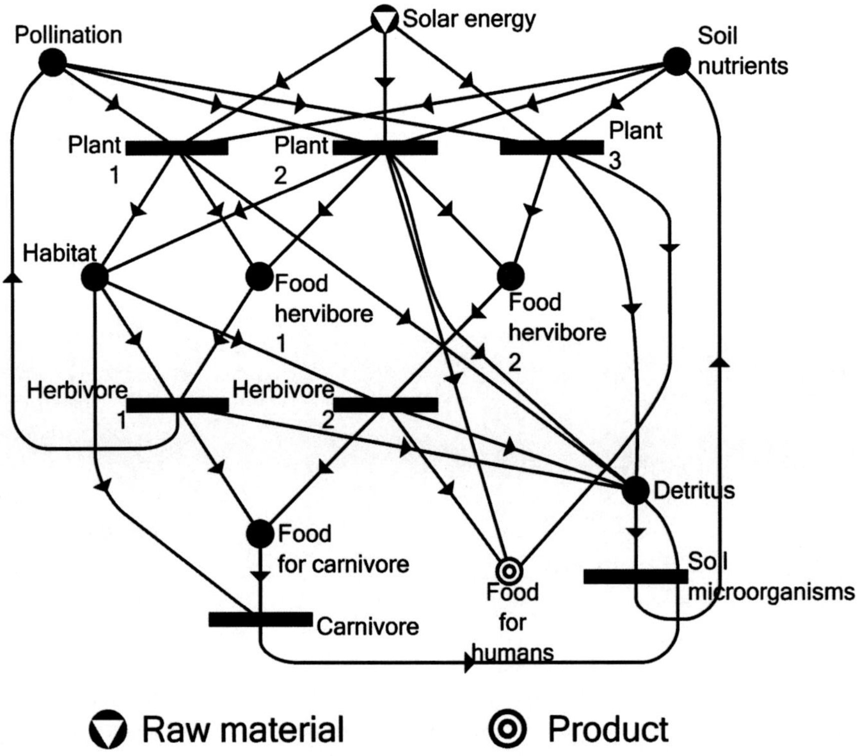

Fig. 2.5 P-graph of an ecological system [2]

Under the assumption of this definition, it can be seen that the graph determined by $(m_1 \cup m_2, o_1 \cup o_2)$ is really a P-graph. It is worth noting that $(m_1 \cup m_2) \cap (o_1 \cup o_2) = \varnothing$ is a necessary condition to obtain P-graph.

The inclusion relation for the P-graphs is given next.

Definition 2.3 Let two P-graphs (m_1, o_1) and (m_2, o_2) be given. (m_1, o_1) is called a *subgraph* of (m_2, o_2) if $m_1 \subseteq m_2$ and $o_1 \subseteq o_2$ are valid. This relation is denoted by $(m_1, o_1) \subseteq (m_2, o_2)$, furthermore, we write $(m_1, o_1) \subset (m_2, o_2)$ if $(m_1, o_1) \subseteq (m_2, o_2)$ and $(m_1, o_1) \neq (m_2, o_2)$.

It is worth noting that this notion differs from the traditional notion of subgraph. Namely, if (m_1, o_1) is a subgraph of (m_2, o_2) in accordance with definition 2.3, then the graph given by (m_1, o_1) is a subgraph in the traditional meaning of the graph given by (m_2, o_2). However, the conversion is not always valid, *i.e.*, a traditional subgraph of the graph given by (m_2, o_2) is not necessarily a P-graph. For example, one can consider the traditional subgraph of the P-graph (m_2, o_2) consisting of the set of nodes $m_2 \cup o_2$ and does not contain any arc. This graph is obviously a traditional subgraph of P-graph (m_2, o_2), but it is not a P-graph.

Summarizing the main idea of this section, it can be stated that the P-graph is such a new form of representation of processing systems that captures not only the syntactic but also the semantic contents of the systems structure. In what follows, the combinatorial properties of the structures of processing systems will be examined for establishing a firmer basis of the mathematical modeling of synthesis and reducing the search space in finding the best solution, or solutions.

References

1. Friedler F, Tarján K, Huang YW, Fan LT (1992) Graph-theoretic approach to process synthesis: Axioms and theorems. Chem Eng Sci 47:1973–1988
2. Lao A, Cabezas H, Orosz Á, Friedler F, Tan R (2020) Socio-ecological network structures from process graphs. PLoS One 15(8):e0232384

Chapter 3
Structural Model of Process Network Synthesis

Abstract In this chapter, the concept of inside-out modelling is introduced. Five combinatorial axioms are presented for reducing the search space of process synthesis problems. It is shown that the structure of each feasible solution of a process synthesis problem must satisfy these axioms. The maximal structure is defined as a rigorous superstructure of process synthesis problems. The structure of the optimal solution, or the structures of the n-best solutions, are always part of the maximal structure; therefore, it is an appropriate basis for the mathematical model.

Keywords P-graph · Structural model · Combinatorial axioms · Solution-structure · Maximal structure · Superstructure · Process synthesis

P-graphs, introduced in the previous chapter, will be the basis of the structural model of synthesis. Stated formaly, let M^* be an arbitrarily fixed set containing every material that can be an input to or an output from any operating unit in a process synthesis problem. Note that M^* can be either finite or infinite. For mathematical rigor, it is supposed that $M^* \cap (\wp'(M^*) \times \wp'(M^*)) = \emptyset$, naturally, no real engineering problem violates this assumption.

A process design problem in general and process network synthesis (PNS) in particular can be structurally specified by three sets, *i.e.*, the set of raw materials, the set of desired products, and the set of available operating units. These sets determine the structure of a PNS problem as a P-graph containing the corresponding interconnections among the operating units. For convenience, let the sets of available raw materials, desired products, and available operating units be denoted by R, P, and O, respectively. Since M^* is the set of all materials, it is natural to asssume that $P \subseteq M^*$, $R \subseteq M^*$, and $O \subseteq \wp'(M^*) \times \wp'(M^*)$. Even if M^* can be infinite, it is assumed that sets P, R, and O are all finite; in addition, for all $u = (\alpha, \beta) \in O$, α and β are regarded as finite sets (α and β represent the input and output materials of operating unit u, respectively). It is a natural assumption that P and O are nonempty and $P \cap R = \emptyset$. The structural model of PNS can now be introduced as follows:

F. Friedler et al., *P-graphs for Process Systems Engineering*,
https://doi.org/10.1007/978-3-030-92216-0_3

Definition 3.1 Triplet $\mathbb{S} = (P, R, O)$ is a *structural model* of PNS where
$P \subseteq M^*$ is a finite nonempty set representing the set of *products*,
$R \subseteq M^*$ is a finite set representing the set of *raw materials*, and
$O \subseteq \wp'(M^*) \times \wp'(M^*)$ is a finite nonempty set representing the set of *operating units*. Moreover, $P \cap R = \emptyset$, and α, β are finite sets, for all operating units $u = (\alpha, \beta) \in O$.

Now, a P-graph can be assigned to a structural model $\mathbb{S} = (P, R, O)$ as follows: The set of materials M belonging to the structural model $\mathbb{S}$ is defined as the set of all materials identified in the inputs and outputs of the operating units in O, altogether with the sets of raw materials R and products P. Formally,

$$M = P \cup R \cup (\bigcup_{(\alpha,\beta \in O)} (\alpha \cup \beta)).$$

Note that M and O are nonempty finite sets, $O \subseteq \wp'(M) \times \wp'(M)$ and $M \cap O = \emptyset$. Thus, (M, O) is a P-graph, its nodes are the elements of $M \cup O$ and its arcs are the elements of $A_1 \cup A_2$ where

$$A_1 = \{(X, (\alpha, \beta)) : (\alpha, \beta) \in O \,\&\, X \in \alpha\}$$

and

$$A_2 = \{((\alpha, \beta), X) : (\alpha, \beta) \in O \,\&\, X \in \beta\}$$

P-graph (M, O) represents all interconnections among the operating units of set O. A subgraph of (M, O) can be assigned to each feasible process of the PNS problem; this subgraph represents the structure, or network, of the process of interest. Since $M \cap O = \emptyset$, any node of P-graph (M, O) belongs to either M or O. The elements of M are *material nodes*, or M-type nodes; and those of O, *operating unit nodes,* or O-type nodes of (M, O). The following example illustrates the notions introduced so far.

Example 3.1 Given the operating units,

$$u_1 = (\{B\}, \{D\}), u_2 = (\{F\}, \{D, E\}),$$
$$u_3 = (\{E\}, \{G\}), u_4 = (\{D\}, \{C, G\}), \text{and } u_5 = (\{A, C\}, \{G\})$$

we are going to synthesize a process to generate material G by considering materials A and B as available raw materials and using all, or some, of the given operating units. Thus, the set of desired products is $P = \{G\}$; the set of raw materials is $R = \{A, B\}$; and the structural model is $\mathbb{S} = (\{G\}, \{A, B\}, \{u_1, u_2, u_3, u_4, u_5\})$. The

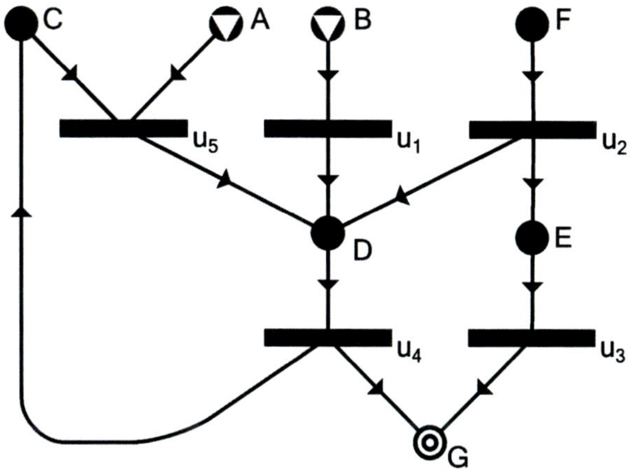

Fig. 3.1 P-graph (M, O) of Example 3.1

set of materials belonging to the available operating units, M, is $\{A, B, C, D, E, F, G\}$. P-graph (M, O) determined by $\mathbb{S}$ is depicted in Fig. 3.1.

There are three feasible solutions for this example provided that no further constraints are imposed. The P-graphs of these solutions are shown in Fig. 3.2. The first solution comprises operating units u_1 and u_4, and the desired product G is generated from raw material B. Fig. 3.2a presents its structure (network) as the subgraph (m, o) of (M, O) with $o = \{u_1, u_4\}$ and $m = \{B, C, D, G\}$. The second solution includes the operating units u_4 and u_5, and the desired product G is obtained from raw material A. The structure of this solution is depicted in Fig. 3.2b, which corresponds to the subgraph (m, o) of (M, O) with $o = \{u_4, u_5\}$ and $m = \{A, C, D, G\}$. The last solution corresponds to the combination of the two previous networks, *i.e.*, it comprises units u_1, u_4 and u_5, and the product G is generated from both raw materials A and B. Its network is depicted in Fig. 3.2c as the P-graph (m, o) where $o = \{u_1, u_4, u_5\}$ and $m = \{A, B, C, D, G\}$.

This example shows that structural model (P, R, O) determines the interconnections among the available operating units. However, this information is insufficient to determine directly those subgraphs of (M, O) that can be the structures of feasible processes. Hence, it is essential to identify the general combinatorial properties of the structures of feasible processes. Such properties, established in [1] from a practical point of view, are necessary conditions for a subgraph of (M, O) to be the structure of a feasible process. In what follows, these necessary conditions are elaborated to precisely define the subgraphs that can be the structures of feasible processes.

Prior to examining the structural conditions that a feasible process is expected to satisfy, let us investigate this problem from a wider perspective. The overall purpose of our effort is to provide theory and tools to find the best solution or solutions of a process synthesis problem. This problem is traditionally solved by engineer's intuition, following heuristic rules, optimizing a mathematical model, or by combinations of them. Engineering failures show that the risk of generating a nonoptimal

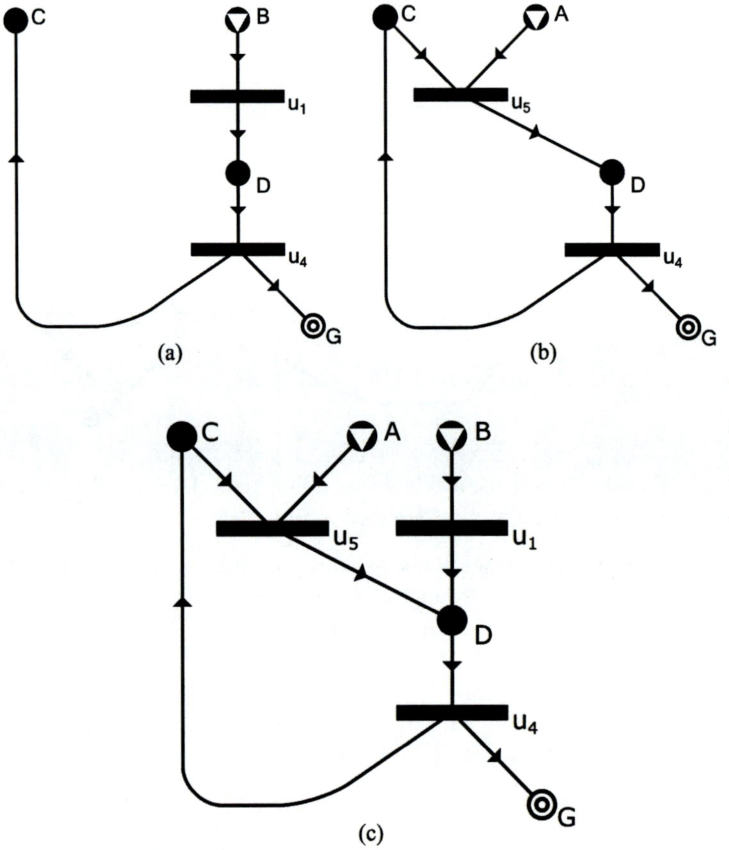

Fig. 3.2 Structures of possibly feasible processes in Example 3.1

solution, or a solution far from the optimal, is very high if nonrigorous tools are used [2–4]. In [5], an example is shown where a nonfeasible solution is given as optimal because of an inappropriate model of scheduling. Therefore, we are forced to solve the problem by mathematical optimization on the basis of a rigorous mathematical model. It is, however, almost impossible because there is no way of verifying the mathematical model conventionally. The reason is simple: the synthesis problem itself is artificial; no real object exists to compare with its model. The structural examination of inherent properties that an appropriate model must satisfy may help to cut this Gordian knot.

By nature, the mathematical model of synthesis is usually a mixed integer nonlinear programming (MINLP) problem of the form given in Fig. 3.3. This formulation is presented as a figure on purpose: it is just an illustration. As far as it has been mentioned, this model is artificial. The reason for setting up an MINLP model is to generate the optimal solution of the synthesis problem as the result of the optimization of this model. The critical question now is who can guarantee that the optimal solution of the MINLP model is really the optimal solution of the original

$$\begin{aligned}&\min f(x,y)\\&\text{s.t.}\\&\qquad g(x,y)\le 0\\&\qquad x\in\mathbb{R}^n\ ,\ y\in\{0,1\}^m\end{aligned}$$

Fig. 3.3 Standard form of an MINLP problem as the most common mathematical model of synthesis

engineering problem. The difficulty escalates if we follow the decision makers' usual requirements to generate the n-best (n > 1, integer) solutions of the original problem instead of a single optimal solution.

To go beyond this issue, let us now assume that there exists a valid MINLP model for the original synthesis problem even though we do not know how to formulate it. Thus, by optimizing this mathematical model, we obtain the optimal solution of the original synthesis problem, if a single best solution is to be found. The same is valid for the n-best solutions, if they are needed.

Naturally, the class of mathematical models of the form of Fig. 3.3 is more general than the class of the mathematical models of synthesis problems, most of the models in the form of Fig. 3.3 do not represent a synthesis problem. Therefore, additional information is embedded in the mathematical model presented in Fig. 3.3 to make it into a valid synthesis model. This information must implicitly be embedded into this model. If this implicit information, or at least a part of it, can be expressed explicitly, this information may serve for the generation of the valid model. This is what we will accomplish next by examining the necessary conditions for the feasibility of process structures.

Since the purpose of process synthesis is to determine a process for producing all products in prespecified amounts, the MINLP model must express the related requirements by a constraint. Consequently, the nodes of the P-graph representing the products must appear in the structure of all feasible processes. This structural condition, or axiom, can be stated as follows.

AXIOM (S1) **Every product must be represented as a material node in the structure of a feasible process.**

If this condition is not satisfied, some of the desired products are not produced by any operating unit of the process, *i.e.*, the condition on the desired products is not satisfied, thus resulting in an infeasibility.

The second condition is not so evident. It expresses the meaning of a raw material. Specifically, a material is regarded as a raw material in a feasible process if it is not to be produced by any available operating unit of the process under consideration. On the other hand, we demand that all the materials other than the raw materials be produced by some operating units of the process. This condition (axiom) can be stated as follows.

AXIOM (S2) **A material node of the structure of a feasible process has no ancestor if and only if it represents a raw material.**

It is worth noting that Axiom (S2) expresses a raw material as an interface between the process and the outside world, *i.e.*, a raw material must not be produced inside the process. This does not serve as a constraint in specifying a PNS problem.

In particular, if the process is allowed to produce internally certain raw material, *e.g.*, by recycling, then we can substitute this with a fictitious material in the operating unit under consideration. This validates axiom (S2) in practice.

In synthesizing a process, only the operating units identified in the outset of the problem can be considered as building blocks of the process, it leads to the following axiom.

AXIOM (S3) **Each operating unit appearing on the structure must be an element of the set of operating units of the problem definition.**

It is expected that each of the operating units contained in a feasible process contributes to the generation of a product. In other words, the appearance of an operating unit node in the structure of a feasible process is forbidden unless the corresponding operating unit participates directly or indirectly in the production of the desired products. This is accomplished by imposing the following necessary condition (axiom).

AXIOM (S4) **For each operating unit node of the structure of a feasible process, there is a path leading from this node to a material node of the structure representing a product.**

As far as the material nodes of the structure of a feasible process are concerned, it is evident that each of the corresponding materials is consumed or produced by at least one of the operating units in the process of interest. This is stated as another necessary condition (axiom) to be satisfied by the structure of a feasible process as follows:

AXIOM (S5) **For each material node of the structure of a feasible process, there exists at least one operating unit in the process, for which the material represented by this material node is an input or an output.**

Now, let us consider the reverse situation. Suppose that a PNS problem is provided with structural model $\mathbb{S} = (P, R, O)$, and the P-graph belonging to it is denoted by (M, O); moreover, let (m, o) be a subgraph of (M, O) satisfying axioms (S1) through (S5). The fact that P-graph (m, o) satisfies these conditions implies that it can be the structure of a feasible solution (process) of the PNS problem for the following reasons. Since (m, o) is a subgraph of (M, O), the process consists of only available operating units; furthermore, all the interconnections are established by these operating units. Axiom (S1) ensures that all the desired products are produced in this process. Axiom (S2) guarantees that all the unproduced materials appearing in the process are raw materials. Axiom (S3) ensures that all operating units in the structure are available to build in the process. Finally, Axioms (S4) and (S5) imply that this process does not contain unnecessary operating units and unnecessary materials. Hence, it can be considered as a feasible process from a structural point of view, *i.e.*, structurally feasible. Thus, these five conditions must be implicitly embedded into the MINLP model to specify it to be a valid model of process synthesis.

Naturally, no P-graph can represent the structure of a feasible process unless it satisfies the five axioms. Some of the structures satisfying these five axioms, however, are not necessarily feasible because of possibly available additional constraints. These constraints are mainly related to the continuous parameters. Therefore, the five axioms are necessary but not sufficient conditions of the feasibility.

Since any of the structures satisfying the five axioms can be optimal under certain circumstances (*e.g.*, because of additional constraints), no additional general structural conditions can be introduced to further reduce the set of structurally feasible solutions without the risk of losing potentially optimal solutions.

On the basis of the conditions elucidated above, we can precisely define the structures (networks) of feasible processes of a given structural model of PNS as follows.

Definition 3.2 Let $\mathbb{S} = (P, R, O)$ be the structural model of PNS and (M, O) be the corresponding P-graph.

P-graph (m, o) is a *solution-structure* or *combinatorially feasible structure* of $\mathbb{S}$ if the following conditions are satisfied.

(S1) $P \subseteq m$,

(S2) $\forall X \in m, X \in R \Leftrightarrow$ there is no (Y, X) arc in P-graph (m, o),

(S3) $o \subseteq O$

(S4) $\forall Y_0 \in o, \exists$ path $[Y_o, Y_n]$ of (m, o) with $Y_n \in P$, and

(S5) $\forall X \in m, \exists (\alpha, \beta) \in o$ such that $X \in \alpha \cup \beta$.

The five conditions will be cited as the axioms of combinatorially feasible process structures. The main consequence of the above examination is that the optimal (or n-best) solution is always among the processes represented by the combinatorially feasible process structures. Furthermore, if we have an algorithm for generating this set of structures and there is an analysis tool to evaluate them, the selection of the optimal or n-best solutions is guaranteed.

This is illustrated by revisiting Example 3.1. Its P-graph (M, O) has exactly three subgraphs (m, o) satisfying (S1) through (S5), *i.e.*, this structural model gives rise to three solution-structures as given in Fig. 3.2. Let us now examine a more complex synthesis problem [6].

Example 3.2 There are seven operating units, they are specified in Table 3.1. The sets of the structural model are $P = \{A\}$, $R = \{E, J, G, K, L\}$, $O = \{O1, O2, O3, O4, O5, O6, O7\}$ [6].

This example has 19 solution-structures that are all shown in Fig. 3.4. This example is simple enough to draw all 19 networks without additional tools. In

Table 3.1 Operating units for Example 3.2

Unit	Inputs	Outputs	Unit	Inputs	Outputs
$O1$	$\{C\}$	$\{A, F\}$	$O5$	$\{G, H\}$	$\{D\}$
$O2$	$\{D\}$	$\{A, B\}$	$O6$	$\{J\}$	$\{F\}$
$O3$	$\{E, F\}$	$\{C\}$	$O7$	$\{K, L\}$	$\{H\}$
$O4$	$\{F, G\}$	$\{C, D]$			

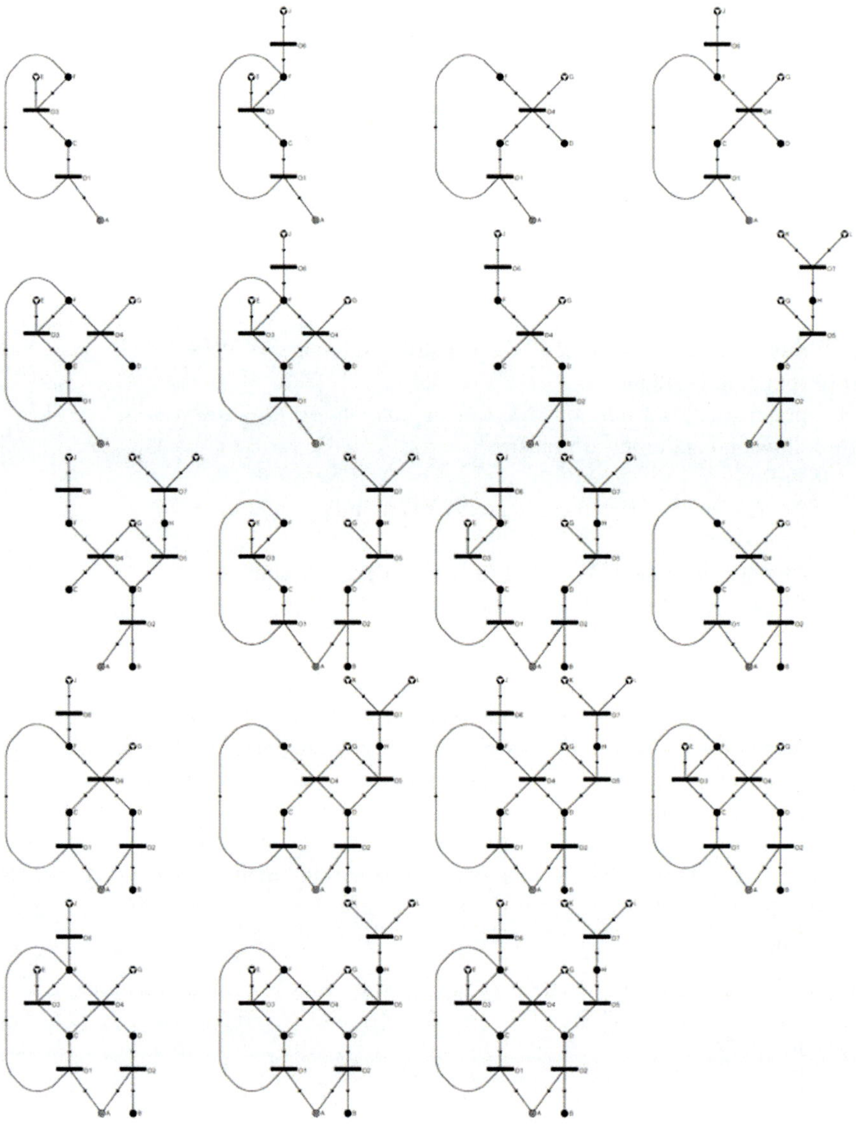

Fig. 3.4 The 19 solution-structures of Example 3.2 [6]

solving industrial problems, however, algorithmic methods are needed. Chapters. 4, 5, and 6 will introduce such algorithms. The industrial synthesis problem given by Example 3.3 has higher complexity; its solution requires the use of these computational tools.

Table 3.2 Operating units for Example 3.3

Unit	Inputs	Outputs	Unit	Inputs	Outputs
op1	$\{A1\}$	$\{A5\}$	$op19$	$\{A14, A27, A28, A29, A30\}$	$\{A41\}$
op2	$\{A2, A3, A4\}$	$\{A9\}$	$op20$	$\{A11, A31\}$	$\{A40, A42\}$
op3	$\{A3, A4, A6, A11\}$	$\{A10\}$	$op21$	$\{A32\}$	$\{A44, A45\}$
op4	$\{A3, A4, A5\}$	$\{A12\}$	$op22$	$\{A33, A34\}$	$\{A46\}$
op5	$\{A3, A4, A5\}$	$\{A13\}$	$op23$	$\{A36\}$	$\{A14, A48\}$
op6	$\{A7, A8, A14\}$	$\{A16\}$	$op24$	$\{A38\}$	$\{A14, A48\}$
op7	$\{A8, A14, A18\}$	$\{A16\}$	$op25$	$\{A41\}$	$\{A50, A51\}$
op8	$\{A9, A11\}$	$\{A21, A22, A24\}$	$op26$	$\{A43, A44\}$	$\{A53\}$
op9	$\{A10, A11\}$	$\{A22, A24, A37\}$	$op27$	$\{A46\}$	$\{A55, A56\}$
op10	$\{A12\}$	$\{A25, A26\}$	$op28$	$\{A47, A48\}$	$\{A5, A57\}$
op11	$\{A13\}$	$\{A25, A31\}$	$op29$	$\{A48, A49\}$	$\{A5, A58\}$
op12	$\{A15, A16\}$	$\{A32\}$	$op30$	$\{A50\}$	$\{A59, A60\}$
op13	$\{A14, A17, A18, A19, A20\}$	$\{A33\}$	$op31$	$\{A51, A54\}$	$\{A61\}$
op14	$\{A6, A21\}$	$\{A35\}$	$op32$	$\{A52, A53\}$	$\{A61\}$
op15	$\{A22, A23\}$	$\{A48\}$	$op33$	$\{A54, A55\}$	$\{A61\}$
op16	$\{A5, A24\}$	$\{A36\}$	$op34$	$\{A59\}$	$\{A62, A63\}$
op17	$\{A5, A11, A25\}$	$\{A37, A38, A39\}$	$op35$	$\{A60\}$	$\{A64, A65\}$
op18	$\{A11, A26\}$	$\{A40, A42\}$			

Example 3.3 Industrial synthesis problem with 35 available operating units [7] specified in Table 3.2 where

$P = \{A61\}$,
$R = \{A1, A2, A3, A4, A6, A7, A8, A11, A15, A17, A18, A19, A20,$
$A23, A27, A28, A29, A30, A34, A43, A47, A49, A52, A54\}$, and
$O = \{op1, op2, \ldots, op35\}$.

This industrial synthesis problem has been solved by the algorithms developed for the generation of the combinatorially feasible structures (see Chap. 5). One of the 3465 solution-structures is shown in Fig. 3.5.

The set of solution-structures of structural model $\mathbb{S} = (P, R, O)$ will be denoted by *S(P,R,O)* or $S(\mathbb{S})$. Note that any element of this set can be a candidate for the optimal solution of the PNS problem. Moreover, no structure can represent a feasible process unless its structure is an element of $S(\mathbb{S})$. Consequently, the n-best solutions can also be firmly determined by simply evaluating all structures in $S(\mathbb{S})$.

Since sets P, R, and O are finite, the corresponding P-graph (M, O) is also finite; therefore, $S(\mathbb{S})$ is also finite. Nevertheless, set $S(\mathbb{S})$ can be either empty or nonempty depending on structural model $\mathbb{S}$. For example, $S(\mathbb{S}) = \{(m, o)\}$ is nonempty in Example 3.1. However, if the set of products is $\{E, G\}$, no solution-structure exists for the resultant structural model $\mathbb{S}'$ defined by $\mathbb{S}' =$

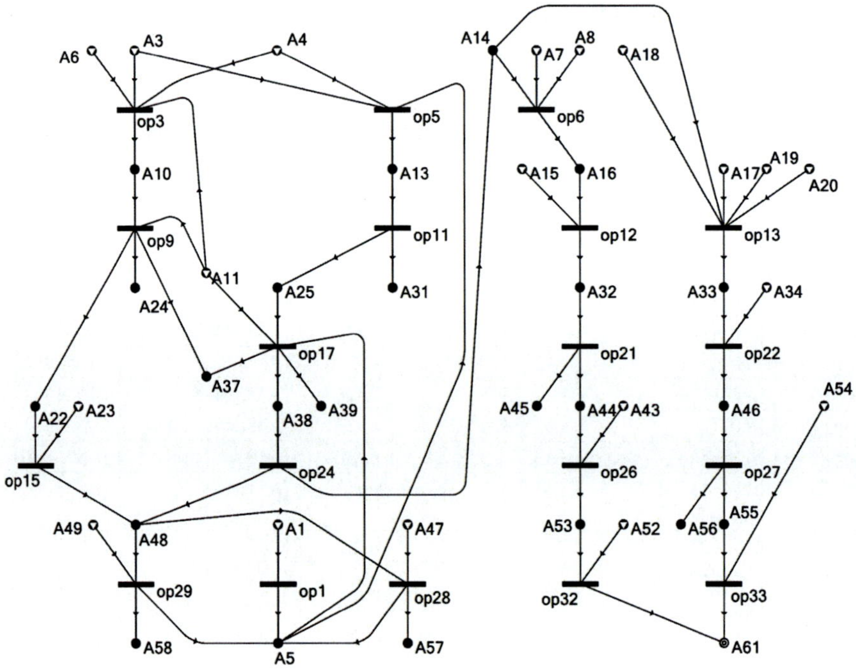

Fig. 3.5 One of the 3465 solution-structures of Example 3.3

$(\{E, G\}, \{A, B\}, \{u_1, u_2, u_3, u_4, u_5\})$, *i.e.*, $S(\mathbb{S}') = \emptyset$, as it is not possible to produce material E under the conditions imposed.

Suppose that there is an evaluation tool capable of determining whether a solution-structure is feasible or not (*i.e.*, whether it satisfies all constraints). Furthermore, if a solution-structure is found feasible, this tool determines the minimal cost function value of the solution-structure. Let $\widehat{S}(\mathbb{S})$ denote the set of feasible solution structures of structural model $\mathbb{S}$, thus, $\widehat{S}(\mathbb{S}) \subseteq S(\mathbb{S})$.

Let function z be defined on set $\widehat{S}(\mathbb{S})$ (*i.e.*, $z : \widehat{S}(\mathbb{S}) \to \mathbb{R}$) where function z expresses the minimal value of the objective function (cost function) of any feasible solution-structure (note that all over this work, the objective function is considered to be minimized). Then, the determination of the optimal solution or solutions of PNS can be stated as the generation of the set of solution structures

$$\left\{s : z(s) \leq z(s') \;\; \& \;\; s, s' \in \widehat{S}(\mathbb{S})\right\}.$$

This set of solution structures can readily be determined by algorithms given in Chaps. 5 and 6 independently of the types of mathematical models and constraints of PNS.

The following theorem states an essential property of the set of solution-structures.

Theorem 3.1 Let $\mathbb{S} = (P, R, O)$ be a structural model of PNS. If (m, o) and (m', o') are solution-structures of $\mathbb{S}$, then $(m, o) \cup (m', o')$ is also a solution-structure of $\mathbb{S}$.

Proof By definition, the union of the solution-structures considered is P-graph $(m \cup m', o \cup o')$. Thus, it is sufficient to prove that $(m \cup m', o \cup o')$ satisfies axioms (S1) through (S5).

Since (m, o) and (m', o') are in $S(\mathbb{S})$, $P \subseteq m$ and $P \subseteq m'$ in light of axiom (S1). Hence, $P \subseteq m \cup m'$ thereby, indicating that (S1) is valid for the union.

To prove axiom (S2), let $X \in m \cup m'$ be an arbitrary material; also assume that $X \in R$. Since both (m, o) and (m', o') are elements of $S(\mathbb{S})$ satisfying (S2), no input arc to X, *i.e.* (Y, X) arc, exists in either (m, o) or (m', o'); therefore, no (Y, X) arc exists in their union. Conversely, if it is assumed that $X \notin R$, we may suppose that $X \in m$ without loss of generality. Then, there exists an arc (Y, X) in (m, o) from axiom (S2). This implies that (Y, X) is an arc in the union as well, thus indicating that axiom (S2) is also valid for the union. Condition (S3) is naturally satisfied by the union of solution-structures.

To verify axiom (S4), let Y_0 be an arbitrary operating unit in $o \cup o'$. Without loss of generality, we may suppose that $Y_0 \in o$. Since $(m, o) \in S(\mathbb{S})$, it satisfies axiom (S4), *i.e.*, a $[Y_0, Y_n]$ path exists in (m, o) where $Y_n \in P$. Since the operation of union keeps each edge of P-graphs (m, o) and (m', o'), path $[Y_0, Y_n]$ in (m, o) is also a path in $(m, o) \cup (m', o')$ with $Y_n \in P$, thereby validating (S4) for the union.

Finally, let $X \in m \cup m'$ be an arbitrary material. Without loss of generality, we may assume that $X \in m$. Then, operating unit $(\alpha, \beta) \in o$ with $X \in \alpha \cup \beta$ exists from axiom (S5) of (m, o). Since $(\alpha, \beta) \in o \cup o'$, axiom (S4) holds for the union. This completes the proof.

Theorem 3.1 indicates that set $S(\mathbb{S})$ of the solution-structures is closed under the operation of finite union. Since $S(\mathbb{S})$ is a finite set, the union of all its elements also yields a solution-structure which is the "maximal" with respect to ordering relation "subgraph". The significance of this structure is revealed in the following definition.

Definition 3.3 Let $\mathbb{S} = (P, R, O)$ be a structural model of PNS. By the *maximal structure* of $\mathbb{S}$, we mean the solution-structure defined as

$$\mu(\mathbb{S}) = \bigcup_{(m,o) \in S(\mathbb{S})} (m, o).$$

If $S(\mathbb{S}) = \emptyset$, then $\mu(\mathbb{S}) = \emptyset$; as such, $\mu(\mathbb{S})$ is termed *degenerate*.

For instance, Fig. 3.6 shows the maximal structure of Example 3.2.

The above definition necessarily leads to the following observation.

Remark 3.1 For any structural model of PNS, the set of solution-structures is nonempty if and only if the maximal structure of this structural model is not degenerate.

Now, by letting o denote an arbitrary finite set of operating units, *i.e.*, α and β are finite sets for all $(\alpha, \beta) \in o$, we can assign the following functions to set o; $mat^{in}(o)$,

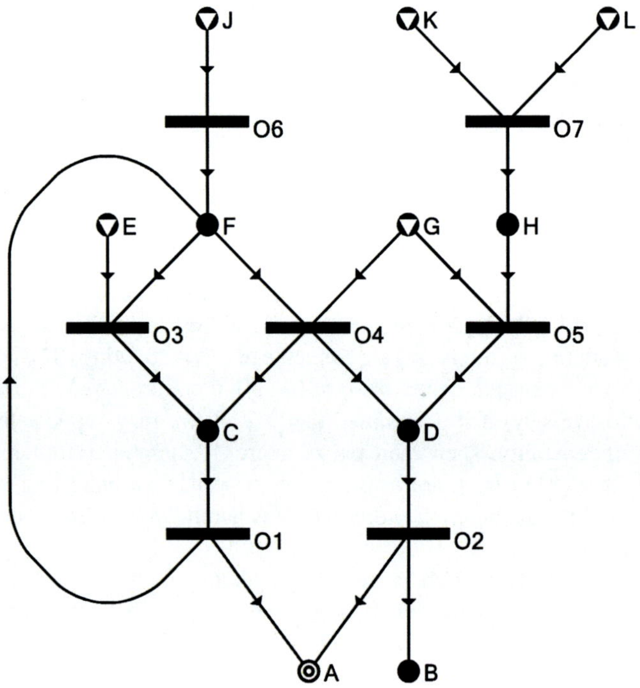

Fig. 3.6 Maximal structure for Example 3.2 [6]

the set of input materials to the operating units of set o; $mat^{out}(o)$, the set of output materials from the operating units of set o; and $mat(o)$, the set of all the materials associated with the operating units of set o.

These functions are defined as follows:

$mat^{in}(o) = \bigcup_{(\alpha,\beta)\in o} \alpha,$

$mat^{out}(o) = \bigcup_{(\alpha,\beta)\in o} \beta,$ and

$mat(o) = mat^{in}(o) \cup mat^{out}(o).$

It follows directly from these definitions that

$mat^{in}(o) \subseteq mat^{in}(o')$, $mat^{out}(o) \subseteq mat^{out}(o')$, $mat(o) \subseteq mat(o')$if $o \subseteq o'$ and $o, o' \subseteq O$. Given below are an application of the functions introduced above and an important property of the solution-structures.

Lemma 3.1 Let $\mathbb{S} = (P, R, O)$ be a structural model of PNS. If $(m, o) \in S(\mathbb{S})$, then $m = mat(o)$.

Proof First, it will be proved that $m \subseteq m'$ if $m' = mat(o)$. For this purpose, let $X \in m$ be an arbitrary material; then, axiom (S5) indicates that operating unit $(\alpha, \beta) \in o$

with $X \in \alpha \cup \beta$ exists since $(m, o) \in S(\mathbb{S})$. Hence, $X \in mat(o) = m'$, and consequently, $m \subseteq m'$.

Second, for verifying the inclusion, $m' \subseteq m$, let X be an arbitrary element of m'. Then, there exists operating unit $(\alpha, \beta) \in o$ with $X \in \alpha \cup \beta$. On the other hand, $(m, o) \in S(\mathbb{S})$, and therefore, (m, o) is a subgraph of (M, O). By definition, any subgraph of a P-graph is also a P-graph, and thus, (m, o) is a P-graph. Hence, $o \subseteq \wp'(m) \times \wp'(m)$, $\alpha \subseteq m$, and $\beta \subseteq m$, thereby resulting in $X \in m$. This completes the proof of the lemma.

It is worth noting that the above lemma has an interesting implication stated by the following remark.

Remark 3.2 Since m is a function of o, *i.e.*, $m = mat(o)$, solution-structure (m, o) of structural model $\mathbb{S}$ is uniquely determined by set o.

At this juncture, let us reexamine the set of solution-structures. If $\mathbb{S} = (P, R, O)$ is a structural model of PNS, then $S(\mathbb{S})$ is determined uniquely; however, the sets of solution-structures of different structural models can be identical. For instance, suppose that an additional operating unit, whose inputs and outputs are not shared by any operating unit in O, is incorporated into structural model $\mathbb{S}$; then, we obtain a new structural model $\mathbb{S}'$ with $S(\mathbb{S}) = S(\mathbb{S}')$. Consequently, a variety of structural models may result in an identical set of solution-structures; nevertheless, the degree of difficulty in determining the optimal structure may vary among these structural models. In practice, therefore, it is essential that the structural model having desirable properties for optimization, *e.g.*, minimal complexity, be identified. To analyze the structural models of PNS, their P-graphs should be explored first.

Let the set of all structural models on material set M^* of PNS be denoted by G_{M^*}. A binary relation is defined on G_{M^*} as follows.

For any structural models $\mathbb{S}, \mathbb{S}' \in G_{M^*}$ with $\mathbb{S} = (P, R, O)$ and $\mathbb{S}' = (P', R', O')$, $\mathbb{S}$ and $\mathbb{S}'$ are defined to be *equivalent* if the sets of products as well as the sets of solution-structures are identical for them, *i.e.*, $P = P'$ and $S(\mathbb{S}) = S(\mathbb{S}')$; this relation is denoted by $\mathbb{S} \sim \mathbb{S}'$. It can be shown that this relation is reflexive, symmetric, and transitive; hence, it is an equivalence relation on G_{M^*}. Thus, partitioning G_{M^*} by this relation into equivalence classes yields that the sets of solution-structures are identical for two structural models if they are involved in the same equivalence class. Nevertheless, the set of solution-structures can be empty for some equivalence classes. Since structural models of these equivalence classes have no practical significance, only equivalence classes for which the sets of solution-structures are nonempty will be considered.

Now, let G' denote an equivalence class of the partition of G_{M^*}, and assume that the set of solution-structures is nonempty for any structural model included in G'. An additional binary relation is defined on G' in the following way. For any pair of $\mathbb{S} = (P, R, O)$, $\mathbb{S}' = (P, R', O') \in G'$, we have $\mathbb{S} \leq \mathbb{S}'$ if $(M, O) \subseteq (M', O')$ and $R \subseteq R'$ where (M, O) and (M', O') are the P-graphs belonging to $\mathbb{S}$ and $\mathbb{S}'$, respectively. Naturally, this relation is reflexive, antisymmetric, and transitive; therefore, it is a partial ordering on G'.

The notions introduced above are illustrated with the following example.

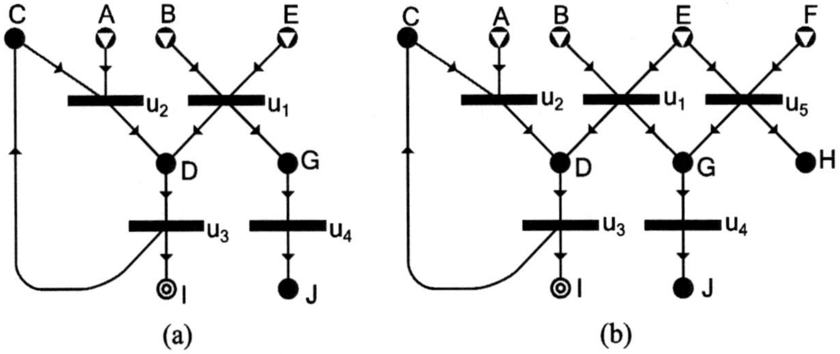

Fig. 3.7 P-graphs corresponding to (**a**) structural model of $\mathbb{S}$; and (**b**) structural model of $\mathbb{S}'$

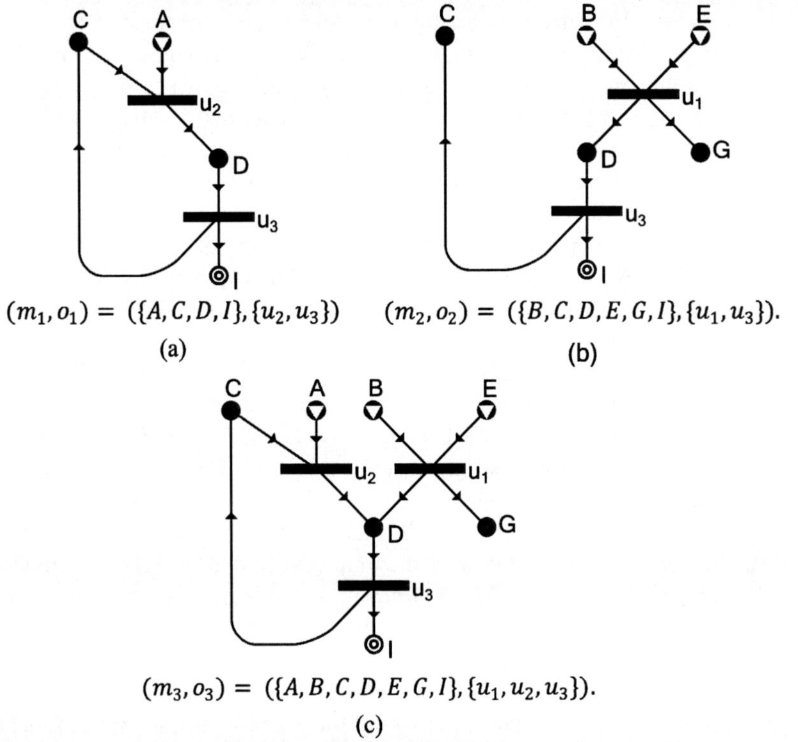

Fig. 3.8 Solution-structures of structural models $\mathbb{S}$ and $\mathbb{S}'$

Example 3.4 Let us consider the two structural models of PNS, $\mathbb{S} = (P, R, O)$ and $\mathbb{S}' = (P, R', O')$, with $P = \{I\}$, $R = \{A, B, E\}$, $R' = \{A, B, E, F\}$, $O = \{u_1, u_2, u_3, u_4\}$, $O' = \{u_1, u_2, u_3, u_4, u_5\}$ where the operating units are defined by $u_1 = (\{B, E\}, \{D, G\})$, $u_2 = (\{A, C\}, \{D\})$, $u_3 = (\{D\}, \{C, I\})$, $u_4 = (\{G\}, \{J\})$, and $u_5 = (\{E, F\}, \{G, H\})$.

P-graphs (M, O) and (M', O') belonging to $\mathbb{S}$ and $\mathbb{S}'$, respectively, are given in Fig. 3.7.

The three solution-structures of structural model $\mathbb{S}$ are the same as those of structural model $\mathbb{S}'$; they are presented in Fig. 3.8

Thus, $\mathbb{S}$ and $\mathbb{S}'$ are equivalent, *i.e.*, $\mathbb{S} \sim \mathbb{S}'$. On the other hand, $R \subseteq R'$, $(M, O) \subseteq (M', O')$, and hence, $\mathbb{S} \leq \mathbb{S}'$.

In connection with the above example, let us observe that since the maximal structure depends on the set of solution-structures, $S(\mathbb{S})$, both structural models have the same maximal structure, *i.e.*, $\mu(\mathbb{S}) = \mu(\mathbb{S}') = (m_3, o_3)$. Based on this common maximal structure, an additional structural model of PNS can be constructed, which is equivalent to structural models $\mathbb{S}$ and $\mathbb{S}'$. Indeed, for structural model $\overline{\mathbb{S}} = (\{I\}, \{A, B, E\}, \{u_1, u_2, u_3\})$, it can be readily seen that $\overline{\mathbb{S}} \sim \mathbb{S}$; moreover, $\overline{\mathbb{S}} \leq \mathbb{S}$ and $\overline{\mathbb{S}} \leq \mathbb{S}'$. By generalizing this idea, we shall construct the smallest element for a class of equivalent structural models of PNS.

Recall that G' denotes a class of equivalent structural models of PNS for which the set of solution-structures is nonempty. Let $\mathbb{S} = (P, R, O)$ be an arbitrary element of G'; then, $S(\mathbb{S}) \neq \varnothing$, and hence $\mu(\mathbb{S})$ is not degenerate in light of Remark 3.1. For $\mu(\mathbb{S}) = \left(\overline{M}, \overline{O}\right)$, let us form the triplet, $\overline{\mathbb{S}} = \left(P, \overline{R}, \overline{O}\right)$, where $\overline{R} = R \cap \overline{M}$. This leads to the following theorem.

Theorem 3.2 $\overline{\mathbb{S}} = \left(P, \overline{R}, \overline{O}\right)$ is a structural model of PNS, and it is equivalent to $\mathbb{S}$; moreover, $\overline{\mathbb{S}} \leq \mathbb{S}$.

Proof First, we prove that $\overline{\mathbb{S}}$ is a structural model of PNS. Since $\mathbb{S} = (P, R, O)$ is a structural model of PNS, $P \subseteq M^*$ and $O \subseteq \wp'(M^*) \times \wp'(M^*)$ are nonempty finite sets, $R \subseteq M^*$ is a finite set, $P \cap R = \varnothing$; furthermore, α and β are finite sets for all $u = (\alpha, \beta) \in O$. Since $\left(\overline{M}, \overline{O}\right)$ is a solution-structure of $\mathbb{S}$, $\left(\overline{M}, \overline{O}\right) \subseteq (M, O)$ and it is a P-graph; thus, $\overline{O}$ is a finite nonempty set. Hence, $\overline{O} \subseteq \wp'(M^*) \times \wp'(M^*)$ is a finite nonempty set; moreover, α and β are finite sets for all $u = (\alpha, \beta) \in \overline{O}$. Since $\overline{R} = R \cap \overline{M}$, $\overline{R} \subseteq M^*$, $\overline{R}$ is a finite set. Finally, $P \cap R = \varnothing$ implies $P \cap \overline{R} = \varnothing$. Consequently, $\overline{\mathbb{S}} = \left(P, \overline{R}, \overline{O}\right)$ is a structural model of PNS by definition.

It will now be proven that $S(\mathbb{S}) = S\left(\overline{\mathbb{S}}\right)$. For this purpose, let $(m, o) \in S\left(\overline{\mathbb{S}}\right)$ be an arbitrary solution-structure of $\overline{\mathbb{S}}$; then, (m, o) is a subgraph of $\left(\overline{M}, \overline{O}\right)$. On the other hand, $\left(\overline{M}, \overline{O}\right) \in S(\mathbb{S})$, and thus, $\left(\overline{M}, \overline{O}\right)$ is a subgraph of (M, O); therefore, (m, o) is also a subgraph of (M, O).

Next, it will be proven that (m, o) satisfies axioms (S1) through (S5) for PNS problem $\mathbb{S} = (P, R, O)$. Axioms (S1), (S3), (S4), and (S5) are obviously valid, because $(m, o) \in S\left(\overline{\mathbb{S}}\right)$. To verify axiom (S2), let $X \in m$ be an arbitrary material. If

$X \in R$, then, $X \in m \cap R \subseteq \overline{M} \cap R = \overline{R}$. Since $(m, o) \in S(\overline{\mathbb{S}})$, (m, o) satisfies axiom (S2) for PNS problem $(P, \overline{R}, \overline{O})$. Thus, $X \in \overline{R}$ implies that there is no arc (Y, X) in (m, o). Conversely, let us suppose that no (Y, X) arc exists in (m, o). Since $(m, o) \in S(\overline{\mathbb{S}})$, $X \in \overline{R} = R \cap \overline{M} \subseteq R$ in light of (S2).

Now, let $(m, o) \in S(\mathbb{S})$ be an arbitrary solution-structure of $\mathbb{S}$. Since $(\overline{M}, \overline{O})$ is the maximal structure of $\mathbb{S}$, $(m, o) \subseteq (\overline{M}, \overline{O})$. What follows show that (m, o) satisfies axioms (S1) through (S5) for PNS problem $\overline{\mathbb{S}} = (P, \overline{R}, \overline{O})$.

Axioms (S1), (S3), (S4), and (S5) are clearly valid, since $(m, o) \in S(\mathbb{S})$. Axiom (S2) will be verified in both directions. Since $\overline{R} \subseteq R$ and $(m, o) \in S(\mathbb{S})$, (m, o) satisfies axiom (S2) for structural model $\mathbb{S}$; this indicates that no (Y, X) arc exists in (m, o) for any element, X, of $\overline{R}$. Conversely, let us suppose that no (Y, X) arc exists in P-graph (m, o) for $X \in m$. Since $(m, o) \in S(\mathbb{S})$, $X \in R$ in light of (S2). Then, $X \in m \cap R \subseteq \overline{M} \cap R = \overline{R}$, implying that (m, o) satisfies (S2) for structural model $\overline{\mathbb{S}}$; therefore, $S(\mathbb{S}) = S(\overline{\mathbb{S}})$. Since the sets of products are identical for structural models $\mathbb{S}$ and $\overline{\mathbb{S}}$, these two structural models are equivalent, *i.e.*, $\mathbb{S} \sim \overline{\mathbb{S}}$. To complete the proof, we simply recall that $(\overline{M}, \overline{O}) \subseteq (M, O)$ and $\overline{R} \subseteq R$, thereby resulting in $\overline{\mathbb{S}} \leq \mathbb{S}$.

With regard to Theorem 3.2, note that structural model $\overline{\mathbb{S}}$ depends only on $\mu(\mathbb{S})$, the maximal structure of $\mathbb{S}$. Since $\mathbb{S}$ is an arbitrary element of equivalence class G' of the structural models and each structural model in class G' shares the same maximal structure, we have the following corollary.

Corollary 3.1 $\overline{\mathbb{S}} \leq \mathbb{S}$ is valid for all structural model $\mathbb{S}$ included in equivalence class G'; consequently, $\overline{\mathbb{S}}$ is the smallest element of G' in terms of the ordering introduced.

Definition 3.4 The structural model of PNS, $\mathbb{S} = (P, R, O)$, is the *reduced structural model of* PNS if $S(\mathbb{S}) \neq \emptyset$, and for any structural model, $\mathbb{S}'$, $\mathbb{S} \leq \mathbb{S}'$ provided that $\mathbb{S} \sim \mathbb{S}'$.

For any class G' of the equivalent structural models of PNS having a common nonempty set of solution-structures, a unique reduced structural model of PNS exists in light of Corollary 3.1. The construction of the reduced structural model can be guided by the preliminaries to Theorem 3.2.

To illustrate, let us revisit Example 3.4. The maximal structure is $(m_3, o_3) = (\{A, B, C, D, E, G, I\}, \{u_1, u_2, u_3\})$, and therefore, the reduced structural model of the example is

$$\overline{\mathbb{S}} = (\{I\}, \{A, B, E\}, \{u_1, u_2, u_3\}).$$

It is of utmost importance to question whether the above theoretical results can be exploited in solving a PNS problem. For the structural model of a given PNS problem, the following items need be accomplished.

(I) To ascertain if $S(\mathbb{S}) = \emptyset$.

(II) To construct the maximal structure of $\mathbb{S}$ to obtain the reduced structural model of $\mathbb{S}$ if $S(\mathbb{S}) \neq \emptyset$.

Chapter 4 aims at dealing with the above items. Specifically, an efficient algorithm is presented therein which either verifies if $S(\mathbb{S}) = \emptyset$ or generates the maximal structure of the structural model of a PNS problem.

References

1. Friedler F, Tarján K, Huang YW, Fan LT (1992) Graph-theoretic approach to process synthesis: Axioms and theorems. Chem Eng Sci 47:1973–1988
2. Kovács Z, Ercsey Z, Friedler F, Fan LT (2000) Separation-network synthesis: global optimum through rigorous super-structure. Comput Chem Eng 24(8):1881–1900
3. Kovacs Z, Friedler F, Fan LT (1993) Recycling in a separation process structure. AICHE J 39(6): 1087–1089
4. Friedler F (2006) Effective modeling for process synthesis. J Chin Inst Chem Engrs 37(5): 481–490
5. Hegyháti M, Majozi T, Holczinger T, Friedler F (2009) Practical infeasibility of cross-transfer in batch plants with complex recipes: S-graph vs MILP methods. Chem Eng Sci 64(3):605–610
6. Friedler F, Varga JB, Fan LT (1995) Decision-mapping: A tool for consistent and complete decisions in process synthesis. Chem Eng Sci 50:1755–1768
7. Friedler F, Tarján K, Huang YW, Fan LT (1992) Combinatorial algorithms for process synthesis. Comput Chem Eng 16(Supplement 1):S313–S320

Chapter 4
Algorithmic Generation of the Maximal Structure

Abstract Because the maximal structure gives an appropriate model of process network synthesis (PNS), the first step of process synthesis based on P-graphs is always the determination of the maximal structure. This chapter presents algorithm MSG, an efficient, i.e., polynomial algorithm, to generate the maximal structure. The operation of the algorithm is illustrated with examples and a software utility devised for demonstration.

Keywords P-graph · Maximal structure · Algorithm MSG · Combinatorial algorithm · Superstructure · Process synthesis

The maximal structure plays an essential role in PNS. The structure of any feasible solution of the PNS problem is a substructure of the maximal structure, and also, it is the smallest structure with this property. In addition, the maximal structure is the basis of the smallest structural model. Therefore, it is essential to have a rigorous and effective algorithm for its generation.

In the preceding chapter, it has been demonstrated that if a structural model of PNS is satisfactory, *i.e.*, at least one solution-structure exists, then the smallest structural model also exists, which is the reduced model with the same set of solution-structures (see Theorem 3.2, Corollary 3.1 and Definition 3.4). Note that smallest structural model means that it cannot be reduced further without excluding some solution-structures from consideration; consequently, the guarantee of optimality is lost with additional reduction in the model. On the other hand, only models for which solution-structures exist should be of interest from a practical point of view. Hence, it is of great importance to decide algorithmically with relatively small computational effort if there is at least one solution-structure for a given structural model of PNS (*i.e.*, $S(\mathbb{S})\neq\emptyset$). Herein, an efficient algorithm conceived by Friedler et al. [1, 2] is presented, which can evaluate if $S(\mathbb{S})$ is empty. If not, the algorithm generates the corresponding maximal structure.

This algorithm consists of two major phases, the reduction phase and the composition phase. The first phase is intended to reduce the set of available operating units by eliminating some or all inappropriate operating units. If no operating unit produces any of the products after this elimination, no solution-structure is available

F. Friedler et al., *P-graphs for Process Systems Engineering*,
https://doi.org/10.1007/978-3-030-92216-0_4

for the structural model of PNS under consideration; consequently, there is no maximal structure. If the maximal structure exists, it is generated in the composition phase of the algorithm as a P-graph from a subset of the operating units remaining from the reduction phase. To elucidate this procedure, the structural model of PNS be given by $\mathbb{S} = (P, R, O)$.

Algorithm for Maximal Structure Generation (Algorithm MSG)

1. Reduction

   Initialization

   Omit every operating unit from set O producing at least one raw material and denote the remaining set by O_0, *i.e.*, let $O_0 = O\{(\alpha, \beta) : (\alpha, \beta) \in O \,\&\, \beta \cap R = \varnothing\}$. Let M_0 be the set of input and output materials of operating units in O_0, *i.e.*, $M_0 = mat(O_0)$. If none of the remaining units in O_0 can produce the desired products, *i.e.*, $P \nsubseteq M_0$, then terminate indicating that no maximal structure exists for structural model $\mathbb{S}$. Otherwise, collect into set T_0 those materials from M_0 that are neither produced by the operating units in O_0 nor are raw materials.

$$T_0 = \{X : X \in M_0 \backslash R \& ((\alpha, \beta) \in O_0 \longrightarrow X \notin \beta)\}.$$

   As these materials are consumed but not produced by any operating unit and they are not defined as raw materials for the PNS, they must be excluded in subsequent steps of the algorithm. Finally, set $r := 0$.

   Iteration

   - *Step 1.1.* Proceed to the initialization of the composition phase if $T_r = \varnothing$. If not, choose one material from T_r; and denote it by X. As material X belongs to the group of materials to be excluded, the operating units that require it must also be removed. Therefore, collect those operating units from O_r, that consume X; denote it by O_X, *i.e.*, $O_X = \{(\alpha, \beta) : (\alpha, \beta) \in O_r \,\&\, X \in \alpha\}$. Let $O_{r+1} = O_r \backslash O_X$ and let $M_{r+1} = mat(O_{r+1})$. It is necessary to verify if the desired products can still be generated, *i.e.*, if $P \nsubseteq M_{r+1}$, then terminate the iteration since there is no maximal structure for $\mathbb{S}$. If $P \subseteq M_{r+1}$, collect into set $T_{r'}$ every material that is consumed by the remaining units (O_{r+1}) but is solely produced by the units in the set of excluded units O_X. In other words,

$$T_{r'} = \{Y : Y \in mat^{out}(O_X) \& Y \notin mat^{out}(O_{r+1}) \& Y \in mat^{in}(O_{r+1})\}.$$

     These materials are to be excluded as their production becomes infeasible once the operating units in O_X are eliminated. Consequently, the new set of materials to be excluded, T_{r+1}, is formed by the union of the elements of T_r that are comprised in the remaining materials (M_{r+1}), and the new set of materials to be omitted, $T_{r'}$. Thus, let $T_{r+1} = (T_r \cap M_{r+1}) \cup T_{r'}$, $r := r+1$. Then, proceed to the following iteration for reduction.

2. Composition

   Initialization

- This stage considers three sets, the set of required materials, W_s; the set of included materials, m_s; and the set of included operating units, o_s. For initialization, let $W_0 = P$; $m_0 = \emptyset$; $o_0 = \emptyset$; and set $s := 0$.

  Iteration
- *Step 2.1.* Terminate if $W_s = \emptyset$. There exists at least one solution-structure for $\mathbb{S}$. In particular, $(\overline{m}, o_s)$ is the maximal structure of $\mathbb{S}$ where $\overline{m} = mat(o_s)$. Proceed to Step 2.2 if $W_s \neq \emptyset$.
- *Step 2.2.* Select one material from the set of required materials W_s; denote it by X; and let $m_{s+1} = m_s \cup \{X\}$. Then, form the set O_X^* of those operating units from O_r, that produce X, *i.e.*, let

$$O_X^* = \{(\alpha, \beta) : (\alpha, \beta) \in O_r \& X \in \beta\}.$$

- These operating units should be incorporated into the set of included units for the next iteration (o_{s+1}), *i.e.*, let $o_{s+1} = o_s \cup O_X^*$. Additionally, update the set W_s, incorporating the materials consumed by the units in O_X^* that are not raw materials and have not been already included in m_s. Let $W_{s+1} = \left(W_s \cup mat^{in}\left(O_X^*\right)\right) \backslash \left(R \cup m_{s+1}\right)$, set $s := s + 1$, and proceed to the succeeding iteration for composition.

Prior to validation, the algorithm is illustrated with the following examples.

Example 4.1 Let us consider structural model $\mathbb{S} = (P, R, O)$ of PNS where $P = \{B\}$; $R = \{F, H, M, T\}$; and $O = \{u_1, u_2, \ldots, u_{11}\}$. The operating units contained in O are listed in Table 4.1 (the example is available in the demonstration software, see Fig. 4.14); the corresponding P-graph is shown in Fig. 4.1.

Table 4.1 Operating units for Example 4.1

unit	inputs	outputs
u_1	*F, C, D*	*A*
u_2	*D*	*B, G*
u_3	*E*	*B, U*
u_4	*F, G*	*C, D*
u_5	*G, H*	*D*
u_6	*H, I*	*E*
u_7	*J, K*	*E*
u_8	*M*	*G*
u_9	*N, Q*	*H*
u_{10}	*T, U*	*I*
u_{11}	*V*	*J*

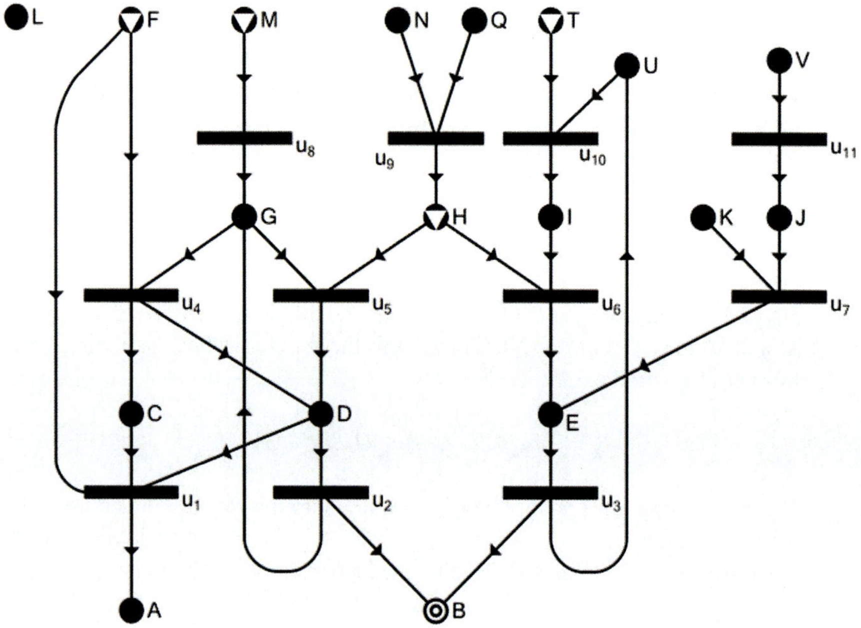

Fig. 4.1 P-graph of Example 4.1[2]

In the first step of the reduction phase, operating unit u_9 is eliminated since it produces a raw material, specifically, H; these yields $O_0 = \{u_1, u_2, u_3, u_4, u_5, u_6, u_7, u_8, u_{10}, u_{11}\}$ and $M_0 = \{A, B, C, D, E, F, G, H, I, J, K, M, T, U, V\}$ (Fig. 4.2).

Let us choose a material, *e.g.*, material V, from set T_0 which is obviously not empty, *i.e.*, $X = V$. Only operation u_{11} consumes this material; thus, $O_V = \{u_{11}\}$. This unit is excluded from the set of operating units for the next iteration, consequently $O_1 = \{u_1, u_2, u_3, u_4, u_5, u_6, u_7, u_8, u_{10}\}$. In addition, material V is also excluded from the P-graph, *i.e.*, $M_1 = \{A, B, C, D, E, F, G, H, I, J, K, M, T, U\}$. In forming set $T_{0'}$, it is recognized to be comprised by material J, because the elimination of u_{11} conveys that no operation renders this material. The set T_1 is formed by $T_1 = (T_0 \cap M_1) \cup T_{0'}$, *i.e.*, $T_1 = (\{K, V\} \cap \{A, B, C, D, E, F, G, H, I, J, K, M, T, U\}) \cup \{J\}$, thereby yielding $T_1 = \{J, K\}$. The result of this step is depicted as a P-graph in Fig. 4.3.

Now, we can proceed to the next iteration step, *i.e.*, $r = 1$. In this step, lets choose material J ($X = J$). The set of units that consumes material J is comprised uniquely by operating unit u_7, therefore $O_J = u_7$. This operation is excluded for the next iteration, thus the set $O_2 = \{u_1, u_2, u_3, u_4, u_5, u_6, u_8, u_{10}\}$, and the set $M_2 = \{A, B, C, D, E, F, G, H, I, M, T, U\}$. From here, $T_{1'}$ is identified as the empty set since material E produced by u_7 can also be generated from u_6; therefore, it does not need to be excluded. Because both materials in $T_1 = \{J, K\}$ are excluded simultaneously with operation u_7, the set T_2 is empty; $T_2 = (\{J, K\} \cap \{A, B, C, D, E, F, G, H, I, M, T, U\}) \cup T_{1'} = \emptyset$. Consequently, the reduction phase of the procedure is completed, and its outcome is

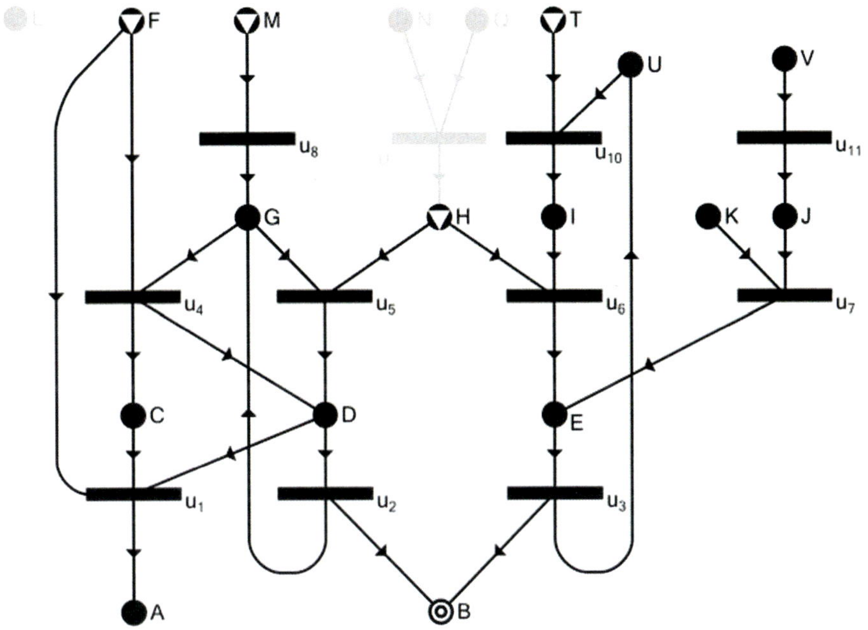

Fig. 4.2 P-graph of the reduction phase's first step of algorithm MSG in Example 4.1

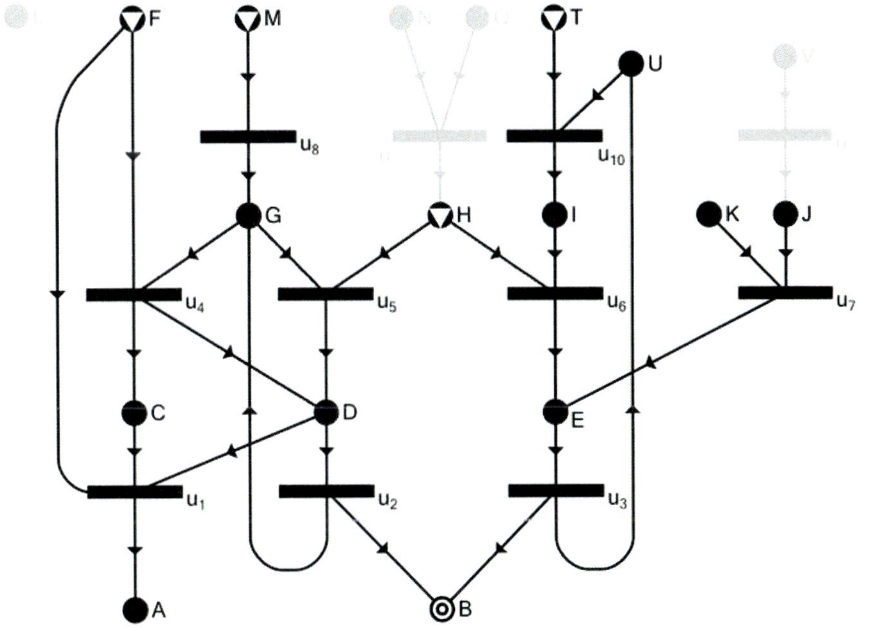

Fig. 4.3 P-graph of the reduction phase's first iteration of algorithm MSG in Example 4.1

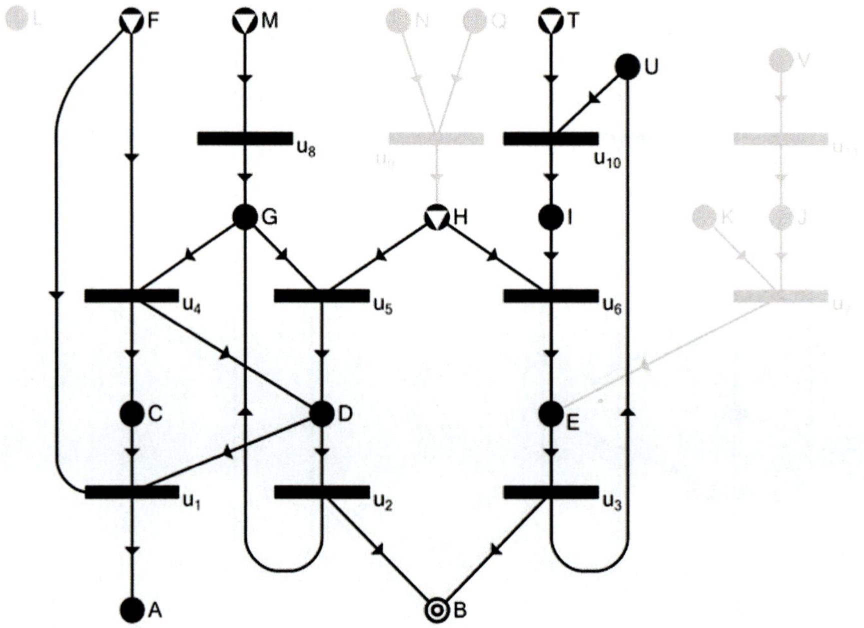

Fig. 4.4 P-graph of the reduction phase's outcome of algorithm MSG in Example 4.1

$$M_2 = \{A, B, C, D, E, F, G, H, I, M, T, U\}$$

and

$$O_2 = \{u_1, u_2, u_3, u_4, u_5, u_6, u_8, u_{10}\}.$$

The sets of nodes (M_2, O_2) are depicted as a P-graph in Fig. 4.4.

By entering in the composition phase of the procedure, we have that in the initialization step, the set of required materials only contains the desired products, and no material or operating unit has been included, *i.e.*, $W_0 = \{B\}$; $m_0 = \varnothing$; and $o_0 = \varnothing$. Figure 4.5 depicts the set of nodes rendered by the reduction phase in grey, from where the maximal structure is constructed.

Naturally, B needs to be selected from W_0. Thus, this material is included in the solution, *i.e.*, $m_1 = \{B\}$. Material B can be produced by units u_2 and u_3, therefore $O_B^* = \{u_2, u_3\}$. These units are included in the solution-structure as $o_1 = \{u_2, u_3\}$. In a subsequent step, the set of required materials is updated by means of $W_{s+1} = \left(W_s \cup mat^{in}\left(O_X^*\right)\right) \setminus (R \cup m_{s+1})$, which includes the materials consumed by u_2 and u_3 that are not raw materials and have not been incorporated yet. Hence, $W_1 = (\{B\} \cup \{D, E\}) \setminus (\{F, H, M, T\} \cup \{B\})$, and $W_1 = \{D, E\}$. The result of the first step of the composition phase is depicted in Fig. 4.6. In this figure, the included nodes are presented in black.

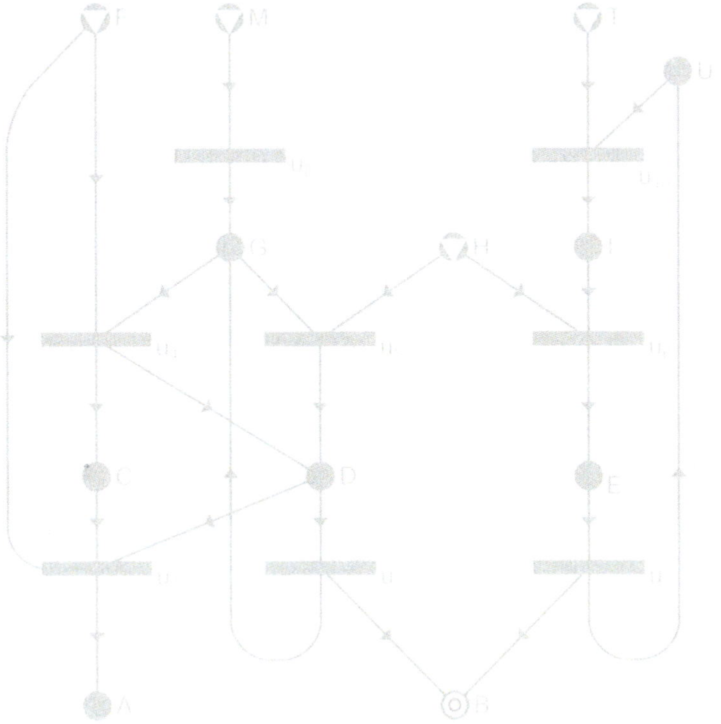

Fig. 4.5 P-graph of the initial nodes for composition phase of algorithm MSG in Example 4.1

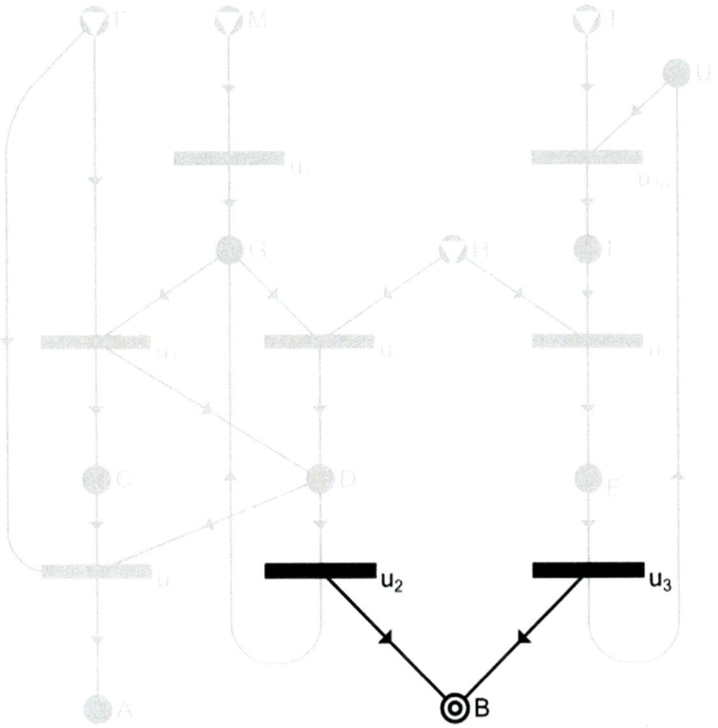

Fig. 4.6 P-graph of the first iteration in the composition phase of algorithm MSG for Example 4.1

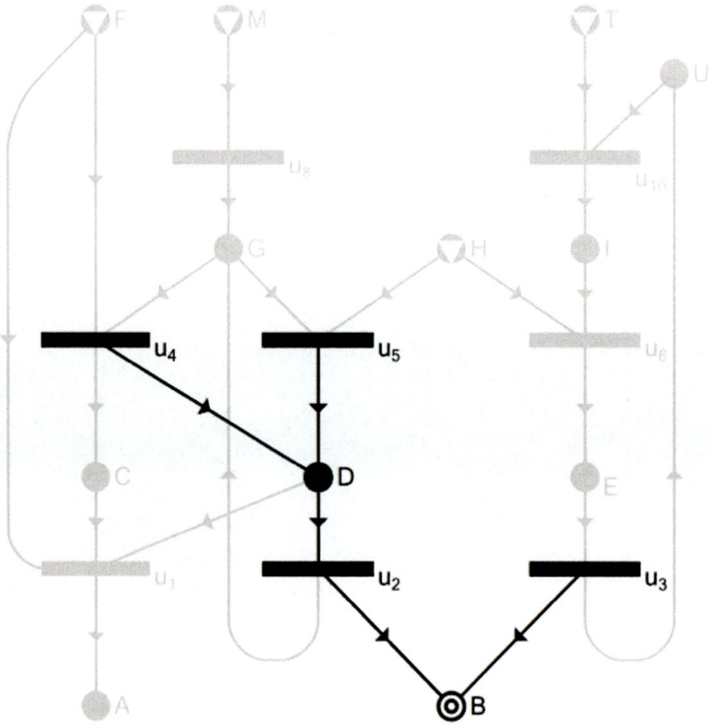

Fig. 4.7 P-graph of the second iteration in the composition phase of algorithm MSG for Example 4.1

In the succeeding iteration of the composition phase, *i.e.*, $s = 1$, by selecting material D from W_1, we have $m_2 = \{B, D\}$. The set of units that can produce D is $O_D^* = \{u_4, u_5\}$; then $o_2 = \{u_2, u_3, u_4, u_5\}$; and only G is included in the set W_2 provided that F and H are raw materials. Therefore, $W_2 = \{E, G\}$. The result of this step is depicted in Fig. 4.7.

Moreover, in the next iteration step, by choosing E from W_2, we obtain $m_3 = \{B, D, E\}$, and $O_E^* = \{u_6\}$. This unit is included in the solution, thereby yielding $o_3 = \{u_2, u_3, u_4, u_5, u_6\}$ and $W_3 = \{G, I\}$. The current state of the solution is depicted in Fig. 4.8.

In a subsequent iteration step, let us select material G from W_3. This leads to $m_4 = \{B, D, E, G\}$. The operating units that can produce G are u_2 and u_8, then $O_G^* = \{u_2, u_8\}$. Only u_8 is included in the set of operations for the solution $o_4 = \{u_2, u_3, u_4, u_5, u_6, u_8\}$ as u_2 had been included in step 1. On the other hand, the new set of required materials is comprised only by I, *i.e.*, $W_4 = \{I\}$. Note that no additional material is included in W_4, as the materials consumed by operations in O_G^* have been either already included in m_4 (material D) or defined as raw materials (material M). The current state of the solutions is illustrated in Fig. 4.9.

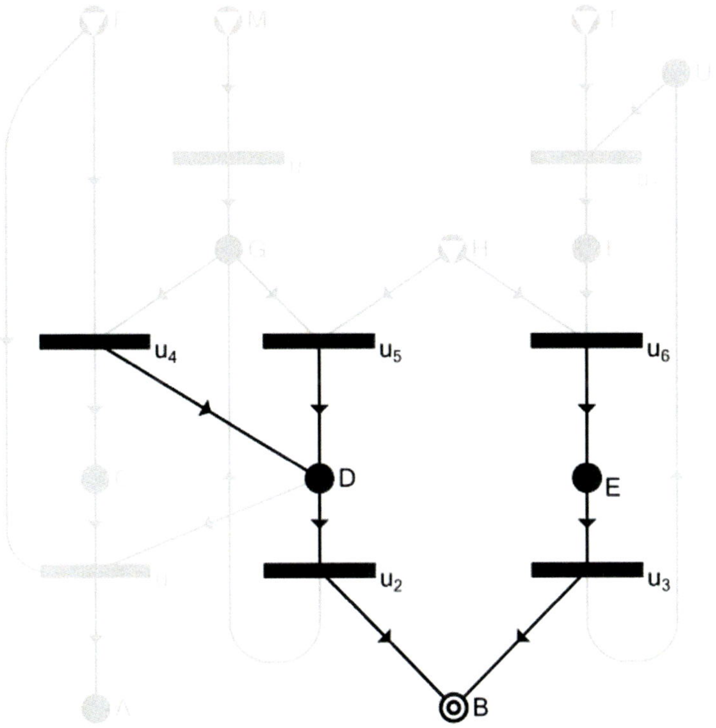

Fig. 4.8 P-graph of the third iteration in the composition phase of algorithm MSG for Example 4.1

Naturally the next iteration step requires to select I from W_4, from here $m_5 = \{B, D, E, G, I\}$; $O_I^* = \{u_{10}\}$; $o_5 = \{u_2, u_3, u_4, u_5, u_6, u_8, u_{10}\}$ and $W_5 = U$. This step renders the structure depicted in Fig. 4.10.

In the next iteration, only U is available to be selected, then $m_6 = \{B, D, E, G, I, U\}$. This is the last iteration as $O_U^* = \{u_3\}$; $o_6 = \{u_2, u_3, u_4, u_5, u_6, u_8, u_{10}\}$ and $W_6 = \varnothing$, thus terminating the procedure. The resulting maximal structure consists of the operating units included in o_6 and their concomitant materials. This structure is displayed in Fig. 4.11.

Since the reduced structural model of a PNS problem is based on the maximal structure, the reduced structural model of the PNS problem of Example 4.1 is $\overline{\mathbb{S}} = (P, \overline{R}, \overline{O})$, where $\overline{R} = \{F, H, M, T\}$ and $\overline{O} = \{u_2, u_3, u_4, u_5, u_6, u_8, u_{10}\}$. The P-graph of this reduced model is the same as the maximal structure of the original PNS problem (see Fig. 4.11).

The following simple example illustrates a structural model yielding no solution-structure.

Example 4.2 For structural model $\mathbb{S} = (P, R, O)$, let $P = \{A_2\}$, $R = \{A_6, A_7, A_8, A_9\}$, and $O = \{u_1, u_2, u_3, u_4\}$, where $u_1 = (\{A_3\}, \{A_1\})$, $u_2 = (\{A_5\}, \{A_2\})$,

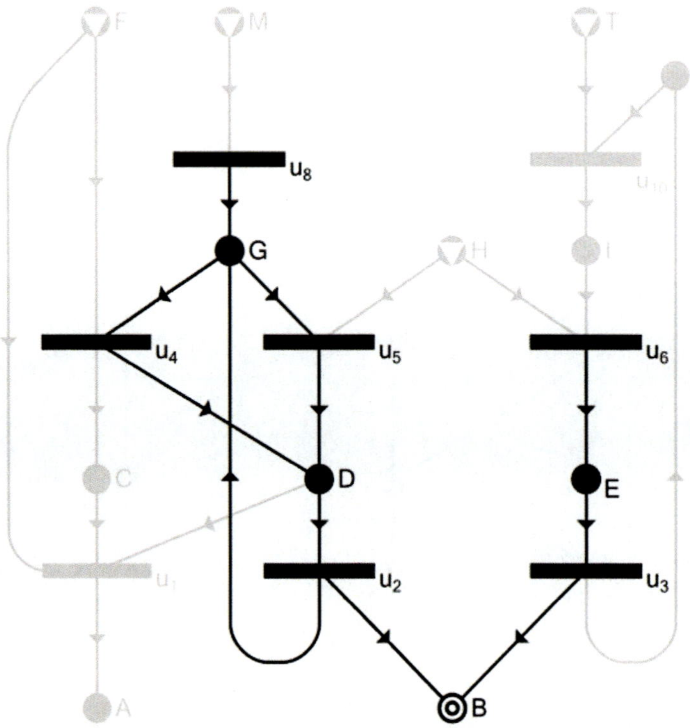

Fig. 4.9 P-graph of the fourth iteration in the composition phase of algorithm MSG for Example 4.1

$u_3 = (\{A_6, A_7\}, \{A_3, A_4\})$, and $u_4 = (\{A_8\}, \{A_5, A_7\})$. The corresponding P-graph is illustrated in Fig. 4.12.

In the first step, operating unit u_4 is eliminated since it produces a raw material, specifically A_7. This gives rise to the set of remaining units O_0 as $O_0 = \{u_1, u_2, u_3\}$, the set of remaining materials M_0 as $M_0 = \{A_1, A_2, \ldots, A_7\}$, and $T_0 = \{A_5\}$. Selecting material A_5 from set T_0 leads to the elimination of operating units that consume it, *i.e.*, sets $O_{A_5} = \{u_2\}$, therefore $O_1 = \{u_1, u_3\}$, and $M_1 = \{A_1, A_3, A_4, A_6, A_7\}$. Since the product A_2 is not included in the set of materials anymore, $P \nsubseteq M_1$, and the procedure terminates; there is no maximal structure for structural model $\mathbb{S}$.

As will be demonstrated in Chap. 10, algorithm MSG is capable of rigorously and efficiently generating the maximal structure independently of the complexity of the problems. As an illustration, Fig. 4.13 depicts the maximal structure generated by algorithm MSG for the industrial problem presented in Example 3.3 of the previous chapter. This example is available in the demonstration software, for details see Fig. 4.14.

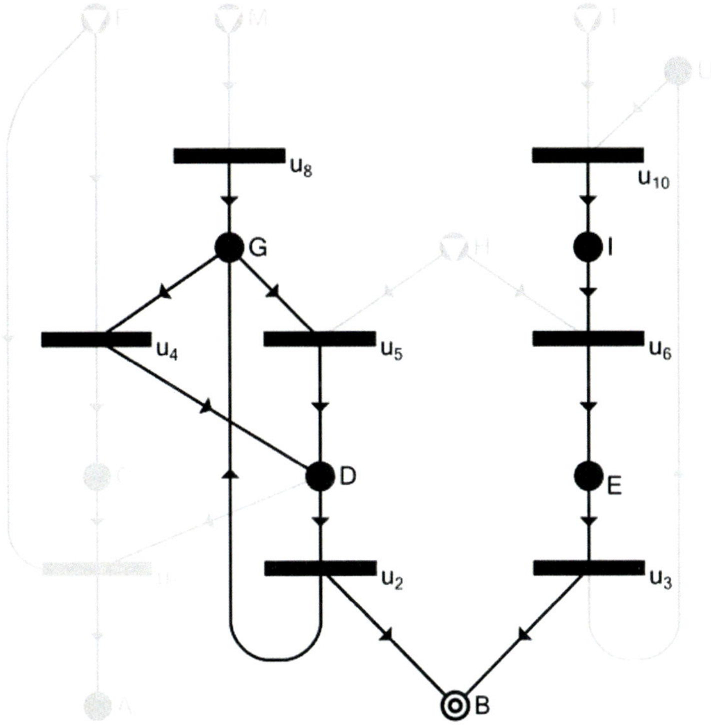

Fig. 4.10 P-graph of the fourth iteration in the composition phase of algorithm MSG for Example 4.1

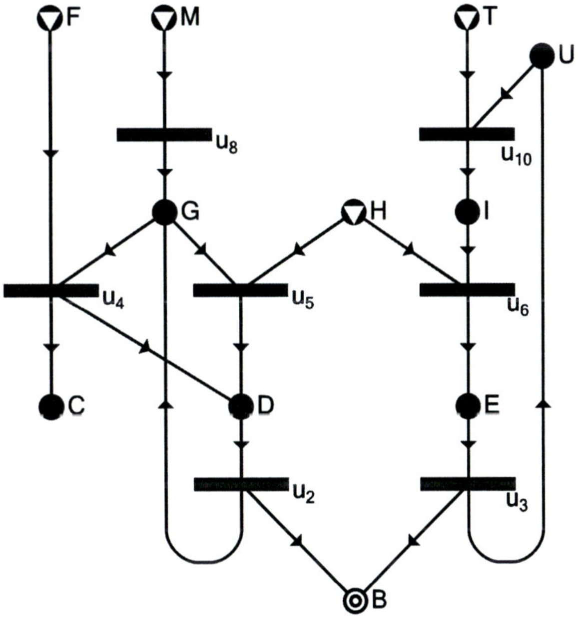

Fig. 4.11 Maximal structure of Example 4.1

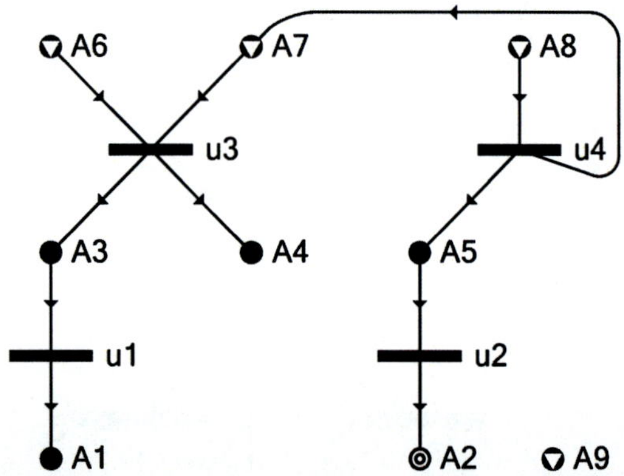

Fig. 4.12 P-graph of Example 4.2

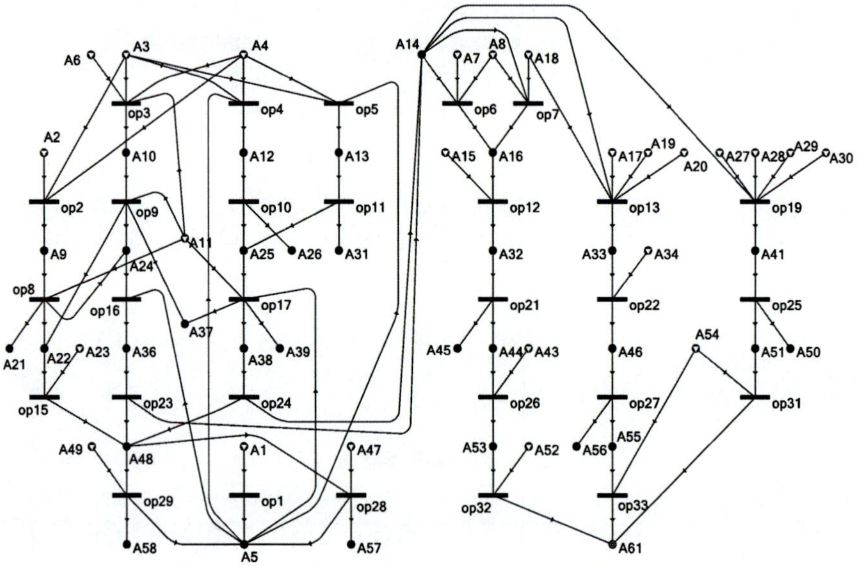

Fig. 4.13 Maximal structure of Example 3.3 generated by algorithm MSG

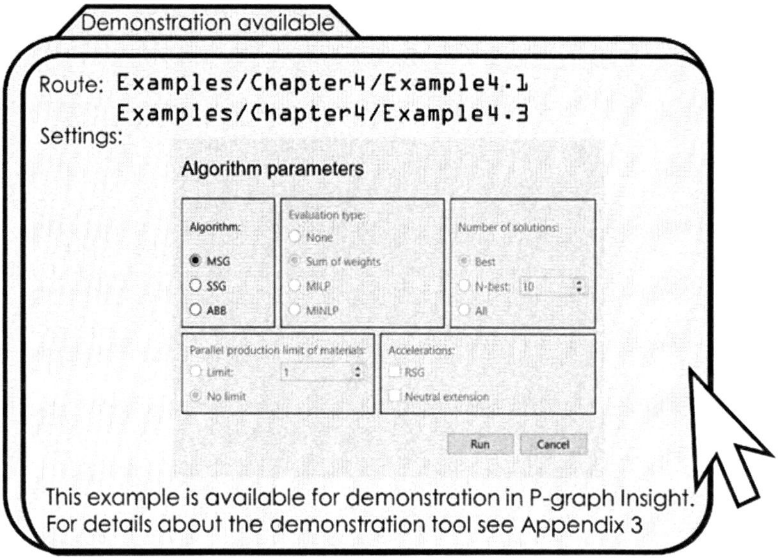

Fig. 4.14 Configuration of demonstration software for generating the maximal structure of Examples 4.1 and 4.3

References

1. Friedler F, Tarján K, Huang YW, Fan LT (1992) Combinatorial algorithms for process synthesis. Comput Chem Eng 16(Suppl):S313–S320
2. Friedler F, Tarján K, Huang YW, Fan LT (1993) Graph-theoretic approach to process synthesis: Polynomial algorithm for maximal structure generation. Comput Chem Eng 17:929–942

Chapter 5
Algorithmic Generation of all Solution-Structures

Abstract The set of all solution-structures of a process network synthesis (PNS) problem always contains the best, or n-best, solutions. If the maximal structure exists for a PNS problem, algorithm SSG, presented in this chapter, is capable of generating all solution-structures exactly once. The best solution, or n-best solutions, of a synthesis problem can always be determined by evaluating all structures generated by SSG. Illustrative examples and the demonstration software show the operation of the algorithm.

Keywords P-graph · Algorithm SSG · Recursive algorithm · Solution-structure · n-best · Process synthesis

One of the ways of determining the globally optimal solution of a PNS problem is to optimize all solution-structures and select the best of them via an enumeration algorithm. Since no process can be feasible unless its structure is a solution-structure of the problem, this enumeration algorithm guarantees the determination of the optimal solution provided that the optimization of the individual solution-structures is correct. Naturally, all commercially or publicly available process analysis tools, including simulation programs, can be used for that purpose. In addition to the generation of the best solution, the guaranteed n-best solutions can also be generated by this enumeration method.

Suppose that the maximal structure of a PNS problem is available, then the objective now is to generate each solution-structure exactly once for any PNS problem, *i.e.*, no limit on the complexity of the structures, number of loops, number of products, *etc.* To satisfy all these requirements, an additional tool, decision-mapping, must be introduced for appropriately managing the systems of decisions. Prior to the formal description of the decision-mapping, examples will illustrate its necessity. Suppose that material A can be produced by three operating units (as seen in Fig. 5.1). Then, for the generation of the whole set of solution-structures, all options for the production of A are to be examined; in this case, the number of options is $(2^3 - 1) = 7$. Figure 5.2 shows the seven options that the structure-

F. Friedler et al., *P-graphs for Process Systems Engineering*,
https://doi.org/10.1007/978-3-030-92216-0_5

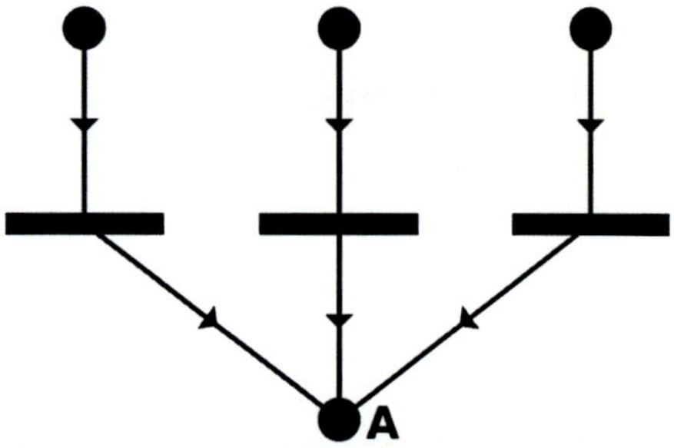

Fig. 5.1 Part of a maximal structure: several options for the production of material *A*

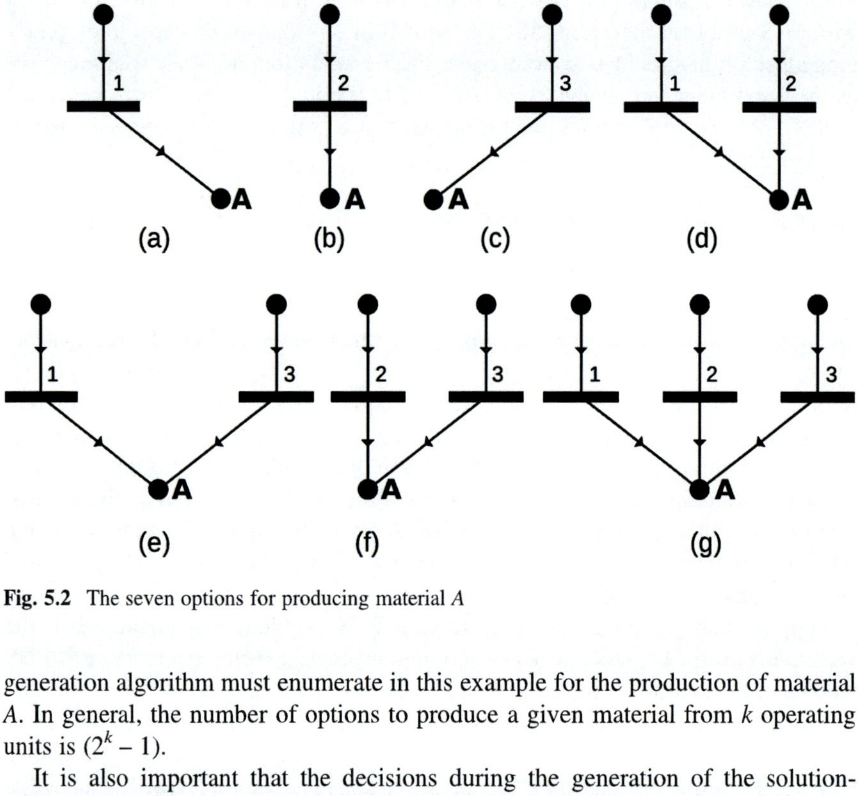

Fig. 5.2 The seven options for producing material *A*

generation algorithm must enumerate in this example for the production of material *A*. In general, the number of options to produce a given material from *k* operating units is $(2^k - 1)$.

It is also important that the decisions during the generation of the solution-structures be made not only exhaustively but also consistently. For illustration, let us analyze the PNS problem given by the maximal structure depicted in Fig. 5.3 [1]. See the end of the chapter for the details on the demonstration software.

Fig. 5.3 Maximal structure of the PNS problem illustrating the decision-system in generating all solution-structures

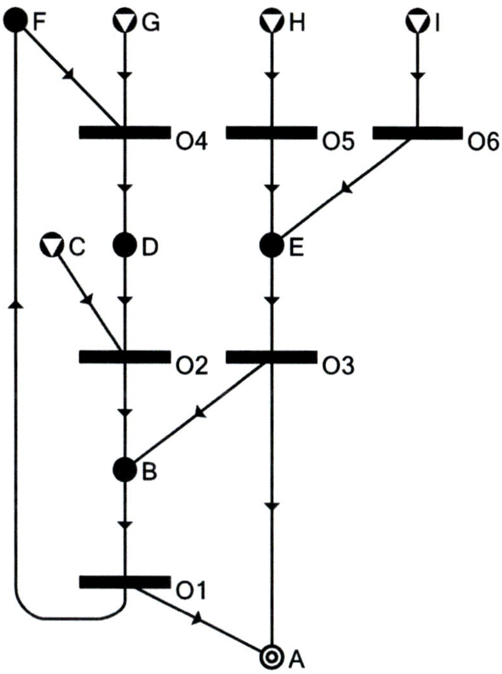

Initially, we have to make a decision on the single product, material *A*. This material can be produced by two operating units ($O1$, $O3$); thus, we have to examine 3 decision options for the production of *A* (see Fig. 5.4), *i.e.*, its production by $O1$, its production by $O2$, and its simultaneous production by $O1$ and $O2$. First, let us suppose that *A* is produced only by operating unit $O1$, *i.e.*, $O3$ is excluded from consideration. By the exclusion of $O3$, units $O5$ and $O6$ become useless, and they are also excluded as shown in Fig. 5.5a. Then, the decision relies on the production of material *B*. Since only one operating unit produces material *B*, this unit must be included in the structure as seen in Fig. 5.5b. Similar cases arise at materials *D* and *F* as indicated in Fig. 5.5c. Please note that material *F* is only produced by $O1$, which had already been included in a previous step. Finally, Fig. 5.5d shows the first generated solution-structure.

Then, back to an unexamined decision at *A*, its production is now by $O3$ and excluding $O1$, the next decision is to be made at material *E* (see Fig. 5.6).

For the production of *E*, we have again three options as shown in Fig. 5.7. Each resultant structure is a solution-structure.

Back again to material *A*, we have one remaining option to produce *A*, *i.e.*, it is produced by $O1$ and $O3$ simultaneously (Fig. 5.8).

Then, the production of *B* and *E* are to be decided. Let us make it first for *B*. In this case, we have only two options even though there are two operating units for its

Fig. 5.4 Decision at material *A*

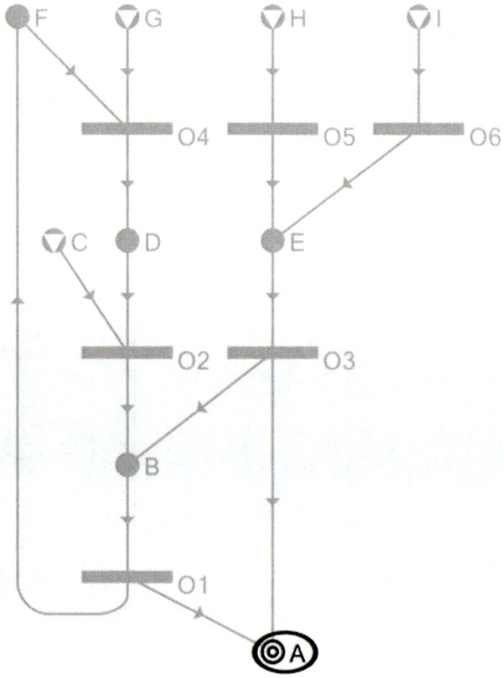

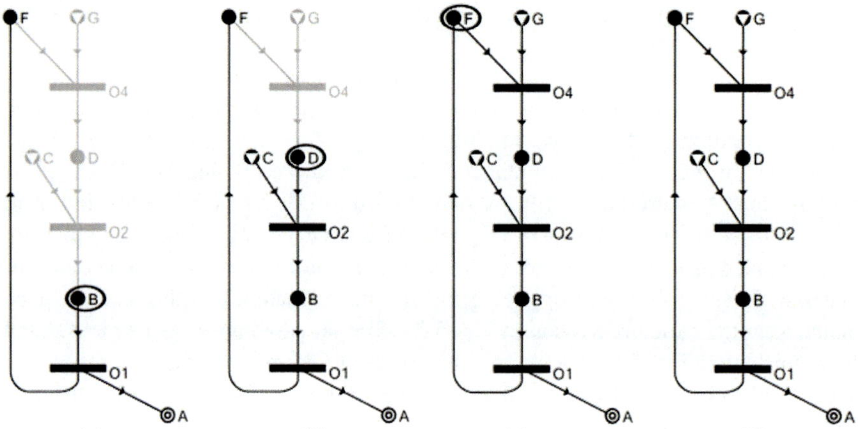

Fig. 5.5 The next four steps are without decision, only one operating unit can produce materials *B*, *D*, and *F* individually

production. Since the involvement of *O*3 is forced by the previous decision, our option is to include *O*2 or exclude it. First, let us exclude *O*2, then the decision is to be made at *E* (see Fig. 5.9). Then we have again three options as shown in Fig. 5.10, each of them represents a solution-structure.

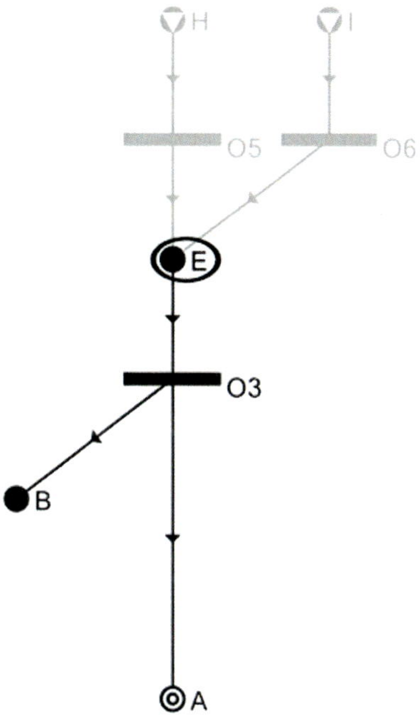

Fig. 5.6 Material *A* is to be produced by *O*3

Then, it is the turn of material *B* for a decision where we have one unexamined option, *i.e.*, the inclusion of *O*2 (Fig. 5.11).

The next steps are performed similarly; the corresponding P-graphs are given in Fig. 5.12.

It is easy to recognize that the system of decisions must be well organized for generating all solution-structures exactly once, which may be difficult even for simple examples. To provide a well established solution procedure for this issue, a new notion, decision-mapping, is introduced. The decision-mapping will greatly facilitate the decision systems in generating all solution-structures or in developing a highly effective branch-and-bound technique for determining the best solution.

Decision-mapping

A decision-mapping can be visualized as a sequence of decisions, each of which is concerned with a single material involved in the process being synthesized; it identifies the set of operating units which are to be used for producing directly the material under consideration. Decisions contained in such a sequence are independent, and each decision is material specific.

The concept of decision-mapping was originally introduced and studied in the work of Friedler et al. [2], a modified version of which is presented here. For the

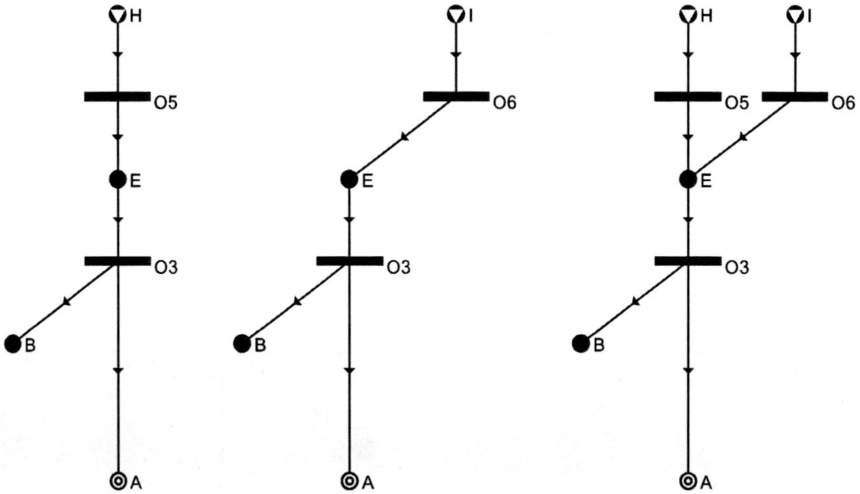

Fig. 5.7 Three options for producing material E

Fig. 5.8 Material A is to be produced by $O1$ and $O3$ simultaneously

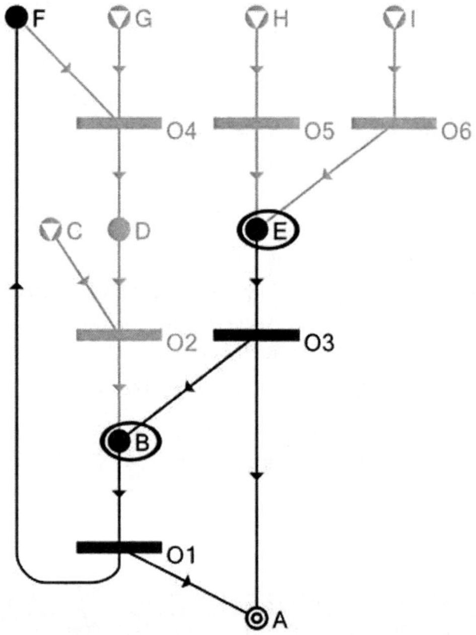

formal definition, let $\mathbb{S} = (P, R, O)$ be a reduced structural model of PNS. Then, the maximal structure, (M, O), of $\mathbb{S}$ determines a function Δ of $M \backslash R$ into $\wp'(O)$ as follows: For any material $X \in M \backslash R$, let

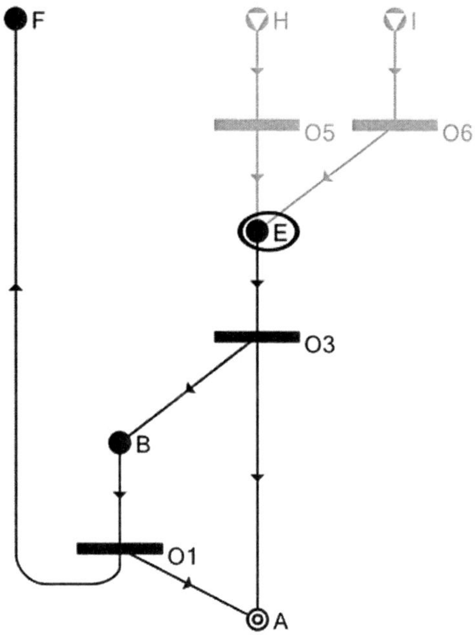

Fig. 5.9 *O*2 is excluded, thus implying the exclusion of *O*4

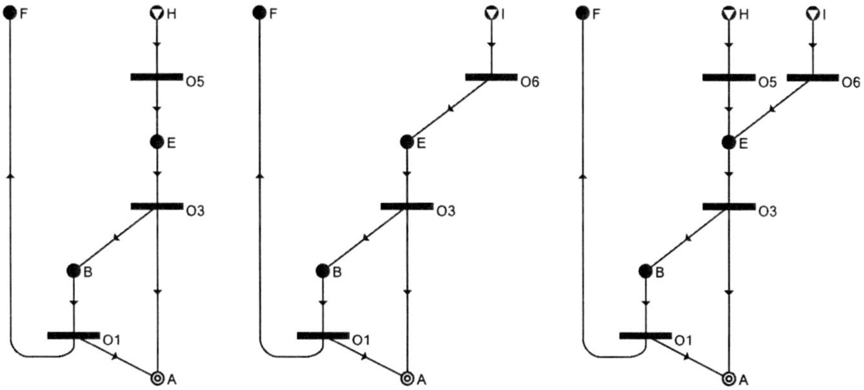

Fig. 5.10 Three options for the production of material *E*

$$\Delta(X) = \{(\alpha,\beta) : (\alpha,\beta) \in O \ \& \ X \in \beta\}.$$

Since (M, O) is the maximal structure of $\mathbb{S}$ satisfying axiom (S2), $\Delta(X) \neq \emptyset$ is valid for all $X \in M \backslash R$. Function $\Delta(X)$ can be visualized in the following way. For any $X \in M \backslash R$, $\Delta(X)$ gives the set of those operating units of the maximal structure, which directly participate in producing X.

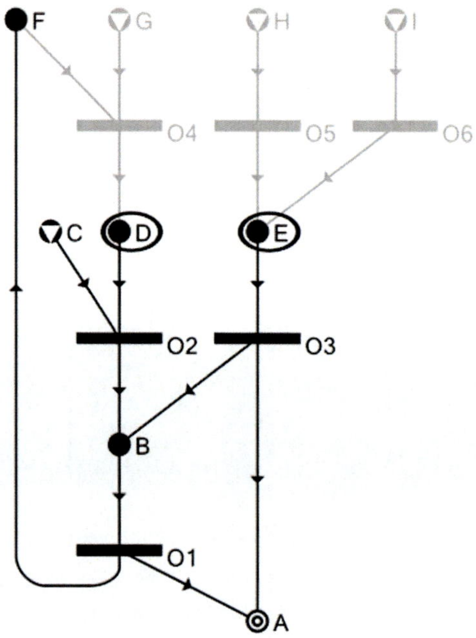

Fig. 5.11 Last decision at B to include $O2$

Definition 5.1 Let m be a subset of $M \backslash R$; furthermore, let $\delta(X)$ be a subset of $\Delta(X)$ for each $X \in m$. Then, mapping δ from set m into the set of subsets of O, $\delta[m] = \{(X, \delta(X)) : X \in m\}$, is called a *decision-mapping belonging* to $\mathbb{S}$. By the *rank* of $\delta[m]$ we mean $|m|$. Provided that no confusion arises, a decision-mapping belonging to $\mathbb{S}$ is simply referred to as a decision-mapping. In practice, a decision-mapping determines the set of operating units of a process to be synthesized. More precisely, if $X \in m$, then the operating units in set $\delta(X)$ are included in the process given by $\delta[m]$ for the production X; however, no operating units in set $\Delta(X) \backslash \delta(X)$ are included in this process even if they also produce material X.

Algorithm SSG [3], the algorithm for generating all solution-structures, is formulated on the decision-mapping. If a maximal structure is given, this algorithm is capable of determining all solution-structures of a PNS problem exactly once independently of the complexity of the maximal structure and the number of products and raw materials. The algorithm is shown in Fig. 5.13; note that it is a recursive algorithm.

This algorithm starts by determining the set of decisions that can be made for generating a product x in set p, *i.e.*, $C = \wp(\Delta(x)) \backslash \emptyset$. For instance, $C = \{\{O1\}, \{O2\}, \{O1, O2\}\}$ for material A in Fig. 5.3. This means that we can decide to produce material A solely by $O1$, solely by $O2$, or by both of them simultaneously. Subsequently, the algorithm examines each separate decision c by validating its consistency; such an assessment consists of two separate tests. First, the algorithm verifies that the operating units to be included by c have not been excluded by any previous decision. Second, the algorithm validates that c does not exclude any unit

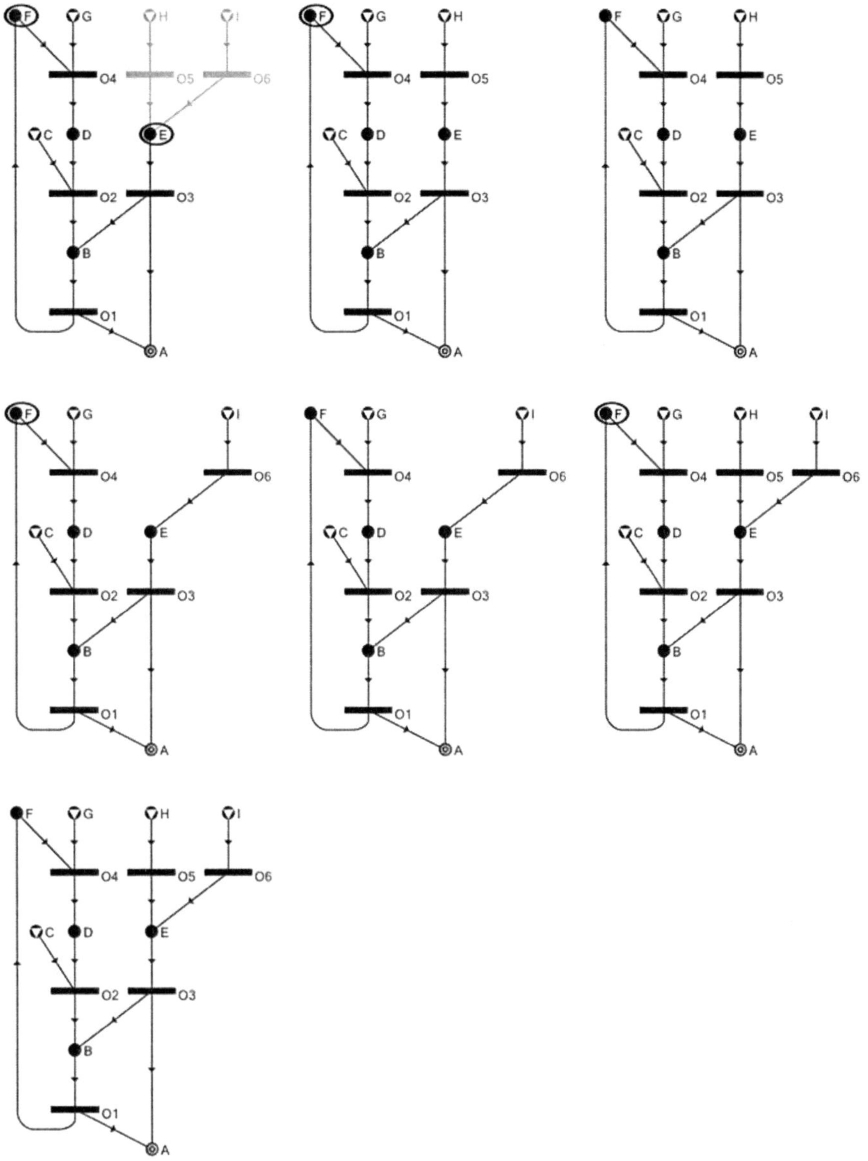

Fig. 5.12 The remaining steps for generating all solution-structures

previously included in the decision-mapping. The first evaluation is performed by the expression $c \cap \bar{\delta}(y) = \varnothing$, which compares the operating units included by c with the set of units excluded in the decisions of the materials previously included (y). This set is represented by the complement of the decision-mapping of y, *i.e.*, $\bar{\delta}(y)$. The second evaluation is performed by the expression $(\Delta(x) \backslash c) \cap \delta(y) = \varnothing$, which first determines the set of units excluded by decision c, *i.e.*, $\Delta(x) \backslash c$, and then

Input: $M, P, R, \Delta[M]$;
Comment: $P, R, \Delta[M]$ belong to synthesis problem $\mathbb{S} = (P, R, O)$, where
$P \subset M, R \subset M, P \cap R = \emptyset$,
$\Delta(x) = \{(\alpha, \beta) | (\alpha, \beta) \in O \,\&\, x \in \beta\}, \Delta(x) = \emptyset \Leftrightarrow x \in R$,
$\Delta[M] = \{(x, \Delta(x) | x \in M\}, \delta[m]$ is a decision-mapping on (M, O);
Output: all solution-structures of synthesis problem $\mathbb{S} = (P, R, O)$;
Global variables: $R, \Delta[M]$;

```
begin
if P = ∅ then stop;
SSG(P, ∅, ∅)
end

procedure SSG(p, m, δ[m]):
begin
if p = ∅ then begin
   write δ[m];
   comment: δ[m] defines a solution-structure;
   return
   end
let x ∈ p;
C := ℘(Δ(x)) \ {∅};
for all c ∈ C do begin
   if ∀y ∈ m, c ∩ δ̄(y) = ∅ & (Δ(x) \ c) ∩ δ(y) = ∅ then begin
      δ[m ∪ {x}] := δ[m] ∪ {(x, c)};
      SSG((p ∪ mat^in(c)) \ (R ∪ m ∪ {x}), m ∪ {x}, δ[m ∪ {x}]);
      end
   end
return
end
```

Fig. 5.13 Algorithm SSG for generating the solution-structures

compares it with the set of units included by the decisions made on the materials previously included (y), *i.e.*, $\delta(y)$. If c is consistent, its operating units are included in the decision-mapping and the algorithm proceeds to the next recursion. Figures 5.4, 5.5, 5.6, 5.7, 5.8, 5.9, 5.10, 5.11 and 5.12 have shown the generation of the solution-structures for the maximal structure given in Fig. 5.3. The same steps are given in Table 5.1. for algorithm SSG. For details on the demonstration see Fig. 5.14.

Table 5.1 Steps of algorithm SSG in solving a PNS problem identified by the maximal structure given on Fig. 5.3

Number of call	Depth of recursion	Parameter p	Parameter m	Parameter $\delta[m]$	Remark
1	0	$\{A\}$	$\emptyset$	$\emptyset$	Initial call
2	1	$\{B\}$	$\{A\}$	$\{(A,\{1\})\}$	
3	2	$\{D\}$	$\{A,B\}$	$\{(A,\{1\}),(B,\{2\})\}$	
4	3	$\{F\}$	$\{A,B,D\}$	$\{(A,\{1\}),(B,\{2\}),(D,\{4\})\}$	
5	4	$\emptyset$	$\{A,B,D,F\}$	$\{(A,\{1\}),(B,\{2\}),(D,\{4\}),(F,\{1\})\}$	Solution #1
6	1	$\{E\}$	$\{A\}$	$\{(A,\{3\})\}$	
7	2	$\emptyset$	$\{A,E\}$	$\{(A,\{3\}),(E,\{5\})\}$	Solution #2
8	2	$\emptyset$	$\{A,E\}$	$\{(A,\{3\}),(E,\{6\})\}$	Solution #3
9	2	$\emptyset$	$\{A,E\}$	$\{(A,\{3\}),(E,\{5,6\})\}$	Solution #4
10	1	$\{B,E\}$	$\{A\}$	$\{(A,\{1,3\})\}$	
11	2	$\{E\}$	$\{A,B\}$	$\{(A,\{1,3\}),(B,\{3\})\}$	
12	3	$\emptyset$	$\{A,B,E\}$	$\{(A,\{1,3\}),(B,\{3\}),(E,\{5\})\}$	Solution #5
13	3	$\emptyset$	$\{A,B,E\}$	$\{(A,\{1,3\}),(B,\{3\}),(E,\{6\})\}$	Solution #6
14	3	$\emptyset$	$\{A,B,E\}$	$\{(A,\{1,3\}),(B,\{3\}),(E,\{5,6\})\}$	Solution #7
15	2	$\{D,E\}$	$\{A,B\}$	$\{(A,\{1,3\}),(B,\{2,3\})\}$	
16	3	$\{E,F\}$	$\{A,B,D\}$	$\{(A,\{1,3\}),(B,\{2,3\}),(D,\{4\})\}$	
17	4	$\{F\}$	$\{A,B,D,E\}$	$\{(A,\{1,3\}),(B,\{2,3\}),(D,\{4\}),(E,\{5\})\}$	
18	5	$\emptyset$	$\{A,B,D,E,F\}$	$\{(A,\{1,3\}),(B,\{2,3\}),(D,\{4\}),(E,\{5\}),(F,\{1\})\}$	Solution #8
19	4	$\{F\}$	$\{A,B,D,E\}$	$\{(A,\{1,3\}),(B,\{2,3\}),(D,\{4\}),(E,\{6\})\}$	
20	5	$\emptyset$	$\{A,B,D,E,F\}$	$\{(A,\{1,3\}),(B,\{2,3\}),(D,\{4\}),(E,\{6\}),(F,\{1\})\}$	Solution #9
21	4	$\{F\}$	$\{A,B,D,E\}$	$\{(A,\{1,3\}),(B,\{2,3\}),(D,\{4\}),(E,\{5,6\})\}$	
22	5	$\emptyset$	$\{A,B,D,E,F\}$	$\{(A,\{1,3\}),(B,\{2,3\}),(D,\{4\}),(E,\{5,6\}),(F,\{1\})\}$	Solution #10

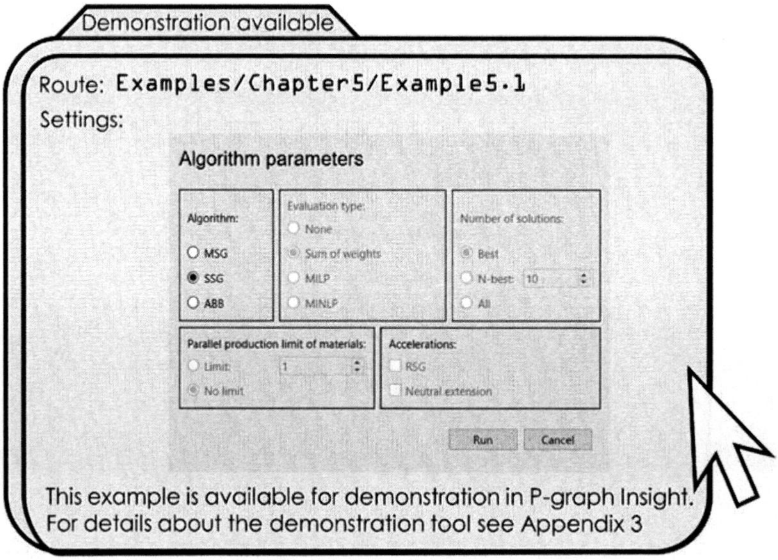

Fig. 5.14 Configuration of the demonstration software for Example 5.1

References

1. Friedler F, Varga JB, Fehér E, Fan LT (1996) Combinatorially Accerelated Branch-and-Bound Method for Solving the MIP Model of Process Network Synthesis. In: Floudas CA, Pardalos PM (eds) Nonconvex Optimization and its Applications. Kluwer Academic Publishers, Norwell, MA, pp 609–626
2. Friedler F, Varga JB, Fan LT (1995) Decision-mapping: A tool for consistent and complete decisions in process synthesis. Chem Eng Sci 50:1755–1768
3. Friedler F, Tarján K, Huang YW, Fan LT (1992) Combinatorial algorithms for process synthesis. Comput Chem Eng 16(Suppl):S313–S320

Chapter 6
Accelerated Branch-and-Bound Algorithm of Process Network Synthesis

Abstract A plausible approach for determining the optimal solution, or the n-best solutions, of a synthesis problem is by evaluating each individual solution-structure generated by algorithm SSG and make the selection as elaborated in Chap. 5. In this chapter, it is shown that the combinatorial properties of feasible process structures can further be exploited for the acceleration of a branch-and-bound type search. This gives rise to algorithm ABB exhibiting three types of acceleration in searching the optimal solution, or n-best solutions, specifically, (i) the number of partial problems is reduced; (ii) the sizes of the partial problems are minimized; and (iii) the bounds related to a partial problem are sharpened. Illustrative examples and the demonstration software show the operation of the algorithm.

Keywords P-graph · Branch-and-bound · Accelerated branch-and-bound · Algorithm ABB · n-best · Process synthesis · Recursive algorithm

This chapter introduces a branch-and-bound algorithm for solving process network synthesis (PNS) problems for which an objective value is assigned to each of the solution-structures in $S(\mathbb{S})$. Branch-and-bound (or B&B in short) methods have been widely published (see *e.g.*, [1–3]). The goal in deploying the algorithm is to find the feasible solution-structure with the best (usually lowest) objective value, or the set of n-best solutions.

In general, branch-and-bound algorithms are a common way to address problems where (i) the search space is either discrete or continuous; and (ii) the best solution is to be found by considering an objective function.

The nature of these problems is usually combinatorial; however, branch-and-bound algorithms have also been utilized for integer (or mixed-integer) programming.

Let us first introduce the basic principles of a branch-and-bound algorithm before explaining the specialized Accelerated Branch-and-Bound algorithm (ABB) of the P-graph framework. For now, let us consider finite discrete problems. Here, all decision variables are discrete, and their values are limited to a certain set of values, and thus, the search space is finite (although it can contain an enormous amount of solutions).

F. Friedler et al., *P-graphs for Process Systems Engineering*,
https://doi.org/10.1007/978-3-030-92216-0_6

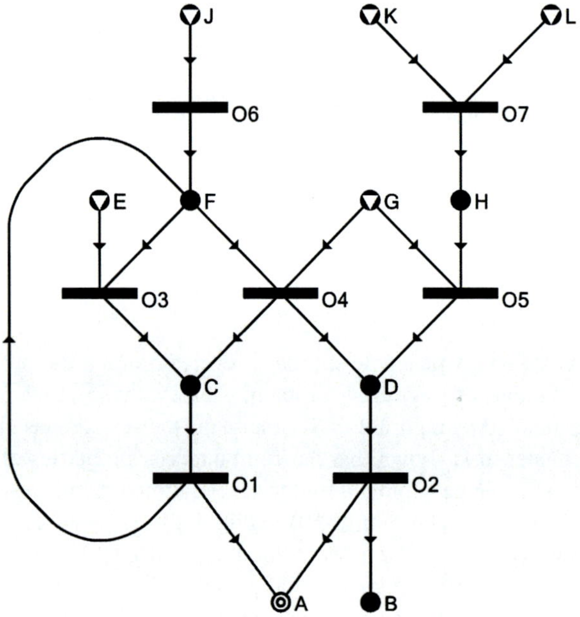

Fig. 6.1 Illustrative example for the branch-and-bound algorithm [4]

Example 6.1 The maximal structure of the example is shown in Fig. 6.1. This example involves a PNS problem where there are seven discrete (in this case, binary) variables, $x_1, x_2, \ldots, x_7$, each related to an operating unit.

As their name indicates, branch-and-bound algorithms consist of two main steps, branching and bounding, and their repeated application builds a search, or branching, tree. The algorithm makes decisions on the values of the discrete variables, and considers all possibilities. The root of this tree represents the whole problem, where no decision has been made, whereas the leaves represent the individual solutions, where a solution is clearly and uniquely defined by a set of decisions. At every step, the algorithm selects a node from the subtree that has not been examined before and makes a decision on one or more of its free discrete variables, thus creating new nodes for the tree.

In Example 6.1, all seven variables are free in the root of the tree. Let us follow this simple branching rule sequence: (i) select the first free variable; (ii) since this variable is binary, there are two possible decisions: set it to 0 or 1; (iii) create the two new nodes for these two cases.

Following this rule sequence, the root node will be selected first where x_1 is the first free variable and its two possible values are 0 and 1. Two new nodes will be created, one for each outcome of this decision, as shown in Fig. 6.2. Following these

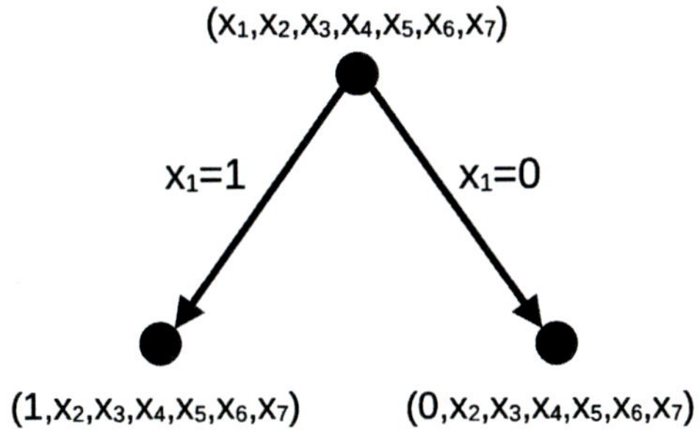

Fig. 6.2 First level of the branching tree, decision on variable x_1

rules, it is possible to build a branching tree of the whole search space that will contain all possible solutions at its leaves (leaf nodes). If the branching rule is determined correctly, all the leaves indicate a separate and unique solution. In Example 6.1, there are $2^7 = 128$ leaves in the complete branching tree (since each of the seven binary variables can take two possible values). Note that this is just an illustrative example for a possible branching tree corresponding to this PNS problem, the actual P-graph algorithms use more sophisticated methods which will be explained later.

However, in some problems, *e.g.*, in PNS, some of the leaves of the branching tree result in infeasible solutions. An important part of the branching algorithm is to detect these infeasibilities. In Example 6.1, there is a node after two decisions in the tree that indicates that the binary variables x_1 and x_2 are both 0. This means that both operating units $O1$ and $O2$ are excluded from the structure. As these are the only units that can manufacture the product, there is no feasible answer to the synthesis problem after this decision. P-graph tools, *e.g.*, algorithm MSG, can determine that no solution-structure can exist without at least one of these two operating units; hence, all leaves in the subtree of this node will be infeasible. As a result, this node can be cut from further examination, thus reducing the computational effort. The complete branching tree of Example 6.1 is shown in Fig. 6.3 where leaf nodes with a dot in the middle indicate the solution-structures (19 structures shown in Fig. 3.4), empty leaf nodes represent feasible structures which contain unneccesary units, and nodes with a horizontal line indicate infeasibility. At each level of the tree, the left branch (child) of a node indicates the inclusion of an operating unit (the binary variable is 1), and the right branch (child) indicates its exclusion (the binary variable is 0).

Now, what happens when the problem has an objective function that should be considered in determining the optimal solution? This is when the bounding part of the algorithm comes in. Of course, it is possible to determine all feasible solutions, determine the objective function value for each, and then select the best one, however, this procedure is usually not effective. Branch-and-bound methods rely on a bounding procedure that can set a bound on all the leaves in a subtree. For now, we will work with minimization problems, but maximization can be done in the same way. In minimization problems, the bounding procedure is supposed to give: (i) for a leaf node (solution), its actual objective value; and (ii) for an internal node, a

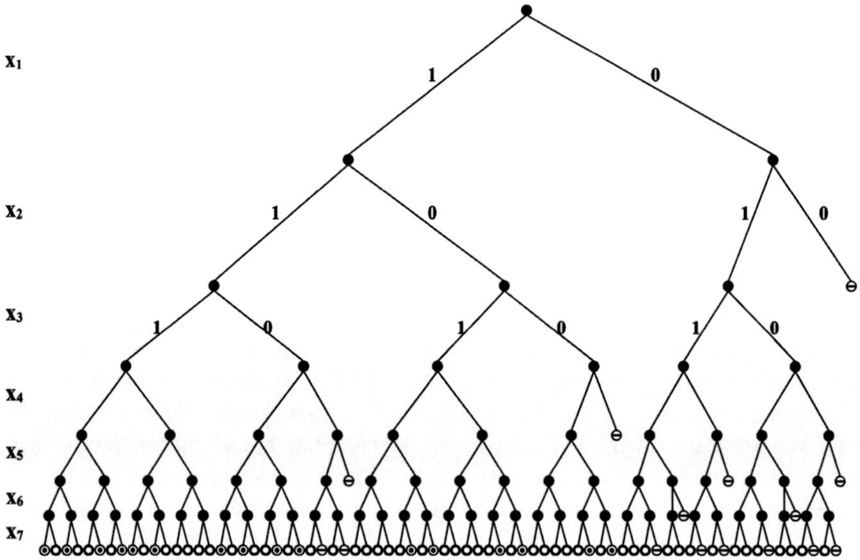

Fig. 6.3 Complete branching tree of the conventional B&B in the worst case for the maximal structure in Fig. 6.1

lower bound that is less than or equal to the objective values of every leaf node derived from it.

With such a bounding procedure, the algorithm can keep track of the best solution it has found so far after the algorithm has identified at least one solution to the problem. Any internal node, whose bound is equal or larger than the value of the current best solution, can be fathomed, *i.e.*, eliminated from future examination. If the bounding procedure gives a sufficiently sharp bound, the branch-and-bound algorithm can reduce the examined branching tree even further by fathoming nodes where the bound indicates that none of the following decisions will lead to a better solution than the current best.

The efficiency of branch-and-bound algorithm depends on several implementation details. The most important of them are the branching strategy which determines the shape of the branching tree, and the bounding procedure which determines the efficiency of the fathoming cuts. Moreover, the order of examination of the nodes can also have a huge impact since it affects how quickly the optimal solution is found. Detection of infeasibility for interior nodes also reduces the size of the branching tree.

The algorithm presented herein is a particular one based on the special properties of the structural model of PNS. For a formal definition, let $\mathbb{S} = (P, R, O)$ be a reduced structural model of PNS with maximal structure (M, O). Suppose that there is an evaluation tool capable of determining whether a solution-structure is feasible or not (i.e., whether it satisfies all constraints). Furthermore, if a solution-structure is found feasible, this tool determines the optimal value of the objective function of the solution-structure. In the current work, the objective function is to be minimized. Considering $\hat{S}(\mathbb{S})$ as the set of feasible solutions structures of structural model $\mathbb{S}$, defined in Chap. 3, let $z : \hat{S}(\mathbb{S}) \to \mathbb{R}$ be a given real-valued function that expresses the minimal value of the *objective function* of any feasible solution-structure. Thus, the problem is to determine the set of optimal solutions, formulated as

$$\{s : z(s) \leq z(s') \;\; \& \;\; s, s' \in \hat{S}(\mathbb{S})\} \tag{6.1}$$

Suppose that $\hat{S}(\mathbb{S})$ is a nonempty finite set, then there exists at least one optimal solution of this model.

Recursive Accelerated Branch-and-Bound Algorithm

Figure 6.4 shows the original Accelerated Branch-and-Bound algorithm for the P-graph framework, which was published by Friedler et al. (1996) [5]. This is a depth-first, search-based recursive algorithm that determines the solution-structure with the lowest objective value.

Next, we describe the steps of Algorithm ABB through the example presented in Fig. 6.1. At the initial call, no decisions have been made, and product A is the only material in set p (see Fig. 6.5), *i.e.*, $x = A$. All units are shown in grey, since no decisions have been made on any of them. The number below each partial problem identifies the generation order, and it is sometimes referenced in the text.

Since product A can be produced by operating units $O1$ and $O2$, *i.e.*, $\Delta(A) = \{O1, O2\}$, the possible decisions are $C = \{\{O1\}, \{O2\}, \{O1, O2\}\}$. All three options will be examined further in recursive function calls. It is worth noting that the first examination performed by ABB is the verification of consistency of the decision similar to that performed by algorithm SSG. The next call will examine the decision that material A is produced only by operating unit $O1$, *i.e.*, $c = \{O1\}$. Because this is the first decision, there are no inconsistencies and $\delta(A) = \{O1\}$. Since $O2$ is not selected here, it will be excluded in all solutions that follow this decision; hence, $m = \{A\}$ and $\overline{\delta}(A) = \{O2\}$. Without $O2$, operating units $O5$ and $O7$ cannot contribute to the production of A; therefore, they will not be considered in the subsequent steps and they will not be shown in subsequent figures. Figure 6.6 shows the graph of the recursive call where the included units (here, only $O1$) are shown in black, and the materials for future decisions (here, set $p = \{C\}$) are circled.

Following the recursive execution, two operating units can produce material C, thus resulting in three options: $\{O3\}$, $\{O4\}$, and $\{O3, O4\}$. The next call will explore decision $c = \{O3\}$. The test of consistency is carried out by comparing decision c to those previously made for every material y included in m.

Naturally, since at this point the set m has only one element, the conditions $c \bigcap \overline{\delta}(y) = \varnothing$ and $(\Delta(x) \backslash c) \bigcap \delta(y) = \varnothing$ are uniquely verified for $y = A$, therefore,

Input: $M, P, R, \Delta(x), (x \in M)$;
Comment: (M, O) is the maximal structure of PNS problem $\mathbb{S} = (P, R, O)$,
$P \subset M, R \subset M, P \cap R = \emptyset, \Delta(x) = \{(\alpha, \beta) | (\alpha, \beta) \in O \;\&\; x \in \beta\}$,
procedure FEASIBLE($\delta[m]$) is a Boolean function, its value is **true** if the partial problem defined by $\delta[m]$ has a feasible solution
procedure BOUND($\delta[m]$) determines a lower bound for the partial problem defined by $\delta[m]$; if $\delta[m]$ is a solution structure, this lower bound is the value of the objective function
Output: optimal solution of synthesis problem $\mathbb{S} = (P, R, O)$;
Global variables: $R, \Delta[M], (x \in M), U, currentbest$;

```
begin
U := ∞; currentbest := anything;
ABB(P, ∅, δ[∅]);
if U < ∞ then print currentbest;
else print 'there is no solution';
end

procedure ABB(p, m, δ[m]):
begin
let x ∈ p;
C := ℘(Δ(x)) \ {∅};
for all c ∈ C do begin
  if ∀y ∈ m, c ∩ δ̄(y) = ∅ & (Δ(x) \ c) ∩ δ(y) = ∅ then begin
    m' := m ∪ {x}; δ[m'] := δ[m] ∪ {(x, c)};
    if FEASIBLE(δ[m']) then begin
      p' := (mat^in(op(δ[m'])) ∪ P) \ (m' ∪ R);
      bound := BOUND(δ[m']);
      if U ≥ bound then begin
        if p' = ∅ then begin U := bound; update currentbest; end
        else ABB(p', m', δ[m']);
        end
      end
    end
  end
return
end
```

Fig. 6.4 Algorithm ABB

$\delta(y) = \{O1\}$ and $\bar{\delta}(y) = \{O2\}$. The consistency of the decision is validated as both conditions are satisfied ($\{O3\} \bigcap \{O2\} = \emptyset$ and $(\{O3, O4\} \backslash \{O3\}) \bigcap \{O1\} = \emptyset$). Subsequently, $\delta(C) = \{O3\}$; $\bar{\delta}(C) = \{O4\}$, see Fig. 6.7. Afterwards, the decision must be made on the production of material F. During this call, two operating units can produce material F: $O1$ and $O6$, and thus, $\Delta(F) = \{O1, O6\}$. This gives three

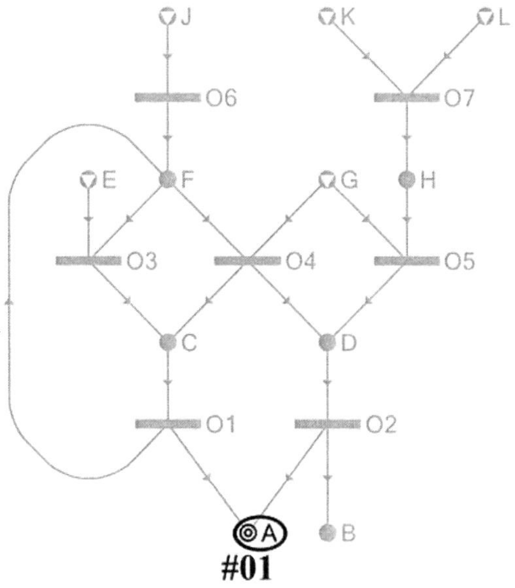

Fig. 6.5 Initial call of Algorithm ABB

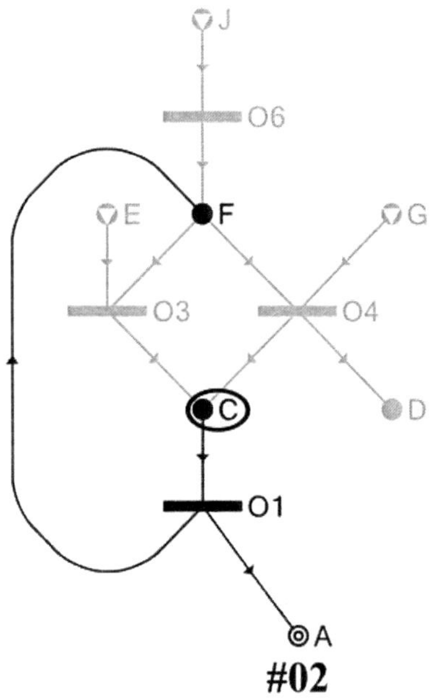

Fig. 6.6 Second call of algorithm ABB

options for c, $\{O1\}$, $\{O6\}$, and $\{O1, O6\}$; however, one of the options, $\{O6\}$, would not give a consistent decision-mapping. The consistency of the decision-mapping is verified considering $y = A$ and $y = C$. For $y = A$, $\delta(A) = \{O1\}$ and $\overline{\delta}(A) = \{O2\}$, here $c \bigcap \overline{\delta}(A) = \{O6\} \bigcap \{O1\} = \varnothing$, however $(\Delta(x)\backslash c) \bigcap \delta(A) = (\{O1,$

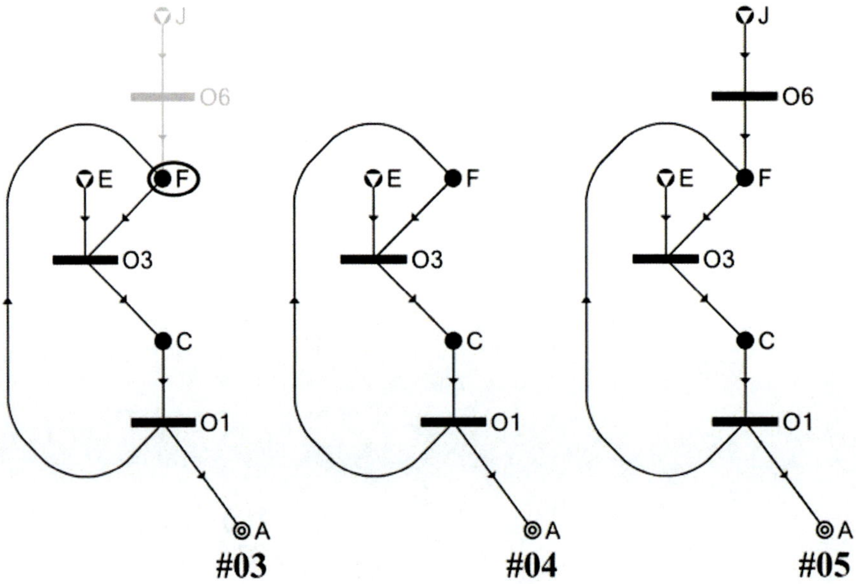

Fig. 6.7 Third call of Algorithm ABB, with the two resulting solutions

$O6\} \setminus \{O6\}) \bigcap \{O1\} = \{O1\}$. Thus, the second part of the validation does not render an empty set, and therefore, the decision $c = \{O6\}$ is not consistent. Note that for $y = C$, the validation would pass: $\delta(C) = \{O3\}$ and $\overline{\delta}(C) = \{O4\}$; hence, $c \bigcap \overline{\delta}(C) = \{O6\} \bigcap \{O4\} = \varnothing$ and $(\Delta(x)\backslash c) \bigcap \delta(C) = (\{O1, O6\} \setminus \{O6\}) \bigcap \{O3\} = \varnothing$. The failed validation for $y = A$ signifies that decision $c = \{O6\}$ implies excluding an operating unit previously included, namely $O1$, thereby rendering it non consistent.

This leaves only the other two options. Both of those decisions will result in a consistent decision-mapping with an empty p set, so they are both solution-structures that do not induce additional recursive calls (#04 and #05 in Fig. 6.7). They are shown in Fig. 6.7. For now, we omit the bounding procedure to show the complete, *i.e.*, worst case, branching tree for the algorithm.

Subsequent branches will generate 4 more solutions where product A is only produced by operating unit $O1$. The corresponding steps are shown in Fig. 6.8. The solution-structures (#07, #08, #10, and #11) can be recognized as they do not have circled materials (in other words, set p is empty).

At this point, the algorithm explored all possibilities where material A is only produced by operating unit $O1$. The next recursive call of the first decision follows, where product A is only produced by operating unit $O2$. Here, since operating unit $O1$ is excluded, operating unit $O3$ is also excluded based on the axioms. The next decision on this branch is about the production of material D, where $\Delta(D) = \{O4, O5\}$. All three options are consistent, the first one, $\delta(D) = \{O4\}$, is explored first. Here, since $O5$ is excluded from the structure, $O7$ should be excluded as well. Here, material F is not yet produced by any operating units, but there is only one unit to

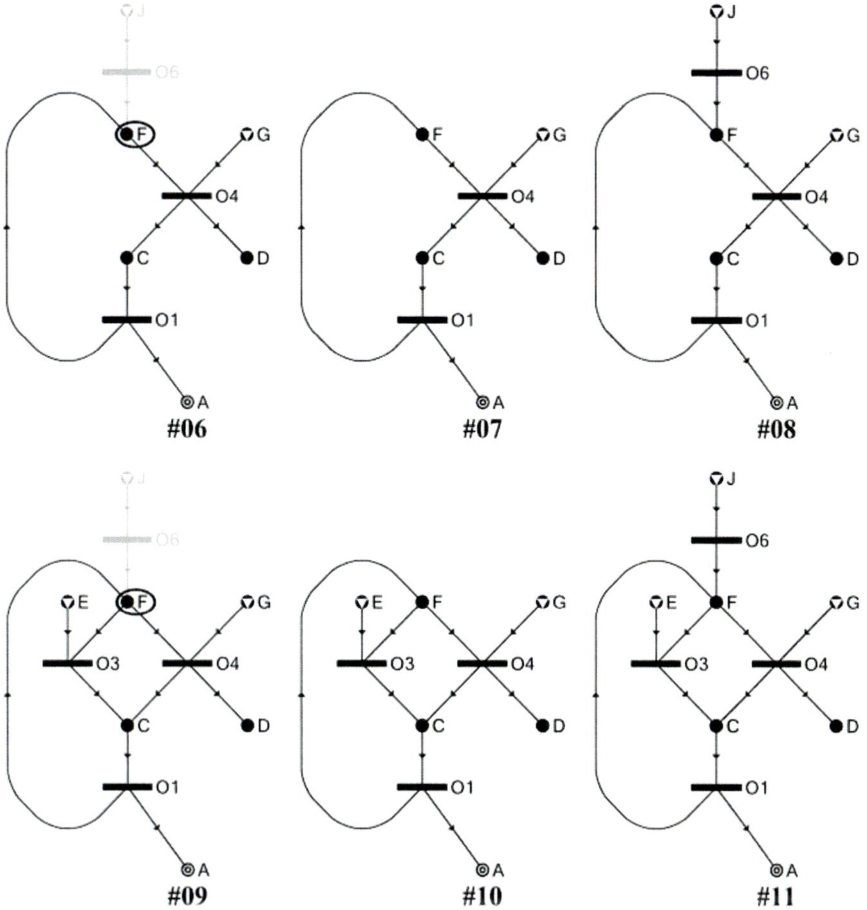

Fig. 6.8 Next steps of generating solutions with Algorithm ABB

choose from, $O6$. Hence, the decision of the next branching step has two options, the inclusion or not of operating unit $O6$, but only one of them leads to a solution-structure. Similarly, if material D is produced by operating unit $O5$ alone, then the next branch after that includes operating unit $O7$, as there is no other feasible option. The same logic is also applied when $\delta(D) = \{O4, O5\}$. These steps are shown in Fig. 6.9.

Next, the branch for $\delta(A) = \{O1, O2\}$ is examined. The algorithm operates on the same principles as before; Figs. 6.10, 6.11, 6.12, 6.13 and 6.14 show the steps. First, both materials C and D are subject to future decisions $p = \{C, D\}$. For the branching step, we need to select one material that will be the basis of the decisions, let this be material C. $\Delta(C) = \{O3, O4\}$; thus, there are three possible decisions, all three are consistent. The next step examines the decision where $\delta(C) = \{O3\}$. Here, after

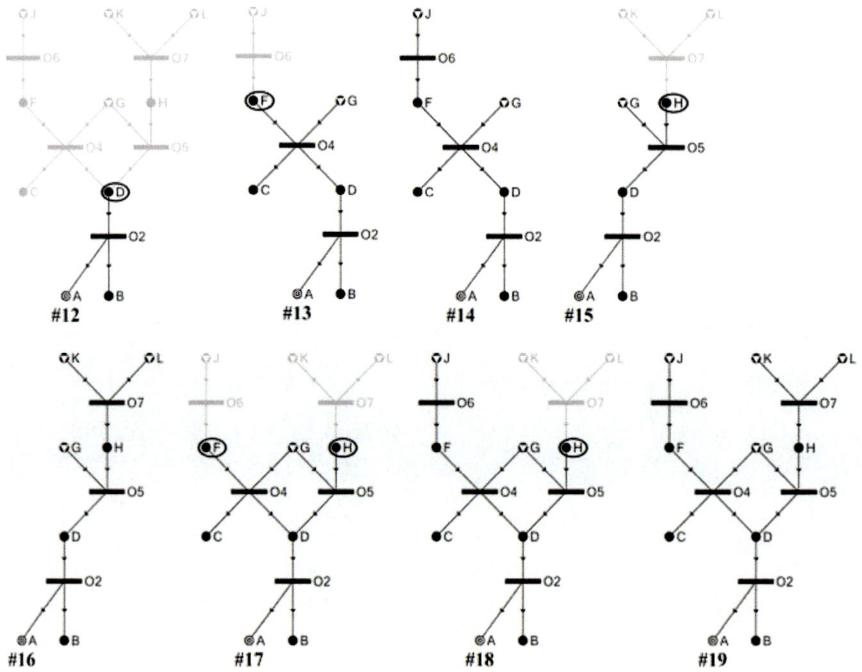

Fig. 6.9 Steps of ABB when $\delta(A) = \{O2\}$

decisions are made on materials *D, F,* and *H*, two new solutions are generated (#24 and #26).

If $\delta(C) = \{O4\}$, at step #27, then material *D* is selected, where $\Delta(D) = \{O4, O5\}$. Unit *O*4 is already included due to previous decisions, so there are only two choices leading to consistent decision-mappings, $\delta(D) = \{O5\}$ and $\delta(D) = \{O4, O5\}$. The former leads to two new solution-structures (#29 and #30 on Fig. 6.11), and the latter leads to two others (see Fig. 6.12). The last few steps of the algorithm are shown in Figs. 6.13 and 6.14.

Algorithm ABB has three types of accelerations compared to the standard branch-and-bound algorithm: (i) it reduces the number of examined nodes in the branch-and-bound tree; (ii) it reduces the size of a node's partial problem, thus resulting in quicker bounding calculations and smaller branch-and-bound subtrees; (iii) it contributes to sharpening the bounding procedure.

These reduction effects are reflected in the step-by-step example discussed earlier. First, it is clear that the branching based on decision-mapping leads to a much smaller branching tree than just considering the binary values. This also results in smaller branch-and-bound subtrees as well as in a more effective bounding procedure.

Figure 6.15 shows the complete branching tree for the example where each node is labelled with the number of the partial problem and the material that is selected for the decision. The leaves without the material are the solution-structures generated by

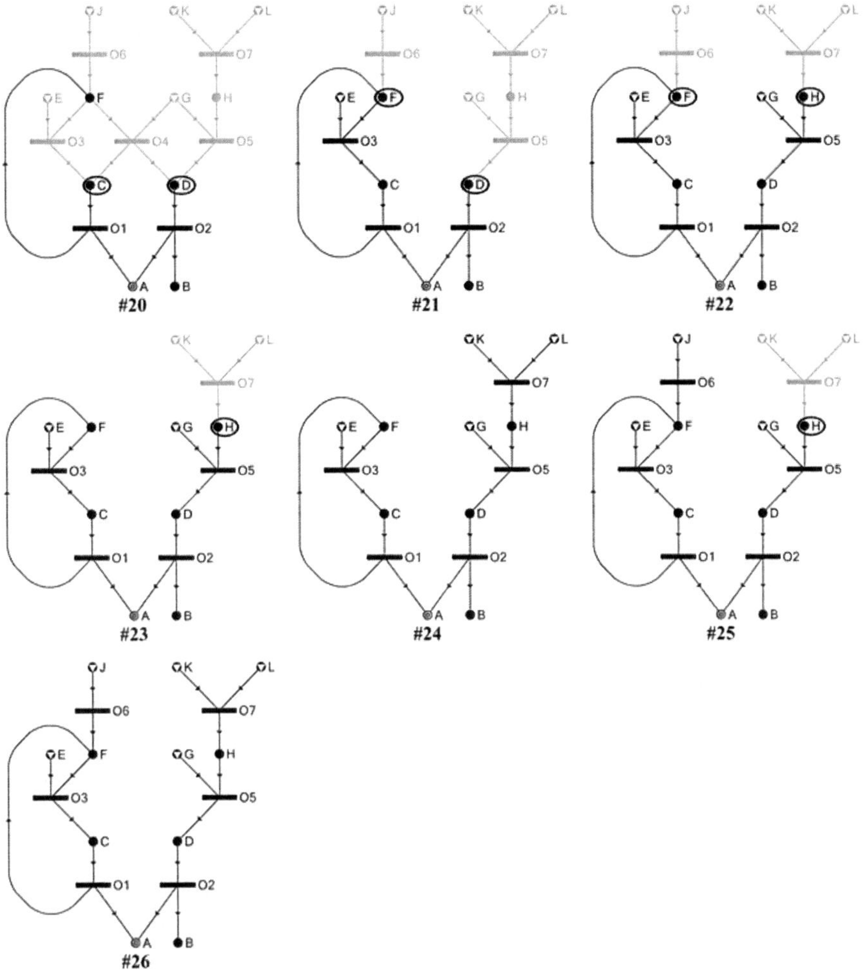

Fig. 6.10 Partial problems #20-#26 of algorithm ABB leading to two more solution-structures

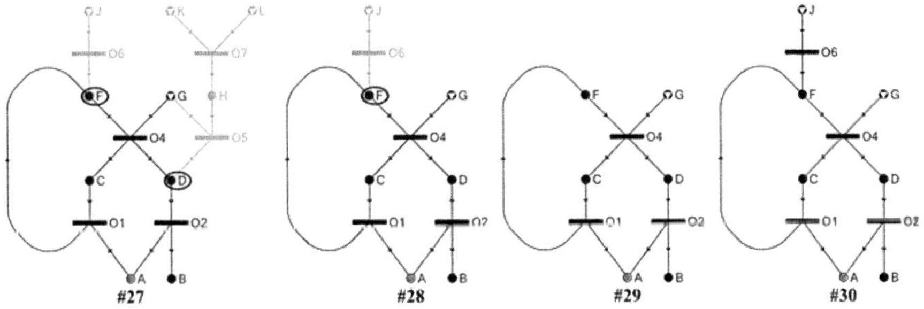

Fig. 6.11 Partial problems #27 to #30 of algorithm ABB

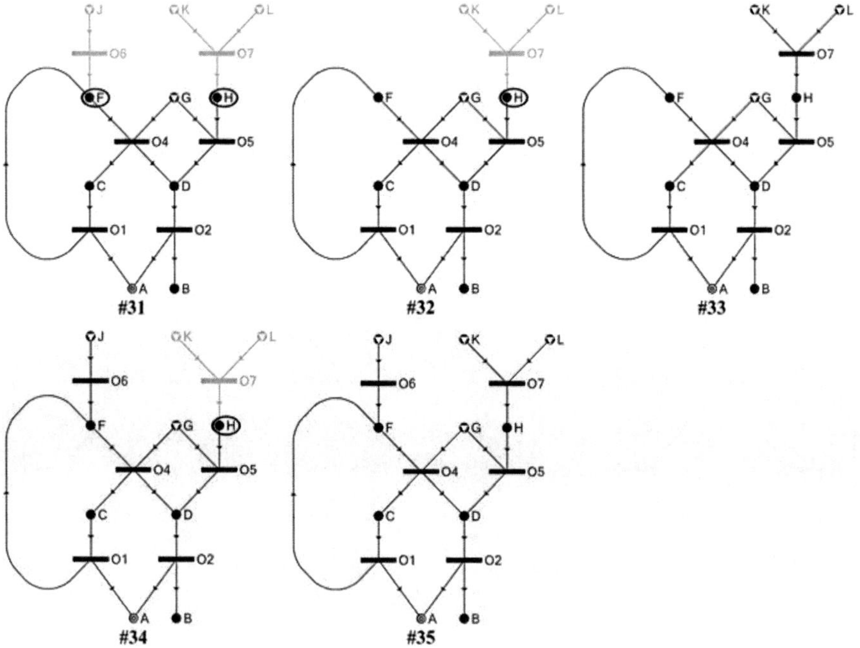

Fig. 6.12 Partial problems #31 to #35 of algorithm ABB

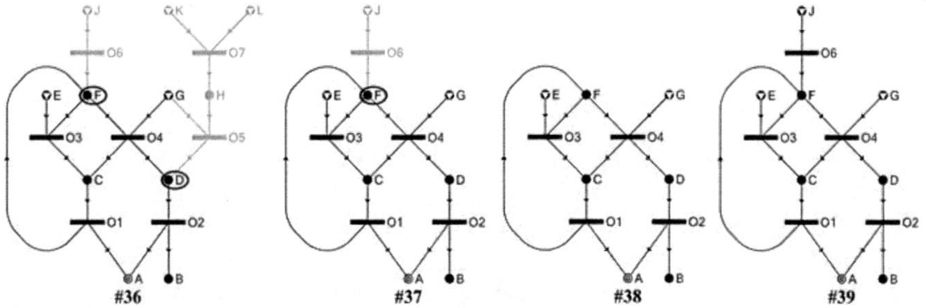

Fig. 6.13 Partial problems #36 to #39 of algorithm ABB

the algorithm. Note that these are the exact same 19 solution-structures that algorithm SSG generates for this maximal structure.

The usual goal of optimization problems is to find the best solution. As seen previously, branch-and-bound algorithms are able to do that. However, there can be other goals, resulting in multiple solutions, which can be easily achieved with branch-and-bound algorithms, but quite difficult to achieve with other types of optimization methods.

For example, a decision maker might want to know not just the best, but the best 10 or 20 solutions, which would allow to make the final decision from a broader set

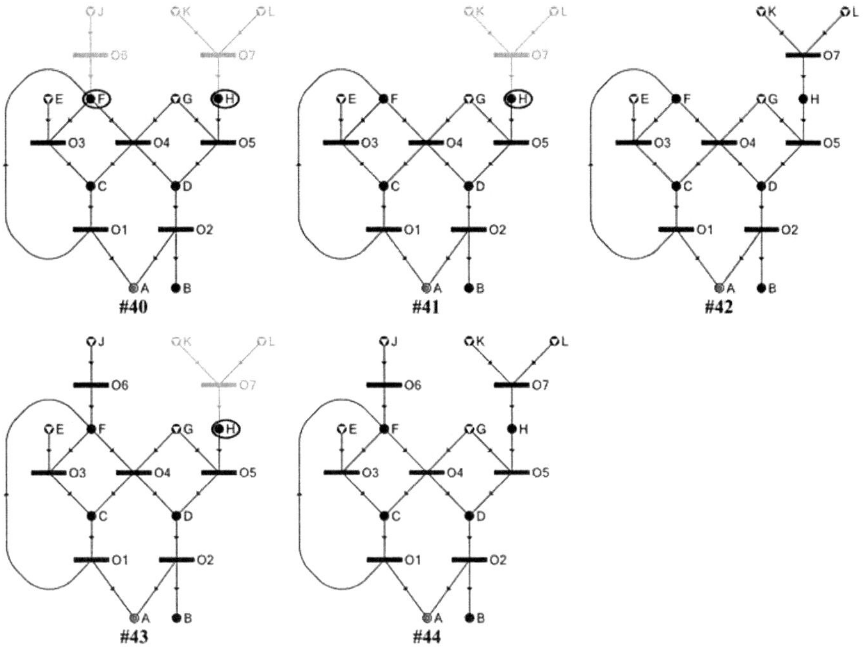

Fig. 6.14 Partial problems #40 to #44 of algorithm ABB

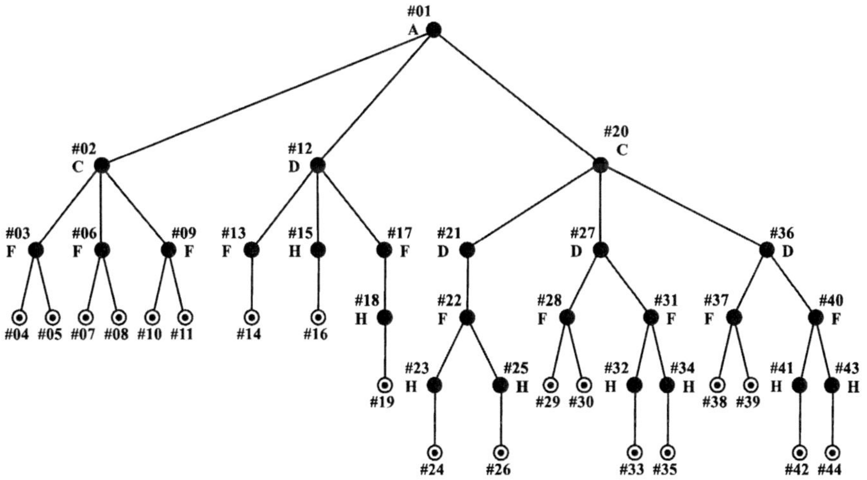

Fig. 6.15 Branching tree of algorithm ABB for the illustrative example (worst case)

of choices. This type of goal is called finding the n-best solutions. If there are solutions with the same objective value, it can happen that, *e.g.*, solutions 18–24 are equivalently good when searching for the best 20 solutions. In this case, it is also possible to extend the set of results to include these additional solutions as well.

```
Input: M,P,R,Δ(x), (x ∈ M), N (number of best networks);
Comment: (M,O) is the maximal structure of PNS problem S = (P,R,O),
P ⊂ M, R ⊂ M, P ∩ R = ∅, Δ(x) = {(α,β)|(α,β) ∈ O & x ∈ β},
procedure FEASIBLE(δ[m]) is a Boolean function, its value is true if the partial
problem defined by δ[m] has a feasible solution
procedure BOUND(δ[m]) determines a lower bound for the partial problem de-
fined by δ[m]; if δ[m] is a solution structure, this lower bound is the value of the
objective function
Output: optimal solution of synthesis problem S = (P,R,O);
Global variables: R,Δ[M], (x ∈ M), U, nbest (a list containing the n-best net-
works ordered by the objective values; LAST(nbest) is the worst network in the
list, i.e., it defines the current bound, SIZE(nbest) gives the size of the list);

begin
U := ∞; nbest is empty;
ABB(P,∅,δ[∅]);
if U < ∞ then print nbest;
else print 'there is no solution';
end

procedure UPDATE(nbest,δ[m])
begin
  insert δ[m] into nbest at the correct place (ordered by objective value)
  if SIZE(nbest) > N then remove last solution from nbest
  if SIZE(nbest) = N then return BOUND(LAST(nbest))
  else return ∞
end

procedure ABB(p,m,δ[m]):
begin
let x ∈ p;
C := ℘(Δ(x)) \ {∅};
for all c ∈ C do begin
  if ∀y ∈ m, c ∩ δ̄(y) = ∅ & (Δ(x) \ c) ∩ δ(y) = ∅ then begin
    m' := m ∪ {x}; δ[m'] := δ[m] ∪ {(x,c)};
    if FEASIBLE(δ[m']) then begin
      p' := (mat^in(op(δ[m'])) ∪ P) \ (m' ∪ R);
      bound := BOUND(δ[m']);
      if U ≥ bound then begin
        if p' = ∅ then U := UPDATE(nbest,δ[m']);
        else ABB(p',m',δ[m']);
        end
      end
    end
  end
return
end
```

Fig. 6.16 Algorithm ABB for finding the n-best solutions

Branch-and-bound algorithms can find the n-best solutions even with equivalent objective values, and it only requires a minor change in the method introduced above. Specifically, instead of remembering the best solution so far, the algorithm remembers the n-best solutions so far. The bounding cuts are only performed after at least n solutions are found, and the currently n^{th} solution determines the bound. The algorithm is given fully in Fig. 6.16.

Branch-and-bound algorithms can take it even further. By removing the bounding cuts completely from the algorithm, it will find all feasible solutions for the optimization problem and rank them based on their objective value. This is generally useful when the number of feasible solutions is not expected to be too high. These solutions are usually forwarded to other algorithms for additional evaluation and testing.

Neutral Extension

If we look at the branching tree shown in Fig. 6.15, it can be seen that some of the nodes have only one child. In these partial problems, there is only one decision-mapping for the selected material which is consistent with the previous decisions, *i.e.*, no decision is needed. To skip such partial problems, algorithm ABB can be extended with the *neutral extension* algorithm. This will be introduced formally in Part III of the book; however, we discuss its conceptual basis herein. A neutral extension can be made for a decision-mapping of a partial problem, if one of the materials of its set p, *i.e.*, from $(mat^{in}(op(\delta[m])) \cup P) \setminus (m \cup R)$, has only one consistent decision.

For example, let us consider partial problem #21 of the illustrative example (see Fig. 6.15). Here, material D has only one consistent decision-mapping where $\delta(D) = \{O5\}$. By applying the neutral extension to the partial problem, this decision with only one possibility is made without a separate branching step and is already included in partial problem #21. Similarly, there is only one way to produce material H; thus, operating unit $O7$ can also be included in the partial problem in the same branching step. As a result, partial problem #21, when extended through netural extension, will already contain decision for two more materials, D and H, and the branching tree of the example becomes smaller.

Figure 6.17 shows the extended ABB algorithm where the maximal neutral extension of a decision-mapping is generated by applying neutral extension repeatedly while it is applicable. Figure 6.18 shows the branching tree of the new version of algorithm ABB for Example 6.1.

Example of Bounding

Now, let us discuss the changes in the branching tree when bounding is also applied. First, let us assume that each solution-structure has an objective value determined by deterministic function z, as given at the beginning of the chapter. For each partial

Input: $M, P, R, \Delta(x), (x \in M)$;
Comment: (M, O) is the maximal structure of PNS problem $\mathbb{S} = (P, R, O)$,
$P \subset M, R \subset M, P \cap R = \emptyset, \Delta(x) = \{(\alpha, \beta) | (\alpha, \beta) \in O \,\&\, x \in \beta\}$,
procedure FEASIBLE($\delta[m]$) is a Boolean function, its value is **true** if the partial problem defined by $\delta[m]$ has a feasible solution
procedure BOUND($\delta[m]$) determines a lower bound for the partial problem defined by $\delta[m]$; if $\delta[m]$ is a solution structure, this lower bound is the value of the objective function
Output: optimal solution of synthesis problem $\mathbb{S} = (P, R, O)$;
Global variables: $R, \Delta[M], (x \in M), U, currentbest$;

begin
$U := \infty$; $currentbest := anything$;
let $\hat{\delta}[\hat{m}]$ be the current maximal neutral extension of $\delta[\emptyset]$;
$p := \left(mat^{in}\left(op(\hat{\delta}[\hat{m}])\right) \cup P\right) \setminus (\hat{m} \cup R)$;
if $p = \emptyset$ **then begin** $currentbest$:=BOUND($\hat{\delta}[\hat{m}]$); $U := currentbest$; **end**
else ABB($p, \hat{m}, \hat{\delta}[\hat{m}]$);
if $U < \infty$ **then print** $currentbest$;
else print 'there is no solution';
end

procedure ABB($p, m, \delta[m]$):
begin
let $x \in p$;
$C := \wp(\Delta(x)) \setminus \{\emptyset\}$;
for all $c \in C$ **do begin**
  if $\forall y \in m, c \cap \bar{\delta}(y) = \emptyset \,\&\, (\Delta(x) \setminus c) \cap \delta(y) = \emptyset$ **then begin**
    $m' := m \cup \{x\}$; $\delta[m'] := \delta[m] \cup \{(x, c)\}$;
    if FEASIBLE($\delta[m']$) **then begin**
      let $\hat{\delta}[\hat{m}']$ be the current maximal neutral extension of $\delta[m']$;
      $p' := \left(mat^{in}\left(op(\hat{\delta}[\hat{m}'])\right) \cup P\right) \setminus (\hat{m}' \cup R)$;
      $bound :=$ BOUND($\hat{\delta}[\hat{m}']$);
      if $U \geq bound$ **then begin**
        if $p' = \emptyset$ **then begin** $U := bound$; update $currentbest$; **end**
        else ABB($p', \hat{m}', \hat{\delta}[\hat{m}']$);
        end
      end
    end
  end
return
end

Fig. 6.17 Algorithm ABB with neutral extension

problem of the branching tree, indicated by decision-mapping $\delta[m]$, let $g(\delta[m])$ be a real valued function that represents the bound calculated for each partial problem. Naturally, for each solution-structure $s = (m, o)$ described by decision-mapping $\delta[m]$, $g(\delta[m]) = z(s)$. For any partial problem represented by decision-mapping $\delta[m]$, $g(\delta[m]) \leq \min\{g(\delta'[m']) : \delta'[m']' \in \omega(\delta[m])\}$, where $\omega(\delta[m])$ is the set of all decision-mappings in the subtree of $\delta[m]$ that represent a solution-structure

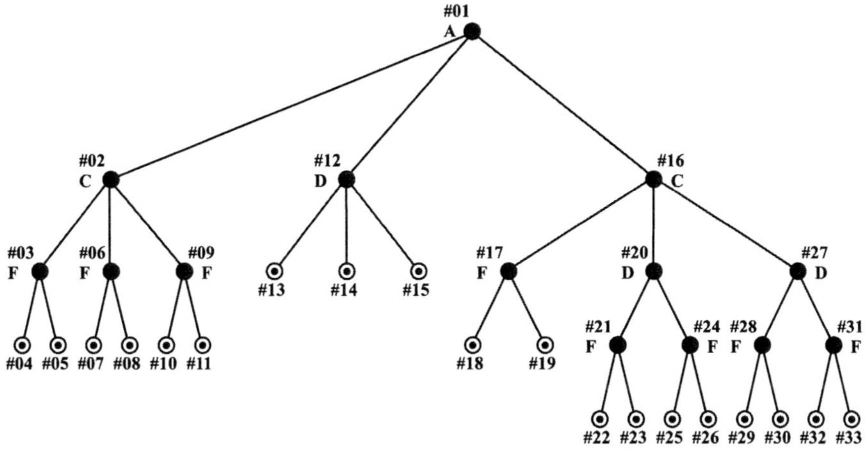

Fig. 6.18 Branching thee of algorithm ABB with neutral extension (worst case)

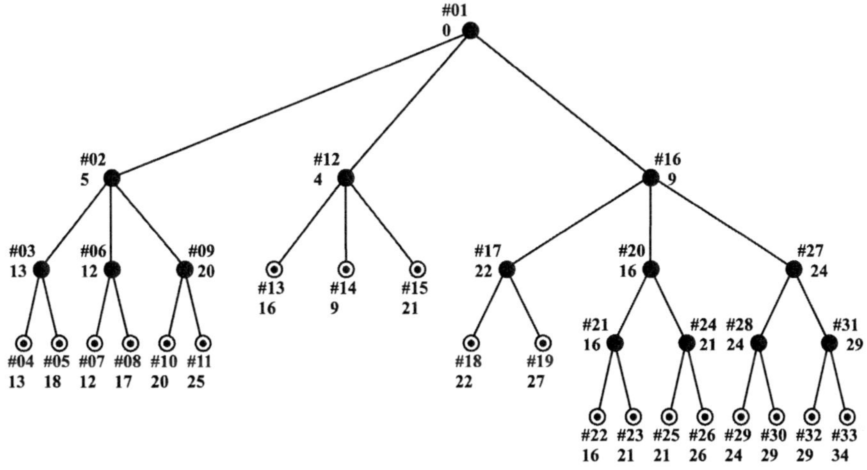

Fig. 6.19 Branching tree with bound values (worst case)

(in other words, these are the extensions of $\delta[m]$ that represent a solution-structure). For a non-leaf node in the branching tree, the $g(\delta[m])$ bounding function gives a lower bound for all solutions in that subtree.

Let us assume that such a bounding function exists for Example 6.1. For illustration, let us assign a weight to each operating unit. Then, the bound for a partial problem is defined as the sum of the weights of the operating units included on the structures by the decisions. Suppose that the weights of operating units $O1, O2, \ldots O7$

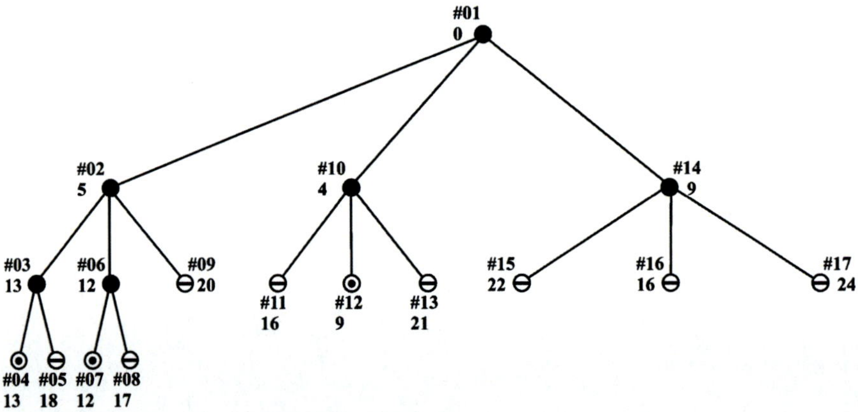

Fig. 6.20 Branching tree with cuts due to bounding

are 5, 4, 8, 7, 3, 5, 2, respectively. Figure 6.19 shows the branching tree where the bounds are assigned to each node. These bound values are not too sharp, but they will be good for demonstration. However, algorithm ABB keeps the generation order after finding a solution, and it is able to perform cuts based on the bound. If only the best solution is to be found, the bound is always given by the best solution found so far. If the bound of a partial problem is larger than the current best solution (or the value of the nth current best solution), that partial problem is cut from the tree, *i.e.*, fathomed. Figure 6.20 shows the modified branching tree where the crossed nodes indicated partial problems that were cut due to their bounds. Note the decrease in the number of steps that the algorithm needs to determine the optimal solution.

Figure 6.20 clearly shows that bounding can reduce the search tree after the first solution is found. However, the displayed search tree is highly dependent on the branching strategy, *i.e.*, the order in which the children of a node are considered. For the current example, the actual branching tree can have as few as 13 nodes, and as many as 31, depending on the branching strategy. We will not discuss the different branching strategies now. For details on finding the best solution of Example 6.1 in the demonstration software, see Fig. 6.21.

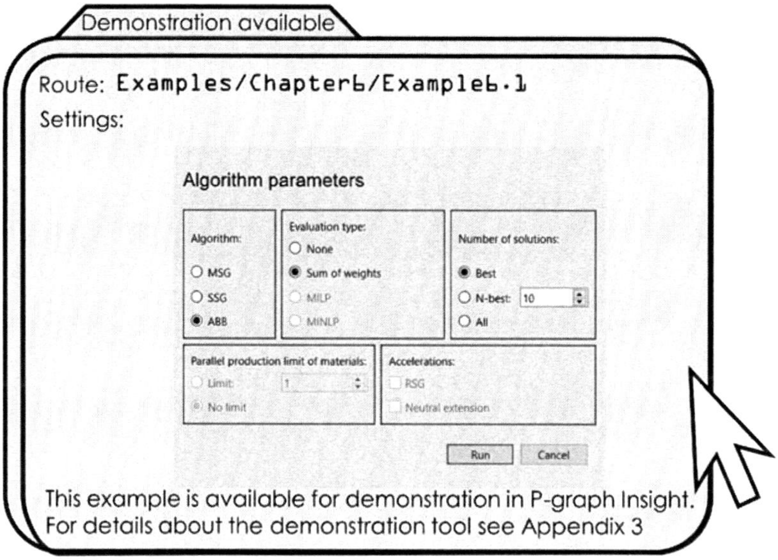

Fig. 6.21 Configuration of the demonstration software for finding the best solution of Example 6.1

References

1. Horst R, Tuy H (2013) Global optimization: Deterministic approaches. Springer Science & Business Media, Cham
2. Ibaraki T (1976) Theoretical Comparisons of Search Strategies in Branch-and-Bound Algorithms. J Assoc Comput Mach 24:315–344
3. Rocktäschel S (2020) A Branch-and-Bound Algorithm for Multiobjective Mixed-integer Convex Optimization. Springer Fachmedien, Wiesbaden
4. Friedler F, Varga JB, Fan LT (1995) Decision-mapping: A tool for consistent and complete decisions in process synthesis. Chem Eng Sci 50:1755–1768
5. Friedler F, Varga JB, Fehér E, Fan LT (1996) Combinatorially accerelated branch-and-bound method for solving the MIP model of process network synthesis. In: Floudas CA, Pardalos PM (eds) Nonconvex optimization and its applications. Kluwer Academic Publishers, Norwell, MA, pp 609–626

Part II
Applications of the P-graph Framework

Chapters 1–6 of Part I introduced the P-graph framework along with its fundamental concepts, definitions, and core algorithms. Because the framework can be applied in a large variety of fields, Chap. 7 of Part II provides a review of literature concerning the P-graph framework, thus showing the versatility of the framework. Chapter 8 discusses the steps of the synthesis procedure of an industrial case study. The scope of the P-graph framework is certainly not restricted: It can be extended to various science and technology fields in new directions. In this regard, Chap. 9 presents the extension of the framework considering process reliability and resilience.

Chapter 7
Literature Review on Research and Applications

Abstract The first paper related to the P-graph framework dates back to the late 1970s. Since then, multiple contributions concerning the framework have been published by numerous researchers worldwide. The main concepts, definitions, and algorithms of the P-graph framework were introduced in Part I of the book. In this chapter, a thorough literature review is given on the research related to P-graph in the past decades, as well as on the diverse applications of the framework.

Keywords P-graph · Reliability · Resilience · Supply chain · Sustainability · P-graph applications · Process synthesis

Part I of this book introduced the theoretical background of the P-graph framework. As elaborated therein, the framework's pillars are a set of axioms and a set of algorithms [1–3] that confer on it mathematical rigorousness as well as computational effectiveness. Such features imbue the framework with a flexibility that facilitates its application for identifying the best possible system structures solving a variety of problems in diverse fields of science and engineering.

The P-graph framework was conceived with the aim of solving problems concerning synthesis of chemical processes, *i.e.*, process network synthesis, or PNS. An inherent trait of these problems is their combinatorial nature as they involve a large number of individual units performing specific functions, or operations, which can be interconnected in myriad ways. Nevertheless, the interconnection of operating units giving rise to complex structures can be observed not only in the chemical industry but also in other areas of science and engineering. Consequently, the P-graph framework can be applied to the solution of problems beyond the chemical industry whose structure and operation are inherently combinatorial.

The origins of what would become the P-graph framework date back to the late 1970s and they can be found in the contribution by Friedler et al. [4]. Subsequently, their findings gradually evolved into the framework's theoretical and operational foundation which was formally introduced in a series of works starting in the early 1990s [5–9].

F. Friedler et al., *P-graphs for Process Systems Engineering*,
https://doi.org/10.1007/978-3-030-92216-0_7

A search on a publication database reveals that the number of contributions on P-graph per year has grown consistently since the framework's introduction; currently, the number of manuscripts is in the neighborhood of 200. These contributions are authored by researchers disseminated in about 20 countries in 5 continents, thereby indicating that the framework has been increasingly accepted by the scientific community at large. A single attempt at categorizing these contributions is somewhat challenging in view of their diversity and scope. Moreover, it can be difficult to categorize a contribution in a single discipline as works usually arise from the intersection of several areas of knowledge. Nevertheless, a number of reviews are available in the literature on current applications of the P-graph framework. In this regard, Lam [10] reviews the framework's application to solving instances that can be represented as process network synthesis (PNS) problems, *e.g.*, effective supply chains. Klemes and Varbanov [11] underlined the achievements of the framework in terms of key applications, geographical coverage, strengths and limitation. Moreover, Varbanov et al. [12] discussed the successes, challenges, and potential applications of the framework concerning the solution of PNS problems. Cabezas et al. [13] reviewed the framework's application in regards the sustainable design of processes and supply chains. In addition, Tan et al. [14] presented a detailed discourse on the future and potential applications of the framework and Friedler et al. [15] discussed the prospects and challenges for automatic PNS with P-graph.

In this chapter, we have exhaustively reviewed the various contributions related to the applications of the P-graph framework published up to 2021 and categorized them into several grand areas that are elaborated below. It is worth pointing out that these contributions clearly reveal that the framework has indeed permeated diverse fields of science and technology since its inception. The contributions related to applications are partitioned in seven distinct groups:

1. Industrial applications: PNS, process integration, and improvement
2. Supply chain, logistics, and production scheduling
3. Sustainability assessment and circular economy
4. Reliability, resilience, and risk assessments
5. Non-conventional applications
6. Extensions of the model and software implementation
7. Novel directions

A total of 181 papers, in addition to those already cited, are reviewed in this chapter. Their distribution among the seven categories can be seen in Fig. 7.1.

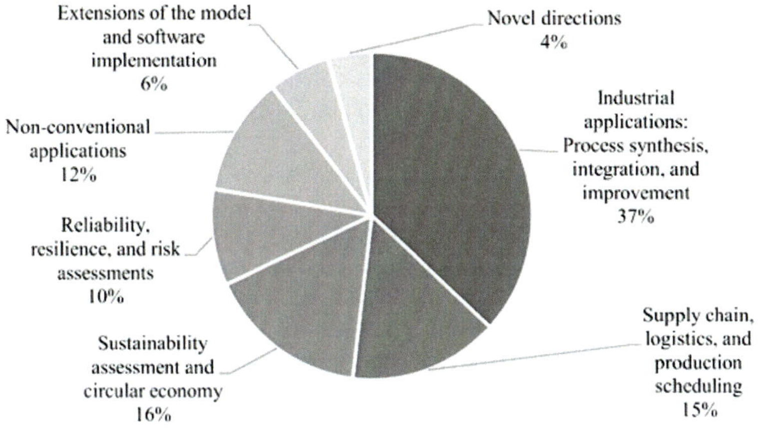

Fig. 7.1 Distribution of contributions found in literature (year 2021)

Industrial Applications: PNS, Integration, and Improvement

The use of the P-graph framework has been continuously extended to various fields where its algorithmic and computational properties can be exploited. As indicated earlier, the framework finds its largest application in problems of combinatorial nature where the framework's algorithms are capable of rapidly reducing the search space in optimizing and delivering the set of n-best system structures.

Clearly, the framework has been extensively employed for PNS as well as the evaluation of productive systems' performance under various circumstances. The framework's application to this sort of problems is naturally straightforward because the framework has been defined in terms of materials and operating units, which are the foundation of processing systems. Instances of these applications abound in the literature. For instance, biorefineries have been optimally synthesized by resorting the framework based on agricultural waste [16, 17], which aimed at generating valuable products. The framework has also been deployed for the technical and economical evaluation of alternatives for processing of pyrolysis-oil and wood [18, 19]. Moreover, some contributions have evaluated the treatment options for lignocellulosic waste considering cost [20] and the trade-off between cost and environmental impact [21]. Such a trade-off is more easily visualized thanks to the set of n-best solutions, which allows making a more rational decision. The framework has also been applied for PNS systematically considering waste minimization [22, 23]; for the synthesis of sugar production processes [24]; the evaluation of target price for processing units under development [25]; the integration of existing industrial parks with new sustainable technologies [26]; in combination with process simulation which enhances the accuracy of the physicochemical models [27], or in Petri-nets for simulation and analysis of complex systems' dynamic behavior [28]; PNS for generation of benzaldehyde from toluene [29]; valorization of lactic acid [30]; and interplant hydrogen integration [31].

Additionally, other contributions have described the employment of the P-graph framework for specific aspects of process systems engineering (PSE) tasks, such as process control and process intensification. Some authors have applied the P-graph framework in plant retrofitting and optimization for utility systems [32, 33]. Also, the framework has been deployed as a design tool in separation-network synthesis [34–36], a special instance of this is the evaluation of alternative designs in azeotropic distillation, where analysis of residual-curve maps is employed to define plausible separations for a particular system [37–39]. The synthesis and analysis of wastewater treatment processes [40, 41] has also been addressed by resorting to the framework.

A fundamental field of application concerns the synthesis of reaction networks as well as the identification of feasible chemical and metabolic reaction pathways. In this regard, the reactions (or reaction steps) to be considered are represented as operating unit type nodes (O-type nodes), thus permitting the solution of these instances as PNS problems by means of the combinatorial algorithms. The works on reaction-pathway identification originate in the work of Fan et al. (2001) [42] who explored the mechanism of ammonia synthesis with the help of the framework. This was further cemented by Fan et al. [43] and Seo et al. (2002) [44] who used the framework for the identification of metabolic pathways in biochemical reactions. Subsequent contributions have addressed the identification of reaction mechanisms in various catalytic and metabolic systems [45–53]. Additional research has focused on the identification and analysis of synthetic reaction routes. For instance, Lakner et al. investigated the feasibility of reaching the reaction network from the initial state of the system, *i.e.*, the startability of the reaction network [54], and Fan et al. [55] employed the framework to generate alternative synthetic routes by assessing their sustainability potential.

Another large field of application related to PSE concerns process intensification problems, which aims at minimizing the resource intake required by a particular process to render it more efficient and sustainable. In this regard, the framework has been employed for energy integration based on heat-exchanger network synthesis (HENS) [56–58]. For instance, Orosz and Friedler [59] discussed HENS by generating the set of n-best heat exchanger network configurations. This work demonstrates the advantages of generating a set of n-best solutions, as it provides additional decision options for the designer, since a network with slightly higher cost might be beneficial according to other criteria, *e.g.*, controllability. Also, the framework has been utilized to explore the total integration of processes [60–66] which, in addition to energy, contemplates mass and water, and power conservation.

The evaluation of process control and the simultaneous synthesis of processes with fault-tolerant control systems have also been explored by means of the P-graph framework [67–69]. For this application, the framework is extended by considering a set of disturbances and actuators that are graphically represented in the P-graphs, in addition to a new axiom of controllability. This is termed as CP-graph when only actuators are contemplated and dCP-graph when disturbances are also considered. These applications related to PSE demonstrate the versatility of the framework and how its capabilities in this field are sought after.

Additionally, the framework has successfully been applied to the evaluation of systems under multi-period conditions, *i.e.*, when the availability of raw materials, processing units, or product requirements varies during the performance of the operation. The implementation of the framework for this type of problems is introduced in detail by Heckl et al. [70, 71]. In these works, the structure of the operating system is depicted with P-graphs, and the resulting structure is replicated for the different periods; thus, the change in the parameters can be evaluated for each period. In addition, the maximum capacity of each unit is considered separately as an O-type node that connects all periods. In this regard, the framework has been employed to derive the best design and the best production policy [72–77]. Moreover, the framework has been applied to the synthesis of polygeneration systems, *i.e.*, networks capable of producing multiple utilities, *e.g.*, heating, electricity, cooling, etc. [78]. For instance, Aviso and Tan [79] developed a methodology combining P-graph with fuzzy optimization for the synthesis of polygeneration systems that simultaneously satisfies multiple product demand targets. Aviso and Tan [80] highlighted such a combination of fuzzy optimization with P-graph to generate the n-best alternative system configurations, which becomes relevant as commercial process units come in discrete sizes and the final network seldom matches the theoretical optimum.

Supply Chain, Logistics, and Production Scheduling

One of the areas with the largest number of P-graph applications in the literature is the design and evaluation of supply chains where the distribution routes, location of manufacture facilities and target sites are optimized in terms of cost or environmental impact. In these cases, the framework is implemented by representing the various steps of the supply chain, such as transportation, loading and manufacture, as O-type nodes. On the other hand, the materials and sites are represented as Material type nodes (M-type nodes) [81, 82]. This concept has been employed for the synthesis and analysis of distribution of hydrocarbons from a refinery [83]; for the enhancement of supply chains of biodiesel [84]; for designing supply chains for distribution and manufacture of biomass [85–88]; design of supply chains with sustainability constraints [89]; and optimal design of supply systems considering uncertainties [90]. A specific instance of this problem, that has attracted significant attention, is the design of systems for supply, distribution, and storage of energy, usually by considering the use of renewable energy and sustainability metrics. In this case, the framework is employed by representing the energy sources and sinks as M-type nodes. Works related to such application have addressed the design of hydrogen energy systems [91–93], the optimal distribution of energy at regional or municipal level with focus in sustainability [94–96], the optimization of energy consumption in production plants [97–99], the generation of a comprehensive superstructure for the analysis of distributed energy systems [100] and the optimization of networks contemplating fatalities during transportation as a safety indicator [101].

Moreover, the framework has been implemented in the management of feedstock and raw materials' scheduling aiming at improving the process' efficiency [102–104]. A recent instance of this type of application is the work of Gu et al. [105] where molecular management is combined with the P-graph framework. The former is employed for optimizing the feedstock structure in ethylene cracking, whereas the latter is deployed to optimize integrated process structures and their parameters. Besides, the framework has been deployed with models that include time constraints, such as the time of earliest availability as well as the time of operation, that permit applications in plant scheduling. Such a model enables the evaluation of numerous alternatives for production considering multiple product constraints, and various policies for storage of materials [106–108].

Sustainability Assessment and Circular Economy

The constant growth of human population and the depletion of resources derived from it have motivated the development of sustainable processes. Such processes rely on the concept that the exploitation of resources should not affect their availability for future generations. Thus, nowadays the concept of sustainability has been extensively disseminated and must be considered since the process' conception. Consequently, various contributions have employed the P-graph framework for the enhancement of processes' environmental performance [196]. In this regard, the framework is usually applied in conjunction to sustainability indicators which can be depicted as M-type nodes of the methodology; these nodes are connected to those operating units that have a pollutant effect. Their representation as M-type nodes facilitates the quantification of the total impact of the units, the assignment of penalties, and the performance of parametric studies by placing constraints on these nodes [109].

Numerous contributions available in the literature employ the P-graph framework to explore sustainability. Some of them focus on the improvement of the process' performance via reduction and management of the waste generated [110–115], while others have dealt with the efficient design of networks for carbon capture [116–120], and negative emission technologies [121]. The framework has also been successfully applied to the evaluation of resource conservation [122, 123] and in life-cycle assessment [124]. In addition, researchers have addressed the sustainability assessment of processes through the raw material management [125, 126] and the synthesis of sustainable supply chains [127–133]. Further research has focused on the evaluation and ranking of clean technologies [134], and the evaluation and optimization of the ammonia cycle as automotive fuel [135].

On the other hand, *Circular economy* (*CE*) is a concept closely related to sustainability. This philosophy intends to renovate the traditional scheme of production into a more ecological system. The traditional production scheme can be regarded as a *linear economy,* which begins with the extraction of resources and finalizes with the generation of waste after consumption. Conversely, in light of

CE, a fraction or the totality of the resources are returned to the productive chain after product consumption. Alternatively, it is proposed to re-use the resources in a second productive process, thereby reducing the generation of waste.

Recent contributions deploy the P-graph framework for the evaluation of such systems. For instance, Fan et al. addressed the integrated design of waste management systems for countries of distinct economic levels in light of *CE*. In this work, the emission of green-house gases is quantified by means of a M-type nodes, and the set of n-best solutions rendered by P-graph offers recommendations for stakeholders with conflicting interests [136]. Furthermore, Yeo et al. synthesized a palm-oil-based biomass network, considering the recycling of materials and energy. In this work, the authors represented structurally the recycle of products to the supply chain by connecting products to operating units designed for resource recovery and return to the initial step of production [137].

Reliability, Resilience, and Risk Assessment

An interesting field of research is the incorporation of reliability considerations into the design of processes. Reliability can be regarded simply as the probability that a process is able to generate all of its desired products. The evaluation of this property of processing systems is extremely important to ensure the correct performance of the process in face to the possible failure of its components. Therefore, it is crucial to determine the reliability of a system since early stages of design. Nevertheless, determination of reliability is not a straightforward task because of the combinatorial nature of the problem, and it requires enumeration of all possible options. However, reliability of a system can be determined resorting to the combinatorial algorithms of P-graph and their capacity of exhaustive enumeration.

In this regard, Holló [138] laid the theoretical foundation of an extension of PNS based on P-graphs where the probability of failure, operating time, and repair time are evaluated besides cost. More recently, Orosz et al. [139] presented a general formula for the evaluation of reliability of processing systems. Such a formula determines the reliability of any process structure on the basis of the individual probabilities of failure of its components. Moreover, Kovacs et al. [140] seized the capability of enumeration of the P-graph framework to establish the algorithmic basis of PNS by considering reliability. In this work, the authors introduce a set of combinatorial algorithms that can estimate the reliability of a complex structure with various types of redundancies.

Furthermore, Orosz et al. [141, 142], have recently introduced a method for the evaluation of resilience of processes in light of the P-graph framework. Resilience is usually considered as the capability of the system to return to its original operation after expected or unexpected failures, and preferably sustain a certain level of production even during the recovery period. Similarly to reliability, the determination of resilience also benefits from the enumeration capacity of the P-graph algorithms. Reliability mainly considers the expected failures in the operation of the components of the system, *e.g.*, due to fatigue. On the contrary, resilience is to address the unexpected failures, where the possible causes are not necessarily known beforehand. For this reason, resilience evaluation is based on the P-graph framework to enumerate over all effects of failures without considering their causes.

Additionally, contributions have been developed in the field of risk analysis as well as criticality analysis [143–150]. Further contributions have examined the effect of redundancy allocation [151], synthesis of heat integrated systems [152], and synthesis of fault-tolerant systems and their operation in abnormal conditions [153, 154].

Non-conventional Applications

The framework's flexibility manifests itself in numerous contributions concerning areas of engineering besides PSE and supply chains. For instance, the framework has been applied to the design and management of vehicle-route programming [155–159], and to information and software science [160]. It has also been utilized in the modelling and synthesis of workflows [161–163]. One particularly interesting application is the evaluation of evacuation routes in buildings [164, 165]. This application is of special interest as the M-type nodes represent a set of people in a particular location in a building structure.

Some contributions address the optimal allocation of human and non-human resources in production lines and institutions [166–171]. Another application where the framework has been deployed deals with the optimization of test sequences during quality assessment in production processes [172] which derived in solutions where testing cost can be reduced. Recently, the framework has been proposed for generating the optimal scheduling of patient appointments in healthcare facilities [173]. A fraction of contributions concerns the results of employing the framework for teaching PSE-related tasks [174, 175].

Extensions of the Model and Software Implementation

In recent years, interesting extensions as well as more accurate models have been implemented into the framework. For instance, Friedler et al. (1992) presented an extension of the axioms that accounts for the manufacture of optional potential products and the purchase and production of raw materials [193]. Holló et al. proposed techniques for enhancing the efficiency of algorithm MSG via combination of operations [194], and for improving the efficiency of algorithm ABB [195]. Tick et al. (2013) proposed a model for PNS where the units have an extended cost termed as *robust PNS* [176]. This model was further investigated by Almási et al. [177]. Also, Kalauz et al. implemented a model with time constraints that allows the effective evaluation of scheduling problems and related processes [178].

Moreover, the methodology has been demonstrated via numerous software realizations the latest version, P-Graph Studio [179], offers a friendly graphical interface for input data and results visualization. The research in software acceleration and improvement is still active, for instance, Bartos and Bertok [180] recently introduced a procedure for the implementation of cooperative parallel computing in the algorithms for PNS.

Depending on the type of mathematical model selected for the operating units of the system, the software realizations have certain level of limitations compared to the general framework. For example, if the mathematical model selected for the operating units is a linear model, the resulting software corresponds to a Mixed-integer linear programming (MILP) solver with additional benefits conferred by the P-graph framework, *e.g.*, n-best solutions. Similarly, if the mathematical models are non-linear the software will correspond to a Mixed-Integer non-linear programming (MINLP) solver, enhanced by the framework's structural capabilities. Furthermore, if the mathematical models of the operating units are given implicitly, *e.g.*, in the form of simulation software's modules, the P-graph framework is able to perform synthesis and optimization of the process despite the fact that there are no explicit equations for the models of the units. It is, therefore, necessary to clearly distinguish between the framework and its software realization. Independently of the fact that P-graph framework is general, its software realization can be limited to certain types of models of the operating units. For instance, software P-Graph Studio employs a fixed-ratio linear model for demonstration purpose.

As previously mentioned, the advantages of the P-graph framework concern the structural part of the problem, which is exploited to confer its capabilities for the enumeration of subnetworks and acceleration of the optimization problem. As the structural properties are invariant for the mathematical model, the advantages of the framework are independent of the model selected to represent the system. This means that any kind of method (linear, non-linear, simulation-based, stochastic, etc.) could be employed to represent the units' performance, materials specifications, objective function, or physical constraints of the system. Some authors imply that the flaws of the linear model with fixed ratio are embedded in the framework [181–183], it is naturally, untrue.

In addition to the structural representations of different models that these contributions propose, How et al. proposed the use artificial neural networks to estimate the performance of the process [184].

Novel Directions

Recent contributions have extended the frontiers of applications of the P-graph framework and have covered recent developments in areas of utmost importance and interest for science and technology. For instance, the optimization of photovoltaic-based microgrids [185], the management of knowledge and data, and organization in industrial facilities [186–188], the analysis of causal networks with desirable and adverse outcomes [189, 190], the analysis networks integrating society and ecology [191] and the analysis and the optimization of industrial production processes by combining the P-graph framework with stochastic modelling [192].

References

1. Friedler F, Tarján K, Huang YW, Fan LT (1993) Graph-theoretic approach to process synthesis: polynomial algorithm for maximal structure generation. Comput Chem Eng 17: 929–942
2. Friedler F, Varga JB, Fan LT (1995) Decision-mapping: a tool for consistent and complete decisions in process synthesis. Chem Eng Sci 50:1755–1768
3. Friedler F, Varga JB, Fehér E, Fan LT (1996) Combinatorially Accerelated branch-and-bound method for solving the MIP model of process network synthesis. In: Floudas CA, Pardalos PM (eds) Nonconvex optimization and its applications. Kluwer Academic Publishers, Norwell, MA, pp 609–626
4. Friedler F, Blicket T, Gyenis J, Tarjáns K (1979) Computerized generation of technological structures. Comput Chem Eng 3:241–249
5. Friedler F, Tarján K, Huang YW, Fan LT (1992) Graph-theoretic approach to process synthesis: axioms and theorems. Chem Eng Sci 47:1973–1988
6. Friedler F, Tarján K, Huang YW, Fan LT (1992) Combinatorial algorithms for process synthesis. Comput Chem Eng 16:S313–S320
7. Friedler F, Fan LT, Imreh B (1998) Process network synthesis: problem definition. Networks 31:119–124
8. Brendel MH, Friedler F, Fan LT (2000) Combinatorial foundation for logical formulation in process network synthesis. Comput Chem Eng 24:1859–1864
9. Friedler F (2006) Effective modeling for process synthesis. J Chinese Inst Chem Eng 37:481–490
10. Lam HL (2013) Extended P-graph applications in supply chain and process network synthesis. Curr Opin Chem Eng 2:475–486
11. Klemeš JJ, Varbanov PS (2015) Spreading the message: P-graph enhancements: implementations and applications. Chem Eng Trans 45:1333–1338
12. Varbanov PS, Friedler F, Klemeš JJ (2017) Process network design and optimisation using P-graph: the success, the challenges and potential roadmap. Chem Eng Trans 61:1549–1554
13. Cabezas H, Argoti A, Friedler F et al (2018) Design and engineering of sustainable process systems and supply chains by the P-graph framework. Environ Prog Sustain Energy 37:624–636
14. Tan RR, Aviso KB, Klemeš JJ et al (2018) Towards generalized process networks: prospective new research frontiers for the p-graph framework. Chem Eng Trans 70:91–96
15. Friedler F, Aviso KB, Bertók B et al (2019) Prospects and challenges for chemical process synthesis with P-graph. Curr Opin Chem Eng 26:58–64
16. Sangalang KPH, Belmonte BA, Ventura J-RS et al (2021) P-graph method for optimal synthesis of philippine agricultural waste-based integrated biorefinery. Chem Eng Trans 83: 103–108
17. Benjamin MFD, Ventura J-RS, Sangalang KPH et al (2021) Optimal synthesis of Philippine agricultural residue-based integrated biorefinery via the P-graph method under supply and demand constraints. J Clean Prod 308:127348
18. Pinheiro Pires AP, Martinez-Valencia L, Tanzil AH et al (2021) Synthesis and techno-economic analysis of pyrolysis-oil-based biorefineries using P-graph. Energy and Fuels 35: 13159–13169

19. Atkins MJ, Walmsley TG, Ong BHY, et al (2016) Application of P-graph techniques for efficient the use of wood processing residues in biorefineries. In: 22nd international congress of chemical and process engineering, CHISA 2016 and 19th conference on process integration, modelling and optimisation for energy saving and pollution reduction, PRES 2016. Czech Society of Chemical Engineering, p 961
20. Van FY, Klemeš JJ, Lee CT (2018) Pre- and post-treatment assessment for the anaerobic digestion of lignocellulosic waste: P-graph. Chem Eng Trans 63:1–6
21. Van FY, Klemeš JJ, Perry S, Lee CT (2019) Anaerobic digestion of lignocellulosic waste: environmental impact and economic assessment. J Environ Manag 231:352–363
22. Friedler F, Fan LT (1992) Process synthesis incorporating in-plant waste treatment: algorithmic approach. In: Session on systematic design approaches to hazardous-waste management at the ACS special symposium on hazardous-waste management. Atlanta, GA, pp. 325–328
23. Friedler F, Tarján K, Huang YW, Fan LT (1991) A systematic approach to waste minimizing synthesis of a chemical process: production of Perchloromethyl Mercaptan. In: Proceedings of the AIChE National Meeting. Pittsburg, PA, pp. 93–98
24. Halasz L, Gwehenberger G, Narodoslawsky M (2007) Process synthesis for the sugar sector - computer based insights in industrial development. Comput Aided Chem Eng 24:431–436
25. Fan LT, Zhang T, Liu J et al (2008) Price-targeting through iterative flowsheet syntheses in developing novel processing equipment: pervaporation. Ind Eng Chem Res 47:1556–1561
26. Halasz L, Ederz M, Sandorz N et al (2010) Optimal integration of sustainable technologies in industrial parks. Chem Eng 19:43
27. Pimentel J, Orosz Á, Tan RR, Friedler F (2020) Automated synthesis of process-networks by the integration of P-graph with process simulation. Chem Eng Trans 81:1171–1176
28. Lakner R, Friedler F, Bertók B (2017) Synthesis and analysis of process networks by joint application of P-graphs and Petri nets. In: van der Aalst W, Best E (eds) Application and theory of Petri nets and concurrency. PETRI NETS 2017. Lecture notes in computer science. Springer, Cham, pp. 309–329
29. Pimentel J, Heckl I, Bertók B et al (2015) Process network synthesis for benzaldehyde production: P-graph approach. Chem Eng Trans 45:1369–1374
30. Pimentel J, Rodríguez G, Gil I (2020) Synthesis of alternative cost-effective process flowsheets for lactic acid valorization by means of the P-graph methodology. Ind Eng Chem Res 59:5921–5930
31. Affery AP, Tan JX, Ong IYB et al (2021) Optimal planning of inter-plant hydrogen integration (IPHI) in eco-industrial park with P-graph and game theory analyses. Process Saf Environ Prot 155:197–218
32. Liu J, Fan LT, Seib P et al (2006) Holistic approach to process retrofitting: application to downstream process for biochemical production of organics. Ind Eng Chem Res 45:4200–4207
33. Halasz L, Nagy AB, Ivicz T et al (2002) Optimal retrofit design and operation of the steam-supply system of a chemical complex. Appl Therm Eng 22:939–947
34. Heckl I, Friedler F, Fan LT (2010) Solution of separation-network synthesis problems by the P-graph methodology. Comput Chem Eng 34:700–706
35. Liu J, Fan LT, Seib PA et al (2004) Downstream process synthesis for biochemical production of butanol, ethanol, and acetone from grains: generation of optimal and near-optimal flowsheets with conventional operating units. Biotechnol Prog 20:1518–1527
36. Xu X, Zhu C, Ma Y, Song H (2015) A robust combinatorial approach based on P-graph for superstructure generation in downstream bioprocesses. Brazilian J Chem Eng 32:259–267
37. Xu W, Diwekar UM (2005) Environmentally friendly heterogeneous Azeotropic distillation system design: integration of EBS selection and IPS recycling. Ind Eng Chem Res 44:4061–4067
38. Feng G, Fan LT, Friedler F, Seib PA (2000) Identifying operating units for the design and synthesis of azeotropic- distillation systems. Ind Eng Chem Res 39:175–184

39. Feng G, Fan LT, Seib PA et al (2003) Graph-theoretic method for the algorithmic synthesis of Azeotropic-distillation systems. Ind Eng Chem Res 42:3602–3611
40. Yenkie KM, Burnham S, Dailey J et al (2019) Generating efficient wastewater treatment networks: an integrated approach comprising of contaminant properties, technology suitability, plant design, and process optimization. In: Kiss AA, Zondervan E, Lakerveld R, Özkan L (eds) 29th European symposium on computer aided process engineering. Elsevier, London, pp 1603–1608
41. Yenkie KM, Pimentel J, Orosz Á et al (2021) The P-graph approach for systematic synthesis of wastewater treatment networks. AICHE J:e17253
42. Fan LT, Bertók B, Friedler F, Shafie S (2001) Mechanisms of ammonia-synthesis reaction revisited with the aid of a novel graph-theoretic method for determining candidate mechanisms in deriving the rate law of a catalytic reaction. Hungarian J Ind Chem 29:71–80
43. Fan LT, Bertók B, Friedler F (2002) A graph-theoretic method to identify candidate mechanisms for deriving the rate law of a catalytic reaction. Comput Chem 26:265–292
44. Seo H, Lee DY, Park S et al (2001) (2001) graph-theoretical identification of pathways for biochemical reactions. Biotechnol Lett 2319(23):1551–1557
45. Díaz-Alvarado FA, Miranda-Pérez J, Grossmann IE (2017) Search for reaction pathways with P-graphs and reaction blocks: methanation of carbon dioxide with hydrogen. J Math Chem 564(56):1011–1102
46. Yun C, Kim TY, Zhang T et al (2013) Determination of the thermodynamically dominant metabolic pathways. Ind Eng Chem Res 52:222–229
47. Barany M, Bertók B, Fan LT, Friedler F (2013) Relationship between extreme pathways and structurally minimal pathways. Bioprocess Biosyst Eng 36:1199–1203
48. Fan LT, Lin Y-C, Shafie S et al (2012) Exhaustive identification of feasible pathways of the reaction catalyzed by a catalyst with multiactive sites via a highly effective graph-theoretic algorithm: application to ethylene hydrogenation. Ind Eng Chem Res 51:2548–2552
49. Lin Y-C, Fan LT, Shafie S et al (2009) Generation of light hydrocarbons through Fischer-Tropsch synthesis: identification of potentially dominant catalytic pathways via the graph-theoretic method and energetic analysis. Comput Chem Eng 33:1182–1186
50. Fan LT, Lin Y-C, Shafie S et al (2008) Graph-theoretic and energetic exploration of catalytic pathways of the water-gas shift reaction. J Chinese Inst Chem Eng 39:467–473
51. Lin Y-C, Fan LT, Shafie S et al (2008) Catalytic pathways identification for partial oxidation of methanol on copper−zinc catalysts: CH3OH + 1/2O2 $\leftrightarrow$ CO2 + 2H2. Ind Eng Chem Res 47:2523–2527
52. Lee DY, Fan LT, Park S et al (2005) Complementary identification of multiple flux distributions and multiple metabolic pathways. Metab Eng 7:182–200
53. Fan LT, Shafie S, Bertók B et al (2005) Graph-theoretic approach for identifying catalytic or metabolic pathways. J Chinese Inst Eng Trans Chinese Inst Eng A/Chung-kuo K Ch'eng Hsuch K'an 28:1021–1037
54. Lakner R, Bertók B, Friedler F (2018) Synthesis of startable reaction pathways. Chem Eng Trans 70:1129–1134
55. Fan LT, Zhang T, Liu J et al (2007) Assessment of sustainability-potential: hierarchical approach. Ind Eng Chem Res 46:4506–4516
56. Nagy AB, Adonyi R, Halasz L et al (2001) Integrated synthesis of process and heat exchanger networks: algorithmic approach. Appl Therm Eng 21:1407–1427
57. Orosz Á, Pimentel J, Friedler F (2019) Simultaneous synthesis of processes with its heat exchanger networks: P-graph approach. Chem Eng Trans 76:1219–1224
58. Orosz Á, How BS, Friedler F (2021) Multiple-solution heat exchanger network synthesis using P-HENS solver. J Taiwan Inst Chem Eng 12
59. Orosz Á, Friedler F (2020) Multiple-solution heat exchanger network synthesis for enabling the best industrial implementation. Energy 208:118330

60. Chin HH, Foo DCY, Lam HL (2019) Simultaneous water and energy integration with isothermal and non-isothermal mixing – a P-graph approach. Resour Conserv Recycl 149: 687–713
61. Tay ZX, Lim JS, Alwi SRW et al (2020) Integrated evaluation framework of the cogeneration energy system via application of fuzzy P-graph. Chem Eng Trans 81:1141–1146
62. Walmsley TG, Jia X, Philipp M et al (2018) Total site utility system structural design using P-graph. Chem Eng Trans 63:31–36
63. Ong BHY, Atkins MJ, Walmsley TG, Walmsley MRW (2017) Total site heat and mass integration and optimisation using P-graph: a biorefinery case study. Chem Eng Trans 61: 727–732
64. Ong BHY, Walmsley TG, Atkins MJ, Walmsley MRW (2017) Total site mass, heat and power integration using process integration and process graph. J Clean Prod 167:32–43
65. Aviso KB, Chiu ASF, Yu KDS et al (2015) P-graph for optimising industrial symbiotic networks. Chem Eng Trans 45:1345–1350
66. Lee S, Park S (1996) Synthesis of mass exchange network using process graph theory. Compurers Chem Engng 20:S201–S205
67. Hangos KM, Friedler F, Varga JB, Fan LT (1994) A graph-theoretic approach to integrated process and control system synthesis. Integr Process Des Control IFAC Postp:61–66
68. Hangos KM, Varga JB, Friedler F, Fan LT (1995) Integrated synthesis of a process and its fault-tolerant control system. Comput Chem Eng 19:465–470
69. Gál IP, Varga JB, Hangos KM (1998) Integrated structure design of a process and its control system. J Process Control 8:251–263
70. Heckl I, Halász L, Szlama A et al (2014) Modeling multi-period operations using the P–graph methodology. Comput Aided Chem Eng 33:979–984
71. Heckl I, Halász L, Szlama A et al (2015) Process synthesis involving multi-period operations by the P-graph framework. Comput Chem Eng 83:157–164
72. Aviso KB, Lee JY, Tan RR (2016) A P-graph model for multi-period optimization of isolated energy systems. Chem Eng Trans 52:865–870
73. Aviso KB, Lee JY, Dulatre JC et al (2017) A P-graph model for multi-period optimization of sustainable energy systems. J Clean Prod 161:1338–1351
74. Bertók B, Bartos A (2018) Algorithmic process synthesis and optimisation for multiple time periods including waste treatment: latest developments in p-graph studio software. Chem Eng Trans 70:97–102
75. Bertók B, Bartos A (2020) Renewable energy storage and distribution scheduling for microgrids by exploiting recent developments in process network synthesis. J Clean Prod 244:118520
76. Friedler F, Varbanov PS, Klemeš JJ (2009) Advanced HENs design for multi-period operation using P-graph. Chem Eng Trans 18:457–462
77. Tan RR, Aviso KB (2016) An extended P-graph approach to process network synthesis for multi-period operations. Comput Chem Eng 85:40–42
78. Tan RR, Cayamanda CD, Aviso KB (2014) P-graph approach to optimal operational adjustment in polygeneration plants under conditions of process inoperability. Appl Energy 135: 402–406
79. Aviso KB, Tan RR (2017) Fuzzy P-graph for optimal synthesis of polygeneration systems. Chem Eng Trans 61:115–120
80. Aviso KB, Tan RR (2018) Fuzzy P-graph for optimal synthesis of cogeneration and trigeneration systems. Energy 154:258–268
81. Kim Y, Park S (2006) Supply network modeling using process graph theory: a framework for analysis. SICE-ICASE Int Jt Conf 2006:1730–1733
82. Fan LT, Kim Y, Yun C et al (2009) Design of optimal and near-optimal enterprise-wide supply networks for multiple products in the process industry. Ind Eng Chem Res 48:2003–2008
83. Kim Y, Fan LT, Yun C et al (2008) Graph-theoretic approach to optimal synthesis of supply networks: distribution of gasoline from a refinery. Comput Aided Chem Eng 25:247–252

84. Bertók B, Kalauz K, Süle Z, Friedler F (2013) Combinatorial algorithm for synthesizing redundant structures to increase reliability of supply chains: application to biodiesel supply. Ind Eng Chem Res 52:181–186
85. How BS, Hong BH, Lam HL, Friedler F (2016) Synthesis of multiple biomass corridor via decomposition approach: a P-graph application. J Clean Prod 130:45–57
86. Lam HL, Klemeš JJ, Varbanov PS, Kravanja Z (2012) P-graph synthesis of open-structure biomass networks. Ind Eng Chem Res 52:172–180
87. How BS, Hong BH, Lam HL, Friedler F (2015) Synthesis of multiple biomass corridor via decomposition approach: a P-graph application. Chem Eng Trans 45:1363–1368
88. Haleem AA, Hashim H, Varbanov PS et al (2015) Case study: optimisation of Federal Land Development Authority Jengka supply chain management using P-graph approach. Chem Eng Trans 45:1375–1380
89. Cabezas H, Heckl I, Bertók B, Friedler F (2015) Use the P-graph framework to design supply chains for sustainability. Chem Eng Prog 111:41–47
90. Süle Z, Bertók B, Friedler F, Fan LT (2011) Optimal design of supply chains by P-graph framework under uncertainties. Chem Eng Trans 25:453–458
91. Lee CH, Chong DYL, Hemmati S et al (2020) A P-graph approach for the synthesis of national-wide bio-hydrogen network from palm oil mill effluent. Int J Hydrog Energy 45: 17200–17219
92. Tan JX, Tan HS, Peter Affery A et al (2021) A P-graph approach for the synthesis of hydrogen networks with pressure and impurity constraints. Int J Hydrog Energy 46:29198–29215
93. Lim JY, How BS, Rhee G et al (2020) Transitioning of localized renewable energy system towards sustainable hydrogen development planning: P-graph approach. Appl Energy 263: 114635
94. Lam HL, Varbanov PS, Klemeš JJ (2010) Optimisation of regional energy supply chains utilising renewables: P-graph approach. Comput Chem Eng 34:782–792
95. Maier S, Narodoslawsky M (2014) Optimal renewable energy Systems for Smart Cities. Comput Aided Chem Eng 33:1849–1854
96. Maier S (2016) Smart energy systems for smart city districts: case study Reininghaus District. Energy Sustain Soc 6
97. Éles A, Halász L, Heckl I, Cabezas H (2018) Energy consumption optimization of a manufacturing plant by the application of the p-graph framework. Chem Eng Trans 70: 1783–1788
98. Lakner R, Varbanov PS, Friedler F (2019) Systems analysis of electricity transmission networks for improved sustainability. Chem Eng Trans 76:619–624
99. Éles A, Halász L, Heckl I, Cabezas H (2019) Evaluation of the energy supply options of a manufacturing plant by the application of the P-graph framework. Energies 12:1484
100. Voll P, Klaffke C, Hennen M, Bardow A (2013) Automated superstructure-based synthesis and optimization of distributed energy supply systems. Energy 50:374–388
101. Ng RLT, Tan RR, Hassim MH (2015) P-graph methodology for bi-objective optimisation of bioenergy supply chains: economic and safety perspectives. Chem Eng Trans 45
102. Cao J, He Y, Zhu Q (2020) Solutions selection based on the P-graph integrated data envelopment analysis for material scheduling in the ethylene production. Can J Chem Eng
103. Cao J, He Y, Zhu Q (2020) Feedstock scheduling optimization based on novel extensible P-graph reasoning in ethylene production. Ind Eng Chem Res 59:18965–18976
104. Mu P, Gu X, Zhu Q (2019) Modeling and optimization of ethylene cracking feedstock scheduling based on P-graph. Huagong Xuebao/CIESC J 70:556–563
105. Gu H, Li J, Mu P, Zhu Q (2020) Improving the operational efficiency of ethylene cracking integrated with refining by molecular management. Ind Eng Chem Res 59:13160–13174
106. Frits M, Bertók B (2021) Routing and scheduling field service operation by P-graph. Comput Oper Res 136
107. Frits M, Bertók B (2020) Scheduling custom printed napkin manufacturing by P-graphs. Comput Chem Eng 141:107017

108. Frits M, Bertók B (2014) Process scheduling by synthesizing time Constained process-networks. In: Klemeš J, Varvanov P, Liew PY (eds) 24th European symposium on computer aided process engineering- ESCAPE 24. Budapest, Hungary
109. Bertók B, Heckl I (2016) Process synthesis by the P-graph framework involving sustainability. In: Ruiz-Mercado G, Cabezas H (eds) Sustainability in the design, synthesis and analysis of chemical engineering processes. Butterworth-Heinemann, Oxford, pp 203–225
110. Halim I, Srinivasan R (2002) Systematic waste minimization in chemical processes. 1. Methodology Ind Eng Chem Res 41:196–207
111. Halim I, Srinivasan R (2006) Systematic waste minimization in chemical processes. 3. Batch operations Ind Eng Chem Res 45:4693–4705
112. Halim I, Srinivasan R (2008) Designing sustainable alternatives for batch operations using an intelligent simulation–optimization framework. Chem Eng Res Des 86:809–822
113. Ali RA, Ibrahim NNLN, Lam HL (2019) Conversion technologies: evaluation of economic performance and environmental impact analysis for municipal solid waste in Malaysia. PRO 7
114. Walmsley TG, Varbanov PS, Klemeš JJ (2017) Networks for utilising the organic and dry fractions of municipal waste: P-graph approach. Chem Eng Trans 61:1357–1362
115. Kollmann R, Maier S, Shahzad K et al (2014) Waste water treatment plants as regional energy cells - evaluation of economic and ecologic potentials in Austria. Chem Eng Trans 39:607–612
116. Aviso KB, Belmonte BA, Benjamin MFD et al (2019) Synthesis of optimal and near-optimal biochar-based carbon management networks with P-graph. J Clean Prod 214:893–901
117. Aviso KB, Arogo JIA, Coronel ALO et al (2018) P-graph approach to planning biochar-based carbon management networks. Chem Eng Trans 70:37–42
118. Tan RR, Aviso KB, Foo DCY (2017) P-graph and Monte Carlo simulation approach to planning carbon management networks. Comput Chem Eng 106:872–882
119. Chong FK, Lawrence KK, Lim PP et al (2014) Planning of carbon capture storage deployment using process graph approach. Energy 76:641–651
120. Varbanov PS, Friedler F (2008) P-graph methodology for cost-effective reduction of carbon emissions involving fuel cell combined cycles. Appl Therm Eng 28:2020–2029
121. Pimentel J, Orosz Á, Aviso KB et al (2021) Conceptual Design of a Negative Emissions Polygeneration Plant for multiperiod operations using P-graph. PRO 9:233
122. Foo DCY, Chew LX, Chong TL et al (2021) A note for the extended P-graph model for the synthesis of batch water network. Process Integr Optim Sustain 2021:1–12
123. Lim CH, Pereira PS, Shum CK et al (2017) Synthesis of resource conservation networks with P-graph approach—direct reuse/recycle. Process Integr Optim Sustain 1:69–86
124. Lim JY, How BS, Teng SY et al (2021) Multi-objective lifecycle optimization for oil palm fertilizer formulation: a hybrid P-graph and TOPSIS approach. Resour Conserv Recycl 166: 105357
125. Mu P, Yang Y, Han Y et al (2020) Raw material management networks based on an improved P-graph integrated carbon emission pinch analysis (CEPA-P-graph) method. Can J Chem Eng 98:676–689
126. Lim CH, Pereira PS, Shum CK et al (2017) Synthesis of material interception networks with P-graph. Process Integr Optim Sustain 1:225–235
127. Heckl I, Cabezas H, Friedler F (2015) Designing sustainable supply chains in the energy-water- food nexus by the P-graph methodology. Chem Eng Trans 45:1351–1356
128. Vance L, Heckl I, Bertók B et al (2015) Designing sustainable energy supply chains by the P-graph method for minimal cost, environmental burden, energy resources input. J Clean Prod 94:144–154
129. Vance L, Cabezas H, Heckl I et al (2012) Synthesis of sustainable energy supply chain by the P-graph framework. Ind Eng Chem Res 52:266–274
130. Wang B, Van FY, Chin HH et al (2020) Emission-cost nexus optimisation and performance analysis of downstream oil supply chains. J Clean Prod 266
131. Hemmati S, Elnegihi MM, Lee CH et al (2020) Synthesis of large-scale bio-hydrogen network using waste gas from landfill and anaerobic digestion: a p-graph approach. Processes 8:505

132. How BS, Lam HL (2019) PCA method for debottlenecking of sustainability performance in integrated biomass supply chain. Process Integr Optim Sustain 3:43–64
133. How BS, Yeoh TT, Tan TK et al (2018) Debottlenecking of sustainability performance for integrated biomass supply chain: P-graph approach. J Clean Prod 193:720–733
134. Low CX, Ng WY, Putra ZA et al (2020) Induction approach via P-graph to rank clean technologies. Heliyon 6
135. Angeles DA, Are KRAG, Aviso KB et al (2017) Optimization of the automotive ammonia fuel cycle using P-graphs. ACS Sustain Chem Eng 5:8277–8283
136. Van FY, Klemeš JJ, Walmsley TG, Bertók B (2020) Implementing circular economy in municipal solid waste treatment system using P-graph. Sci Total Environ 701:134652
137. Yeo JYJ, How BS, Teng SY et al (2020) Synthesis of sustainable circular economy in palm oil industry using graph-theoretic method. Sustain 12:1–29
138. Holló C (2013) Reliability considerations for process network synthesis problems. Cent Eur J Oper Res 21:885–897
139. Orosz Á, Kovács Z, Friedler F (2018) Synthesis of processing systems taking into account reliability. Chem Eng Trans 70:1111–1116
140. Kovács Z, Orosz Á, Friedler F (2019) Synthesis algorithms for the reliability analysis of processing systems. Cent Eur J Oper Res 27:573–595
141. Orosz Á, Pimentel J, Friedler F (2020) General formulation for the resilience of processing systems. Chem Eng Trans 81:859–864
142. Orosz Á, Pimentel J, Friedler F (2021) Processing systems design considering resilience. In: Türkay M, Gani R (eds) 31st European symposium on computer aided process engineering. Elsevier, pp 807–812
143. Polgár B, Selényi E (2003) Probabilistic diagnostics with P-graphs. Acta Cybern 16:279–291
144. Polgár B, Selényi E, Bartha T (2005) On the extension and applicability of the P-graph modeling paradigm to system-level diagnostic problems. Scalable Comput 6:45–57
145. Tan RR, Benjamin MFD, Cayamanda CD et al (2016) P-graph approach to optimizing crisis operations in an industrial complex. Ind Eng Chem Res 55:3467–3477
146. Benjamin MFD, Cayamanda CD, Belmonte BA et al (2016) A risk-based criticality analysis in bioenergy parks using P-graph method. Chem Eng Trans 52:1243–1248
147. Benjamin MFD, Cayamanda CD, Tan RR, Razon LF (2017) P-graph approach to criticality analysis in integrated bioenergy systems. Clean Techn Environ Policy 19:1841–1854
148. Benjamin MFD (2018) Multi-disruption criticality analysis in bioenergy-based eco-industrial parks via the P-graph approach. J Clean Prod 186:325–334
149. Süle Z, Baumgartner J, Dörgő G, Abonyi J (2019) P-graph-based multi-objective risk analysis and redundancy allocation in safety-critical energy systems. Energy 179:989–1003
150. Benjamin MFD (2017) P-graph approach to criticality analysis in bioenergy parks under uncertainty. Chem Eng Trans 61:619–624
151. Süle Z, Baumgartner J, Abonyi J (2018) Reliability - redundancy allocation in process graphs. Chem Eng Trans 70:991–996
152. Orosz Á, Kovács Z, Friedler F (2019) Synthesis of heat integrated processing systems taking into account reliability. Energy 181:214–225
153. Orosz Á, Friedler F (2019) Synthesis technology for failure analysis and corrective actions in process systems engineering. In: Kiss AA, Zondervan E, Lakerveld R, Ozkan L (eds) Proceedings of the 29th European Symposium on Computer Aided Process Engineering. Eindhoven, pp. 1–6
154. Aviso KB, Marfori I.A.V. III, Tan RR, Ubando AT (2020) Optimizing abnormal operations of off-grid community utility systems with fuzzy P-graph. Energy 202:
155. Ercsey Z, Nagy A, Tick J, Kovács Z (2021) Bus transport process networks with arbitrary launching times. Acta Polytech Hungarica 18:2021–2125
156. Konig E, Bertók B (2019) Process graph approach for two-stage decision making: transportation contracts. Comput Chem Eng 121:1–11

157. Nagy A, Ercsey Z, Tick J, Kovács Z (2019) Bus transport process network synthesis. Acta Polytech Hungarica 16:25–43
158. Friedler F, Barany M, Bertók B et al (2011) Solving vehicle assignment problems by process-network synthesis to minimize cost and environmental impact of transportation. Clean Techn Environ Policy 13:637–642
159. Adonyi R, Heckl I, Olti F (2013) Scheduling of bus maintenance by the P-graph methodology. Optim Eng 14:565–574
160. García-Ojeda JC, Bertók B, Friedler F et al (2015) A preliminary study of the application of the P- graph methodology for organization-based multiagent system designs: assessment. Acta Polytech Hungarica 12:103–122
161. Tick J (2007) P-Graph-based Workflow Modelling Acta Polytech Hungarica 4:75–88
162. Tick J (2013) Potential application of P-graph-based workflow in logistics. In: Madarász L, Živčák J (eds) Aspects of computational intelligence: theory and applications. Topics in intelligent engineering and informatics, vol 2. Springer, Berlin, Heidelberg, pp 293–303
163. Tick J (2011) Visualisation and simulation of P-graph based workflow systems. In: 2011 6th IEEE international symposium on applied computational intelligence and informatics (SACI). pp 231–234
164. García-Ojeda JC, Bertók B, Friedler F (2012) Planning evacuation routes with the P-graph framework. Chem Eng Trans 29:1531–1536
165. García-Ojeda JC, Bertók B, Friedler F, Fan LT (2013) Building-evacuation-route planning via time-expanded process-network synthesis. Fire Saf J 61:338–347
166. Aviso KB, Cayamanda CD, Solis FDB et al (2015) P-graph approach for GDP-optimal allocation of resources, commodities and capital in economic systems under climate change-induced crisis conditions. J Clean Prod 92:308–317
167. Aviso KB, Cayamanda CD, Mayol AP, Yu KDS (2017) Optimizing human resource allocation in organizations during crisis conditions: a P-graph approach. Process Integr Optim Sustain 1:59–68
168. Aviso KB, Demeterio FPA, Lucas RIG, Tan RR (2018) P-graph approach to planning human resource expansion for universities in transition. Chem Eng Trans 70:277–282
169. Aviso KB, Chiu ASF, Demeterio FP et al (2019) Optimal human resource planning with P-graph for universities undergoing transition. J Clean Prod 224:811–822
170. Bartos A, Bertók B (2020) Production line balancing by P-graphs. Optim Eng 21:567–584
171. Tan RR, Aviso KB, Yu KDS et al (2015) P-graph approach to allocation of inoperability in urban infrastructure systems. Chem Eng Trans 45:1339–1344
172. Baumgartner J, Süle Z, Bertók B, Abonyi J (2019) Test-sequence optimisation by survival analysis. Cent Eur J Oper Res 27:357–375
173. Süle Z, Baumgartner J, Vathy-Fogarassy Á (2020) Managing alternative patient appointments using P-graph methodology. Stud Health Technol Inform 271:57–64
174. Promentilla MAB, Lucas RIG, Aviso KB, Tan RR (2017) Problem-based learning of process systems engineering and process integration concepts with metacognitive strategies: the case of P-graphs for polygeneration systems. Appl Therm Eng 127:1317–1325
175. Lam HL, Tan RR, Aviso KB (2016) Implementation of P-graph modules in undergraduate chemical engineering degree programs: experiences in Malaysia and the Philippines. J Clean Prod 136:254–265
176. Tick J, Imreh C, Kovács Z (2013) Business process modeling and the robust PNS problem. Acta Polytech Hungarica 10:193–204
177. Almási D, Imreh C, Kovács T (2014) Heuristic algorithms for the robust PNS problem. Acta Polytech Hungarica 11:169–181
178. Kalauz K, Süle Z, Bertók B et al (2012) Extending process-network synthesis algorithms with time bounds for supply network design. Chem Eng Trans 29:259–264
179. P-graph community (2015) P-graph web page. http://p-graph.org. Accessed 13 Jun 2020
180. Bartos A, Bertók B (2019) Parameter tuning for a cooperative parallel implementation of process-network synthesis algorithms. Cent Eur J Oper Res 27:551–572

181. Ong BHY, Walmsley TG, Atkins MJ et al (2016) Approximation of non-linear cost functions in P-graph structures. Chem Eng Trans 52:1093–1098
182. Szlama A, Heckl I, Cabezas H (2016) Optimal design of renewable energy systems with flexible inputs and outputs using the P-graph framework. AICHE J 62:1143–1153
183. Éles A, Heckl I, Cabezas H (2021) Modeling technique in the P-graph framework for operating units with flexible input ratios. Cent Eur J Oper Res 29:463–489
184. How BS, Teng SY, Leong WD et al (2019) Non-linear programming via P-graph framework. Chem Eng Trans 76:499–504
185. Mah AXY, Ho WS, Hassim MH et al (2021) Optimization of photovoltaic-based microgrid with hybrid energy storage: a P-graph approach. Energy 233:121088
186. Cao J, Mu P, Gu X, Zhu Q (2019) Automatic generation method of process knowledge based on P-graph. Huagong Xuebao/CIESC J 70:467–474
187. Cao J, He Y, Zhu Q (2020) Maximal structure generation of superstructure for semantic triple generated by DEVS ontology in the process industry. Chemom Intell Lab Syst 205:104119
188. Cao J, He Y, Zhu Q (2021) An ontology-based procedure knowledge framework for the process industry. Can J Chem Eng 99:530–542
189. Tan RR, Aviso KB, Lao AR, Promentilla MAB (2021) P-graph causality maps. Process Integr Optim Sustain:1–16
190. Tan RR, Aviso KB, Lao AR et al (2021) Modelling vicious networks with P-graph causality maps 1:3
191. Lao AR, Cabezas H, Orosz Á et al (2020) Socio-ecological network structures from process graphs. PLoS One 15
192. Bernini R, Bondavalli A, Lollini P, Montecchi L (2016) Combining SAN and P-graphs for the analysis and optimization of industrial processes. In: Proceedings - 2016 12th European dependable computing conference, EDCC 2016. Institute of Electrical and Electronics Engineers Inc., Gothenburg, Sweden, pp 197–207
193. Friedler F, Tarjan K, Huang YW, Fan LT (1992) Combinatorial foundation of process synthesis. In: 42nd Canadian chemical engineering conference. Canadian society for chemical engineering, Toronto, Canada, pp 207–208
194. Holló CS, Blázsik Z, Cs I, Kovács Z (1999) On a merging reduction of the process network synthesis problem. Acta Cybernetica 14:251–261
195. Holló CS (2000) A look ahead branch-and-bound procedure for solving PNS problems. Pure Mathematics and Applications 11(2):265–279
196. Klemes J, Friedler F, Bulatov I, Varbanov P (2011) Sustainability in the process industry: integration and optimization. McGraw-Hill Education, New York

Chapter 8
Case Study: Synthesis of a Production Process for Adipic Acid

Abstract This chapter illustrates the use of the P-graph framework for the conceptual design of a process, *i.e.*, process network synthesis (PNS). Herein, a methodology for PNS is discussed and applied to the identification of alternative process networks (flowsheets) yielding adipic acid, one of the most employed precursors in the chemical industry. Two types of mathematical models are described which can be employed to solve the synthesis problems in distinct stages. This demonstrates some of the framework's capabilities for algorithmically synthesizing process alternatives when applied to a realistic problem.

Keywords P-graph · P-graph applications · Adipic acid · n-best · Process synthesis

Process Network Synthesis by Resorting to the P-graph Framework

In general, PNS comprises numerous steps in which the P-graph framework can be employed as evaluation and design tool. Figure 8.1 depicts the major steps to be followed in algorithmically identifying process alternatives as per the framework. The first step, termed Reaction-Network Synthesis (RNS), concerns only processes involving chemical transformations. RNS entails the selection of chemical reactions that can yield the product, or products, of interest from a set of raw materials. To this end, information in the available literature, patents, and/or experience is thoroughly reviewed to find out the reaction routes of industrial significance as well as those still being investigated for the process of concern. Upon identifying such routes, the P-graph framework is employed for the systematic evaluation of the plausible chemical routes that the designer could select. This evaluation may be based on various criteria such as environmental impacts, safety indicators, gross profit, etc. Thus, the framework renders a set of different reaction networks from which it is possible to select the most convenient options to be assessed in the next step of synthesis. Instances of RNS can be found in the available literature [1].

Once the identification of plausible chemical reactions is completed, the next step for synthesizing the process is the definition of the set of operating units that can

F. Friedler et al., *P-graphs for Process Systems Engineering*,
https://doi.org/10.1007/978-3-030-92216-0_8

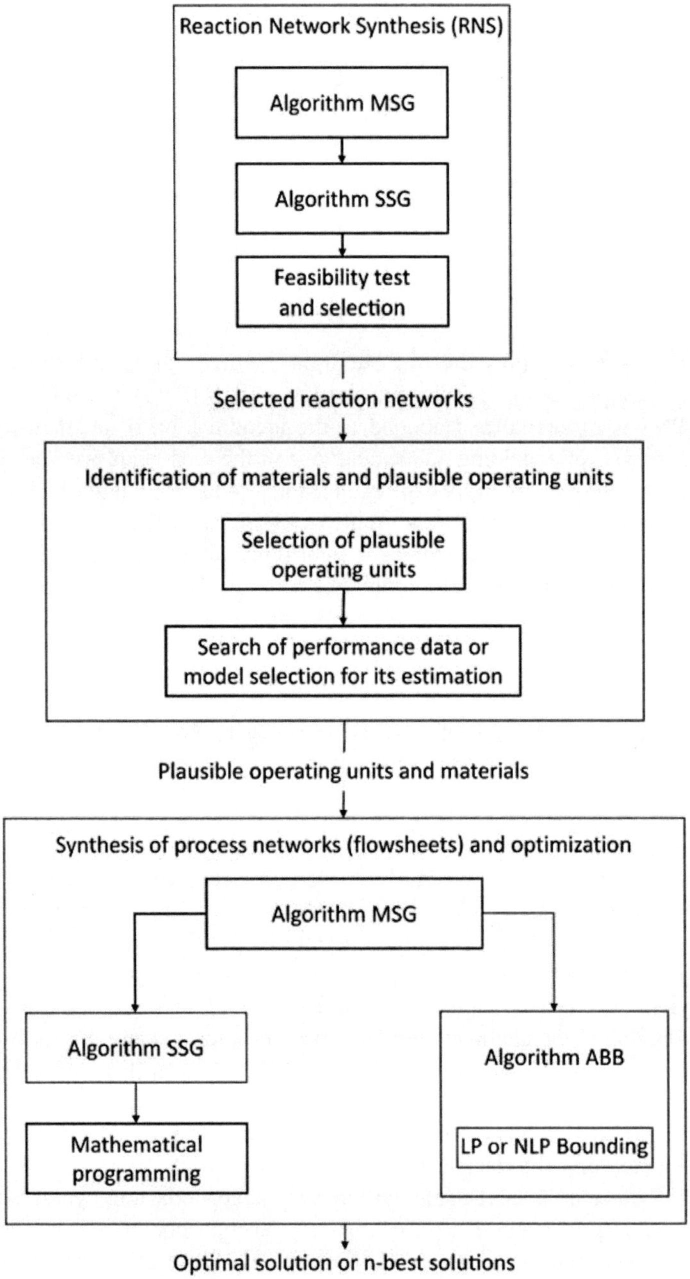

Fig. 8.1 Methodology for process network synthesis in light of the P-graph framework

perform the transformations and separations required, *e.g.*, reactors, liquid-liquid separation systems, filters, etc. Such a definition comprises the identification of not only operating units, but also the concomitant mathematical models that allow the estimation of their performance according to the designer's criteria. Instances of these mathematical models are cost estimation relations, mass and energy balances, thermodynamic behavior, etc. For example, if the designer desires to evaluate the networks of operating units in terms of environmental impact, in addition to mass and energy balances, then models for emissions [2], carbon footprint, or sustainable process index [3] of each operating unit should be incorporated. The selection of operating units and their models can be effected by resorting to the information available in the literature [4–7], simulation software, or engineering knowledge.

Afterwards, the set of material streams and that of operating units are represented as material type nodes (M-type nodes) and operating unit type nodes (O-type nodes), respectively, as delineated by the P-graph framework. Subsequently, algorithm Maximal Structure Generation (MSG) [8] given in Chap. 4 is employed for constructing the maximal (or global) structure, a rigorous superstructure of the problem. The maximal structure comprehensively contains all the plausible processes (flowsheets) arising from the combination of sets of materials, operating units, and the relationships among them specified by the designer. Algorithm MSG guarantees the construction of such a comprehensive structure as it excludes the materials and operating units that violate any of the axioms supporting the P-graph framework [9].

Based on the maximal structure, two distinct approaches can be deployed for identifying the set of best, or most convenient, process designs (flowsheets) that can effectively perform the transformation of raw materials into desired products. Naturally, regardless of the approach taken, the calculations carried out and the results obtained are valid only for the specified input data, constraints, model, and evaluation tools employed. The first approach enumerates all the solution-structures, *i.e.*, the process flowsheets with a logical structure according to the framework's axioms. This is followed by their individual evaluation in light of feasibility and optimality criteria. Such an enumeration is accomplished by resorting to algorithm Solution-Structure Generation (SSG) [10] given in Chap. 5, thereby systematically identifying the individual plausible flowsheets contained in the maximal structure. Subsequently, each plausible flowsheet is optimized by linear or nonlinear programming based on the objective function defined, as well as on the mathematical models of the operating units contained in the plausible flowsheet. This approach allows the systematic comparison of all the alternative designs, upon which the most convenient structures can be selected. The second approach determines and ranks the n-best alternative designs rapidly without generating all feasible flowsheets by deploying algorithm Accelerated Branch-and-Bound (ABB) [11] given in Chap. 6. This algorithm reduces the sizes and the number of subproblems solved during the optimization and improves the effectiveness in bounding steps of the procedure. In solving the synthesis problem, algorithm ABB can be deployed with linear or nonlinear programming for its bounding, depending on the nature of the problem and the requirements of the designer. This approach is particularly useful when the number of solution-structures identified by algorithm SSG is exceedingly large, thereby rendering their optimization extremely time consuming.

Synthesis of a Process for Production of Adipic Acid

Adipic acid (AA), also known as hexanedioic acid, is a white crystalline powder recognized as one of the most used chemicals in the world [12]; in 2018, AA's global production was about 3.9 million metric tons [13]. AA has a low hygroscopicity; it is slightly soluble in cyclohexane, benzene, and water but it is highly soluble in methanol and ethanol. AA is primarily used to produce nylon 66 [14], which accounts for 90% of the acid's world consumption. The remaining 10% is consumed mostly in the production of esters and polyesters that are employed as plasticizers and solvents. In addition, polyurethane resins can be also obtained from AA and employed in the manufacture of lacquers, adhesives, foams, and surface coatings. In the food industry, AA is employed as gelling aid, acidulant, flavoring agent, leavening, and buffer reagent [15].

Selection of Reaction

AA can be produced via two main chemical routes: oxidation of cyclohexanone and carbomethoxylation of butadiene. Additionally, biochemical routes that use strains of *Acinetobacter*, *Pseudomonas*, and *Xanthobacter* for the oxidation of cyclohexanol to AA have been gaining relevance in recent years [12].

The first chemical route involves the use of air or concentrated nitric acid (HNO_3) [16] to effect the oxidation of cyclohexanone obtained from the hydrogenation of benzene. This reaction gives rise to KA oil, a mixture of cyclohexanol and cyclohexanone, which is then oxidized to AA through various reaction pathways catalyzed by copper and vanadium [14, 17]. The oxidation of KA oil generates CO_2, NO_x, and glutaric and succinic acids as main byproducts. A simplified version of the oxidation of KA oil with HNO_3 [18] is depicted in Fig. 8.2.

The second chemical route, *i.e.*, the butadiene-based route, can be carried out in various systems; however, none of these systems has reached significant commercial implementation [14, 16]. For instance, carbon monoxide and methanol are used for the carbomethoxylation of butadiene generating methyl 3-pentenoate that is converted to dimethyl adipate, an ester of AA. Subsequently, AA is obtained from this ester. In another instance, the reaction of hydrocarboxylation from carbon monoxide, water, and 1,3 butadiene is catalyzed by palladium, rhodium, or iridium to yield 3-pentenoic acid, which is further reacted to form AA, 2-methylglutaric acid, and 2-ethylsuccinic acid.

The case study under consideration will focus on the synthesis of alternative processes for AA production based on the first chemical route, *i.e.*, the oxidation of

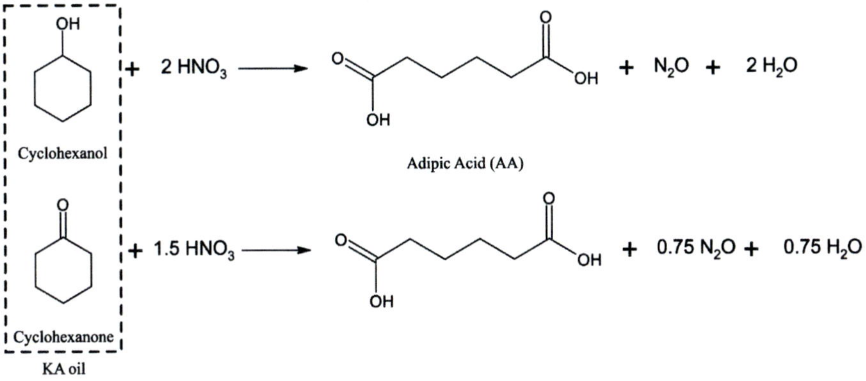

Fig. 8.2 Simplified chemical reactions involved in the production of AA from a mixture of cyclohexanol and cyclohexanone, or KA oil

KA oil with HNO_3 catalyzed by vanadium [V(V)] and copper [Cu(II)]. Because the chemical route has been selected from the outset, the design step concerning Reaction-Network Synthesis (RNS) will not be considered; as a result, the methodology detailed in Fig. 8.1 will be implemented from the identification of plausible operating units onward.

Specification of Materials and Selection of Operating Units

The designer defines a set of operating units capable of performing the process of concern based on the available knowledge. As expected, the production of AA involves various types of operations, such as separation, reaction, and auxiliary operations. Naturally, this set can be extended to include additional operations, reactions, materials, or connections among them. It is also possible to evaluate alternatives where one or more operating units of a particular stage of the process are not selected. This is accomplished by defining bypass units as O-type nodes inside the P-graph representation. The set of plausible operating units considered in this case study is described in detail in what follows; the operating units' names in the process flowsheet will be indicated in italic font. On the basis of this set of plausible units, the set of raw materials is specified as HNO_3, KA oil, catalyst, water, and air. Clearly, adipic acid is the process' main product and all other reaction products are considered as byproducts to be extracted for further processing in a separate facility. Byproducts are represented as intermediate materials.

Reaction Units

Three distinct stages of reaction are selected as plausible operations for the process of concern, specifically, low-temperature oxidation, high-temperature oxidation, and NO_x oxidation. Low-temperature oxidation refers to the reaction of KA oil with HNO_3 at temperatures between 70 and 90 °C. In this regard, two alternative operations are included in the PNS problem, *React_LT_A* and *React_LT_B*. The former considers the oxidation reaction in a continuous stirred tank reactor (CSTR) whereas the latter performs the same reaction in a plug flow reactor (PFR). Conversely, high-temperature oxidation refers to the reaction of KA oil with HNO_3 between 90 and 130 °C; this stage is carried out in a CSTR (*React_HT*) [14]. Finally, a PFR unit (*Ox_React*) is evaluated for performing the oxidation of NO_x generated in the previous stages to recover it as HNO_3 in subsequent steps.

Bleacher Unit

A bleaching operation (*Bleacher*) is included to remove a fraction of the NO_x gases dissolved in the liquid stream that leaves the high-temperature reactor by using air as a desorption agent.

Compressor

The air needed for certain operations, such as the bleacher or the NO_x oxidation reactor, is supplied by a compressor (*Compressor*); this operation could be regarded as an auxiliary unit similar to a pump. However, a compressor's impact on the process cost is usually significant; herein, therefore, it will be considered as a main unit.

Absorption Units

Two absorption units, *Absorber_1* and *Absorber_2*, are included to recover HNO_3. In these units, the NO_x leaving the bleacher (*Bleacher*) and/or the high-temperature reactor (*Reactor_HT*) is absorbed in water to regenerate some acid that can be recirculated into the process.

Membrane

This operation is included as an alternative for the NO_x recovery after the high-temperature reactor (*React_HT*). The liquid stream from this reactor can be passed through a membrane (*Membrane*) instead of the bleacher, thereby enhancing the removal of NO_x from the system. The stream rich in NO_x is subsequently processed in the PFR unit (*Ox_React*).

Flash Separator and Concentrators

After the removal of NO_x from the high-temperature reactor's outlet stream and the recovery of HNO_3 in the absorption units, the excess water needs to be extracted. To this end, a flash operation (*Flash*) is included where the reactor stream is concentrated by removing water in a single vapor-liquid equilibrium stage. This operation can alternatively be performed in multiple-stage separators; thus, two such units are included in the PNS problem, which are modeled as distillation columns (*Concent_1* and *Concent_2*). As a consequence of the use any of these operations, non-condensable gases are also removed.

Crystallizers

Four crystallization units are considered in the problem. The first three (*Crystallizer_1*, *Crystallizer_2*, and *Crystallizer_3*) are included to recover the high-purity adipic acid from the streams leaving the bleacher (*Bleacher*), membrane (*Membrane*), and concentrators (*Flash*, *Concent_1*, and *Concent_2*). After this initial recovery, the solid adipic acid is dissolved again and recrystallized in a fourth crystallization unit (*Recrystallizer*) to enhance the purity of the final product.

Operations for Solids Recovery

Recovery of solids after crystallization or recrystallization can be performed by means of solid-liquid separation units. In this case study, three filtration units (*Filter_1*, *Filter_2*, and *Filter_3*) are evaluated in the PNS problem by considering them as continuous filters. Another option for solids recovery is centrifugation; thus, two centrifuges (*Centrifuge_1* and *Centrifuge_2*) are included in the initial set of operating units.

Catalyst Recovery

The streams rich in HNO_3 leaving the solids recovery stage, or the concentration stage, are usually referred to as nitrogen mother liquor (NML). Such a material is commonly recirculated in the reaction step to recover valuable reactants. In addition, NML also contains fractions of byproducts, such as succinic and glutaric acid, as well as catalyst; therefore, a portion of this stream is purged out of the system and treated to recover the catalyst in a unit designated for it (*Cat_Recov*).

Auxiliary Units

A set of additional units are also included to perform various auxiliary operations, such as mixing, *e.g.*, *Mixer_A*, feed pumping (*Pump*), bypassing, stream conditioning, final drying (*Dryer*), and purging, *e.g.*, *Purge_1*.

Selection of Mathematical Models and Solution Approaches

In this case study, the performance of an individual alternative design will be assessed by means of its total annualized cost (TAC). Selection of the mathematical equations that estimate the process' performance, *i.e.*, mathematical models, depends on the designer's specifications and the accuracy required for the estimation. Such specifications might vary in each stage of design as well as with the involvement of additional stakeholders. For instance, simplified models could suffice for the description and evaluation of process alternatives in early stages of conceptual design whereas superior models for estimating cost, kinetic or thermodynamic properties could be required in advanced stages.

It is worth pointing out that the implementation and benefits of the P-graph framework are independent of the type of model selected and the nature of the equations, *i.e.*, linear or nonlinear, that the model comprises. Moreover, the framework's capability of rendering the n-best alternatives can be exploited by initially generating a large number of process networks and then refining the number of alternative designs in subsequent stages by deploying models of increasing accuracy. For illustration, two mathematical models are employed herein to show the P-graph framework's flexibility in incorporating diverse mathematical models.

Selection by Algorithm ABB with LP or NLP Bounding

The first model relies on the sizes of the operating units, which is utilized to evaluate the TAC of the process considered. The mass balance around the individual units, based on a production of adipic acid of 4000 kg/h, is determined by means of simulation software and data available in the literature. The cost of the operating units is estimated as a function of their sizes. This is accomplished by varying the mass flow processed by the units and computing their cost in the simulation via Aspen Economic Analyzer®. The set of simulated data is regressed to obtain fix-charged cost linear functions for both capital and operating cost of each unit. Table 8.1 lists the mass balance around each of the considered units as well as the coefficients of the linear model used for the cost estimations. In this case, algorithm ABB is selected as the solution approach for identifying the set of alternative designs based on this model.

Table 8.1 Mass balances for the set of plausible operating units and cost parameters for mathematical model 1 in manufacturing adipic acid

Operating unit	Input Material	Input Flow (kg/h)	Output Material	Output Flow (kg/h)	Proportional operating cost (USD/h)	Proportional capital cost (USD)	Fixed capital cost (USD)
Mixer_A	M1	106.7	M2	9877.6	0.0	0.0	0.0
	NML	4579.0					
	HNO3_raw	5191.9					
React_LT_A	KA_Oil	3966.1	M3	13,593.4	13.75	19,861.0	55,381.0
	M2	9877.6	NOx	250.4			
React_LT_B	KA_Oil	3966.1	M3	13,593.4	18.98	13,505.5	42,089.6
	M2	9877.6	NOx	250.4			
React_HT	M3	13,593.4	M4	13,486.1	15.23	23,361.0	70,614.0
			NOx	107.3			
Bleacher	M4	13,486.1	M5	12,681.9	0.0	23,636.0	42,702.0
	Air	5770.1	NOx	6574.3			
Absorber_1	M6	300.0	M7	1351.8	0.0	13,666.0	25,228.0
	NOx	6932.0	Off_gas	5880.2			
Mixer_B	M7	1351.8	M8	14,033.6	0.0	0.0	0.0
	M5	12,681.9					
Membrane	M4	13,486.1	M9	12,955.6	24.27	30,000.0	15,233.0
			M10	530.4			
Ox_React	M10	530.4	M11	5723.5	2.75	49,065.0	24,184.0
	Air	5193.1					
Absorber_2	M11	5723.5	M12	437.4	0.01	13,508.0	23,250.0
	M6	300.0	Off_gas	5586.1			
Mixer_C	M12	437.4	M8	13,393.1		0.0	0.0
	M9	12,955.6					

(continued)

Table 8.1 (continued)

Operating unit	Input		Output		Proportional operating cost (USD/h)	Proportional capital cost (USD)	Fixed capital cost (USD)
	Material	Flow (kg/h)	Material	Flow (kg/h)			
Cristallyzer_1	M8	14,033.6	M13	13,906.4	3.15	659,324.3	386,941.3
Filter_2	M13	13,906.4	Crystal_1	4430.0	3.99	107,570.0	86,056.0
			M14	9476.4			
Concent_1	M14	9476.4	Water_Res	3291.9	37.37	99,034.0	117,601.0
			M15	6184.5			
Flash	M8	14,033.6	Water_Res	4723.6	21.61	7114.7	16,027.0
			M16	9310.0			
Crystallizer_2	M16	9310.0	M17	9310.0	1.15	391,831.0	229,956.0
Concent_2	M8	14,033.6	Water_Res	4725.6	173.04	712,864.0	464,698.0
			M18	9308.0			
Crystallizer_3	M18	9308.0	M17	9308.0	1.15	394,105.0	231,218.0
Filter_1	M17	9308.0	Crystal_1	4487.9	2.67	72,000.0	57,600.0
			M15	4820.0			
Centrifuge_1	M17	9308.0	Crystal_1	4487.9	2.36	120,123.0	356.0
			M15	4820.0			
Purge_1	M15	4820.0	NML	4579.0	0.0	0.0	0.0
			M19	241.0			
Dissolv_tank	M24	5792.2	M20	10,730.1	0.90	247,398.0	145,098.0
	M6	450.0					
	Crystal_1	4487.9					
Recrystallizer	M20	10,730.1	M21	10,730.1	1.15	508,730.6	298,561.5
Filter_3	M21	10,730.1	M22	4294.0	3.08	83,000.3	66,400.3
			M23	6436.1			
Centrifuge_2	M21	10,730.1	M22	4294.0	2.36	120,123.0	356.0

			M23	6436.1			
Purge_2	M23	6436.1	M24	5792.2	0.0	0.0	0.0
			Waste	644.0			
Dryer	M22	4294.0	Adipic_Acid	4079.3	1.76	77,184.0	7786.0
Cat_Recov	M19	241.0	M25	96.4	13.50	3500.0	15,233.0
			To_process_2	144.6			
T1	NOx	1.0	M10	1.0	0.0	0.0	0.0
Pump	Water_feed	450.0	M6	450.0	0.0	4430.0	2840.0
Compressor	Air_feed	5770.1	Air	5770.1	14.5	129,013.0	573,725.0
Cat_Mixer	M25	96.4	M1	106.7	0.0	0.0	0.0
	Fresh_Catalyst	10.3					

Selection by the Integration of Algorithm SSG and Simulation Program

The second model relies on rigorous phenomenological equations that are solved in a commercial simulator, *i.e.*, it is an implicit model given by a simulation program. Thus, the thermodynamic models, physical properties, and block modules available in the software can be used to compute mass and energy balances and determine operation parameters. These parameters are retrieved from the simulator for estimating the cost of the units, which enhances the accuracy of the evaluation. The solution approach for this model involves algorithm SSG, thus enumerating the combinatorially feasible structures, *i.e.*, solution-structures or flowsheets, that are automatically constructed in a commercial simulator for its subsequent optimization. The P-graph framework will perform independently of the nature of the simulator or the selected method of calculation, and it can be carried out with any estimation technique. Various evaluation tools such as MATLAB® can be employed for process assessment, besides to commercial simulators. In this case study, the evaluation is performed by exploiting the feature "ActiveX control" of Aspen Plus® [19, 20], which enables the manipulation of the simulation environment from external software. MATLAB® is employed to automatically create each solution-structure in the simulator and retrieve the relevant parameters for the TAC estimation. Once a solution-structure is created in the simulator, it can be automatically optimized by resorting to the optimization method selected by the designer, thereby yielding the set of operating conditions and optimized structures. Table 8.2 presents the block modules used in the simulator for the main operating units and the decision variables considered for their optimization.

Because the working principle varies among the operating units involved in a process, the estimation of costs based on the parameters of the simulation requires particular models for each operation. Commonly, key aspects related to the operating units' dimensions as well as the units' process variables are employed to estimate the investment cost of the equipment required for the transformation. On the other hand, the units' operating cost is usually related to expenses, such as industrial utilities, labor costs, consumables, etc. Table 8.3 lists the variables used for the estimation of investment and operating costs for the main operating units included in the simulation.

Further details of the cost models and their relationship to the cost variables can be found in the references cited in Table 8.3. The model parameters for the simulation of the adipic acid' system, such as solubilities and rates of reaction, are estimated or retrieved from the literature [17, 18, 25, 26]. Additionally, it is assumed that the cost of the PFR with heat exchange can be estimated as the cost of a shell-and-tube heat exchanger.

Table 8.2 Block models employed in the simulation of plausible operations in Aspen Plus® and decision variables utilized for optimization

Name of operation	Module in Aspen Plus®	Decision variables
React_LT_A	RCSTR	Temperature (T_RLT_A) Volume (V_RLT_A)
React_LT_B	RPlug	Temperature (T_RLT_B) Length (L_RLT_B)
React_HT	RCSTR	Temperature (T_RHT) Volume (V_RHT)
Ox_React	RPlug	Temperature (T_OR) Length (L_OR)
Bleacher	RadFrac	Air Flow (F_Bleach)
Compressor	Compr	Air flow (F_Comp)
Absorber_1	RadFrac	Water flow (WF_A1)
Absorber_2	RadFrac	Water flow (WF_A2)
Membrane	Sep	–
Flash	Flash2	Vapor frac (VF_Flash) Pressure (P_Flash)
Concent_1	RadFrac	Distillate to feed ratio (D:F_C1) Reflux ratio (R_C1)
Concent_2	RadFrac	Distillate to feed ratio (D:F_C2) Reflux ratio (R_C2)
Crysztallizer_1 Crystallizer_2 Crysttallizer_3	Crystallizer	Temperature (T_Cr1) (T_Cr2) (T_Cr3)
Re-crystallizer	Crystallizer	Temperature (T_RCr)
Filter_1 Filter_2 Filter_3	Filter	–
Mixers	Mixer	–
Pumps	Pump	–

Evaluation of the Flowsheets

Figure 8.3 illustrates the maximal structure generated by algorithm MSG from the set of materials and operating units described earlier. No material or operating unit is eliminated; thus, all the units specified can be included in at least one solution-structure that yields adipic acid. As described earlier, this chapter presents two solution approaches for the evaluation of process design alternatives.

Results of the Selection by Algorithm ABB

In the first approach, which relies on algorithm ABB, the data listed in Table 8.1 are entered in software P-Graph Studio [27]. Additionally, a set of parameters is required

Table 8.3 Variables related to cost estimation of operating units

Type of operation	Variables considered for cost estimation	References
Reactors as CSTR	Reactor volume Heating requirement	[21]
Reactor as multitube PFR	Number of tubes Reactor volume Heat exchange area Heating requirement	[7, 22]
Bleacher and absorbers	Weight of column Diameter Number of stages	[7]
Compressor	Power required Electricity cost	[22]
Membrane	Treated mass flow	[23]
Flash separator	Temperature Pressure Weight of vessel	[7]
Concentrators as distillation towers	Weight of column Diameter Number of stages Condenser and reboiler areas Condenser and reboiler heat	[7, 24]
Crystallizers	Diameter Mass flow Heating requirement	[21, 22]
Filters	Filtration area	[22]
Centrifuges	Treated mass flow	[22]

concerning the costs of raw materials, utilities, waste treatment, and various process expenses. Such information can be estimated from technical literature [4–6, 28], industrial or market data, or simulator databases. Table 8.4 presents the cost parameters estimated herein in connection with the process under consideration.

Algorithm ABB identifies the set of n-best designs by considering a pay-out period of 10 years, a total of 8000 operational hours per year, and a production target of 35,000 t/year of adipic acid. The price of "waste treatment + catalyst recovery" is represented in the software by the proportional operating cost of unit *Cat_Recov*; no additional constraints are specified. Upon execution, the algorithm identifies the process structure depicted in Fig. 8.4 as the best alternative under the conditions imposed by the model. This structure has a TAC of 56.6 million USD/year (7079.2 USD/h). Moreover, 39 additional process alternatives are generated with TACs between 7079.6 USD/h and 7408.3 USD/h.

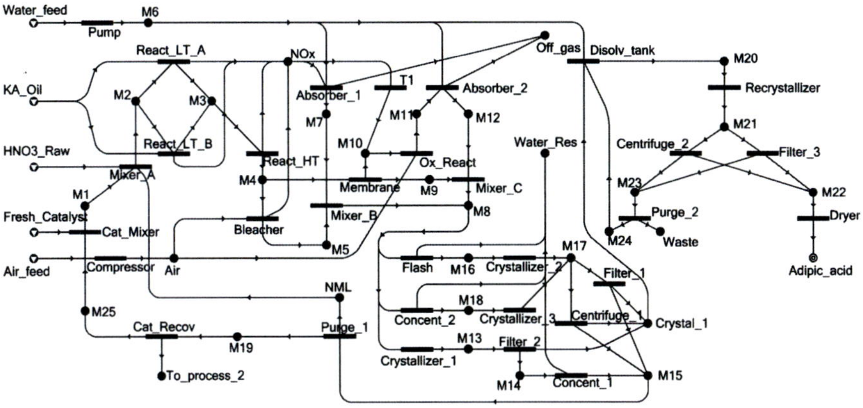

Fig. 8.3 Maximal structure generated by algorithm MSG for synthesizing processes to produce adipic acid

Table 8.4 Parameters used for the evaluation of alternative designs in adipic acid production

Parameter	Value
Cold utility cost	7.66×10^{-7} USD/kJ
Hot utility cost	1.95×10^{-6} USD/kJ
Electricity cost	2.15×10^{-5} USD /kJ
Catalyst price	8.3 USD/kg
Waste treatment + catalyst recovery	0.056 USD/kg
Waste treatment cost	0.043 USD/kg
Gas treatment cost	0.041 USD/kg
KA oil price[a]	900 USD/t
HNO_3 price	250 USD/t
Fresh water cost	1.4 USD/t

[a] Estimated as a fraction of the cost of production of cyclohexanone

Results of the Selection by Algorithm SSG Integrated with Simulation Program

In the second solution approach, the structures enumerated by algorithm SSG can be automatically simulated with the modules listed in Table 8.2 and independently optimized; for this case study, algorithm SSG enumerated 1,512 solution-structures. For illustration, three of these structures (denoted as solution-structure A, B and C) are selected to be shown as examples of the simulation and optimization. The P-graph representation of the solution-structure A selected for simulation is the same as the one depicted in Fig. 8.4; its corresponding conventional representation is shown in Fig. 8.5. The P-graph and conventional representations of solutions structures B and C selected are presented from Figs. 8.6, 8.7, 8.8 to Fig. 8.9. In the conventional representations, the material "Water_Res" is combined with material "Waste" into a single stream that is considered for further treatment.

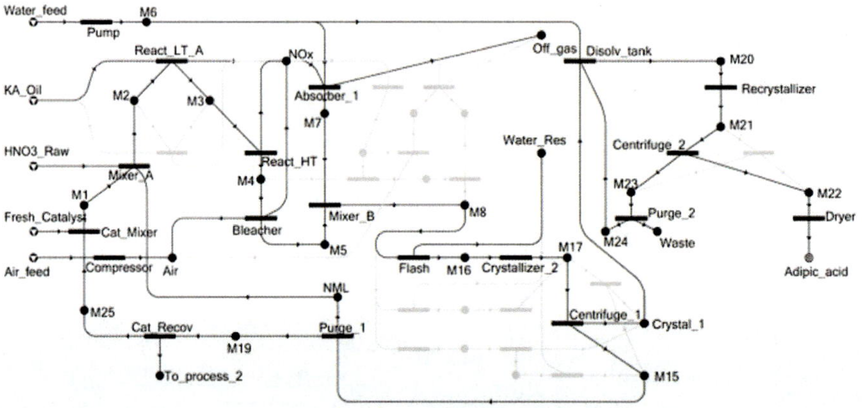

Fig. 8.4 Best solution generated by algorithm ABB for adipic acid production according to mathematical model 1 (solution-structure A)

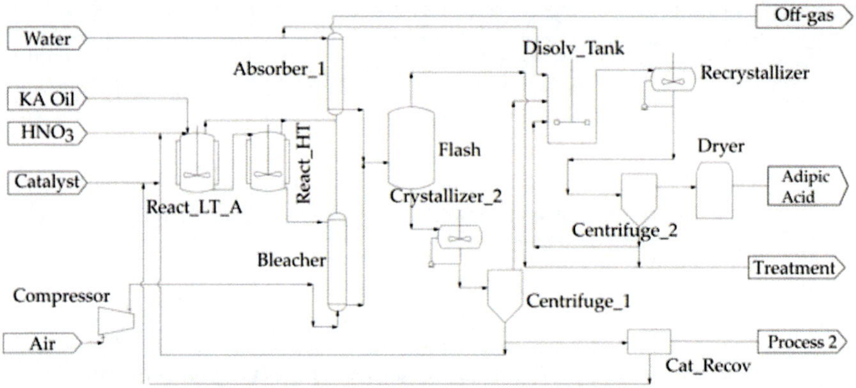

Fig. 8.5 Conventional representation of Fig. 8.4, best structure for mathematical model 1 (solution-structure A)

All these structures can be automatically optimized in addition to their simulation. In this regard, the designer must select the decision variables that will be modified during the optimization. This set of variables can involve all operating units comprised in the structure as well as any number of process conditions or parameters. Naturally, the complexity of the optimization problem and the optimality of the result will depend on the decision variables selected; moreover, they might change according to the accuracy desired. For simplicity, only the variables of the main process operations; *i.e.*, reactors, crystallizers, flash, and concentrators; are selected as decision variables. In addition, the cost estimation of catalyst recovery is included in the treatment of the stream leaving the Purge_1 (see Table 8.4), to reduce the simulation effort. In addition, the cost of catalyst makeup is estimated on the basis of

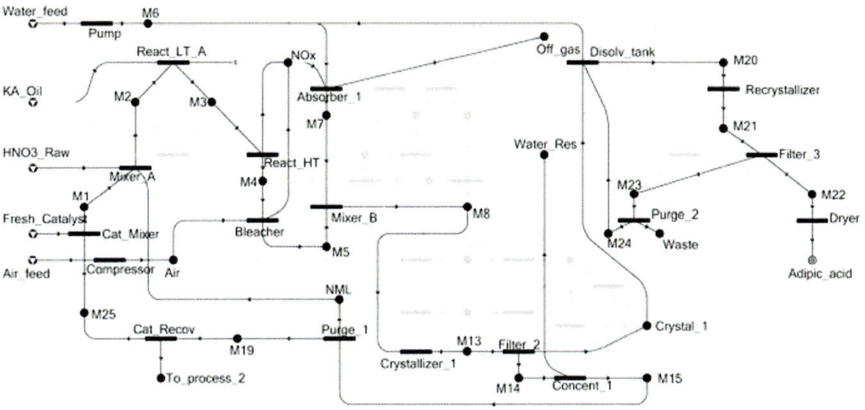

Fig. 8.6 P-graph representation of solution-structure B

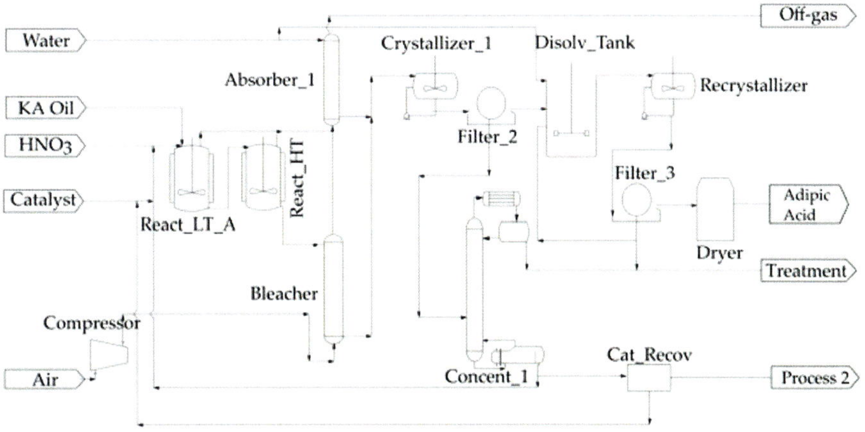

Fig. 8.7 Conventional representation of solution-structure B

an assumed recovery of 90%. Moreover, the flows of raw materials and the split ratios in the purges are also included as optimization variables.

The optimization is performed in MATLAB® linked to Aspen Plus® via ActiveX control. A real coded genetic algorithm with simulated binary crossover and mutation is employed for the optimization [29]. As indicated earlier, the problem assumes a production targets of 35,000 t/year and 8000 h/year, and it also considers the parameters presented in Table 8.4. Table 8.5 lists the optimized values of the decision variables for each of the process structures selected.

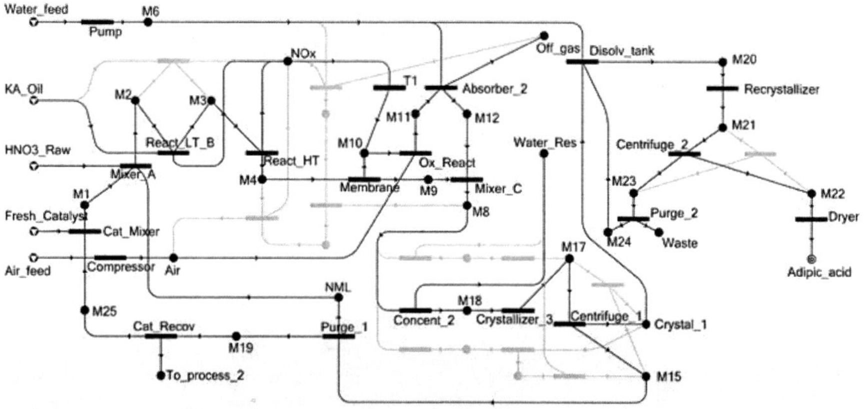

Fig. 8.8 P-graph representation of solution-structure C

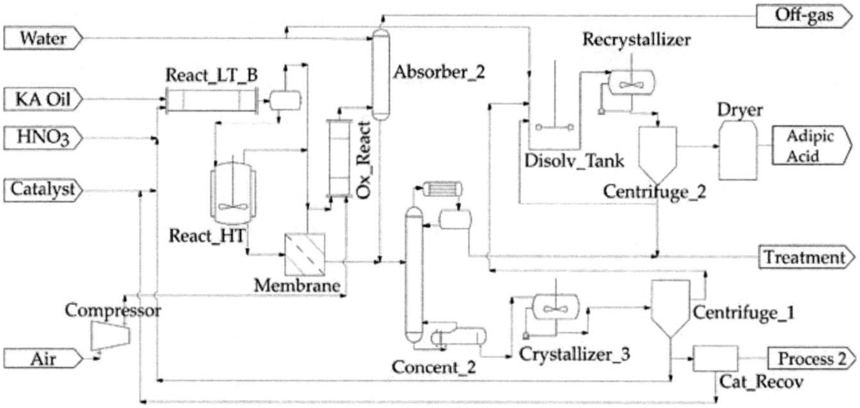

Fig. 8.9 Conventional representation of solution-structure C

Concluding Remarks

This chapter illustrates the use of the P-graph framework for the conceptual design of chemical processes for the production of adipic acid. It should be bear in mind, however, that the methodology followed in this chapter can be extended to any synthesis problem because the P-graph axioms and algorithms are not limited by the nature of the operations defined or the models selected for their evaluation. The P-graph framework allows not only the identification of the best alternative design for generating the desired products but also permits the systematic evaluation of the set of n-best designs. This feature provides the designer and stakeholders with valuable information for decision making in PNS and plant construction.

Table 8.5 Final values of the decision variables for the optimization of the selected solution-structures

Variable	Solution-structure A (Fig. 8.4)	Solution-structure B (Fig. 8.6)	Solution-structure C (Fig. 8.8)
T_RLT_A (K)	373.15	372.87	–
V_RLT_A (m^3)	1.50	1.11	–
T_RLT_B (K)	–	–	372.83
L_RLT_B (m)	–	–	8.21
T_RHT (K)	383.43	389.87	400.15
V_RHT (m^3)	61.00	42.10	60.46
T_OR (K)	–	–	381.03
L_OR (m)	–	–	7.09
T_Cr1 (K)	–	295.96	–
T_RCr (K)	295.41	301.82	301.15
T_Cr2 (K)	293.15	–	–
T_Cr3 (K)			293.153
VF_Flash	0.10	–	–
P_Flash (Pa)	13,340.13	–	–
D:F_C1	–	0.09	–
D:F_C2	–	–	0.15
R_C1	–	2.08	–
R_C2	–	–	2.15
Flow KA oil (kg/h)	0.86	0.98	0.86
Flow HNO_3 (kg/h)	1.70	1.91	1.66
Air flow (kg/h)	1.12	1.13	1.12
Total water flow (kg/h)	0.12	0.14	0.11
Split ratio purge 1	0.01	0.01	0.01
Split ratio purge 2	0.11	0.02	0.03
TAC (Million USD/year)	39.41	45.66	41.20

References

1. Fan LT, Zhang T, Liu J et al (2007) Assessment of sustainability-potential: hierarchical approach. Ind Eng Chem Res 46:4506–4516
2. Allen DT, Shonnard DR (2002) Green engineering: environmentally conscious design of chemical processes. Prentice Hall, New York
3. Krotscheck C, Narodoslawsky M (1996) The sustainable process index. A new dimension in ecological evaluation. Ecol Eng 6:241–258
4. Towler G, Sinnott R (2008) Chemical engineering design: principles, practice and economics of plant and process design. Butterworth-Heinemann, Amsterdam
5. Peters MS, Timmerhaus KD, West RE (2003) Plant design and economics for chemical engineers, 5th edn. McGraw-Hill, New York
6. Turton R, Shaeiwitz JA, Bhattacharyya D, Whiting W (2018) Analysis, synthesis and design of chemical processes, 5th edn. Pearson Education, Boston

7. Seider WD, Lewin DR, Seader JD (2003) Product and process design principles: synthesis, analysis, and evaluation, 2nd edn. Wiley, New Jersey
8. Friedler F, Tarján K, Huang YW, Fan LT (1993) Graph-theoretic approach to process synthesis: polynomial algorithm for maximal structure generation. Comput Chem Eng 17:929–942
9. Friedler F, Tarján K, Huang YW, Fan LT (1992) Graph-theoretic approach to process synthesis: axioms and theorems. Chem Eng Sci 47:1973–1988
10. Friedler F, Varga JB, Fan LT (1995) Decision-mapping: a tool for consistent and complete decisions in process synthesis. Chem Eng Sci 50:1755–1768
11. Friedler F, Varga JB, Feher E, Fan LT (1996) Combinatorially accelerated branch-and-bound method for solving the MIP model of process network synthesis. In: Floudas CA, Pardalos PM (eds) State of the art in global optimization. Springer, Boston, pp 609–626
12. Goldberg I, Rokem JS (2009) Organic and fatty acid production, microbial. In: Schaechter M (ed) Encyclopedia of microbiology, 3rd edn. Academic Press, Oxford, pp 421–442
13. Statista (2019) Market volume of adipic acid worldwide in 2018 and 2023. https://www.statista.com/statistics/1113587/global-market-size-adipic-acid/. Accessed 30 Aug 2021
14. Musser MT (2000) Adipic acid. In: Ullmann's encyclopedia of industrial chemistry. Wiley, New Jersey
15. Dziezak JD (2016) Acids: natural acids and acidulants. In: Caballero B, Finglas PM, Toldrá F (eds) Encyclopedia of food and health. Academic Press, Oxford, pp 15–18
16. Bart JC, Cavallaro S (2015) Transiting from adipic acid to bioadipic acid. 1, petroleum-based processes. Ind Eng Chem Res 54:1–46
17. Qu Y, Fang C, Duan M, Wang J (2014) Investigations on the reaction kinetics and the catalytic effect of Cu(II) and V(V) in the oxidation of cyclohexanol by nitric acid. React Kinet Mech Catal 112:209–226
18. Li G, Liu S, Dou X et al (2020) Synthesis of adipic acid through oxidation of K/A oil and its kinetic study in a microreactor system. AIChE J 66:e16289
19. Aspen Technology Inc (2000) Aspen Plus®: Aspen Plus User Guide. Author, Cambridge
20. Pimentel J, Orosz Á, Tan RR, Friedler F (2020) Automated synthesis of process-networks by the integration of P-graph with process simulation. Chem Eng Trans 81:1171–1176
21. Matches matches' 275 equipment cost estimates. https://www.matche.com/equipcost/Default.html. Accessed 31 Aug 2021
22. Couper JR, Penney WR, Fair JR, Walas SM (2005) Chemical process equipment: selection and design, 2nd edn. Elsevier, Amsterdam
23. Chellam S, Serra CA, Wiesner MR (1998) Estimating costs for integrated membrane systems. J AWWA 90:96–104
24. Kern DQ (1965) Process heat transfer. Mc-Graw Hill, Tokyo
25. Tsukahara H, Ishida T, Mayumi M (1999) Gas-phase oxidation of nitric oxide: chemical kinetics and rate constant. Nitric Oxide 3:191–198
26. Glarborg P, Johnsson JE, Dam-Johansen K (1994) Kinetics of homogeneous nitrous oxide decomposition. Combust Flame 99:523–532
27. P-graph community (2015) P-graph web page. http://p-graph.org. Accessed 13 Jun 2020
28. Dai SB, Lee HY, Chen CL (2018) Design and economic evaluation for the production of ethyl lactate via reactive distillation combined with various separation configurations. In: Eden MR, Ierapetritou M, Towler G (eds) Proceedings of the 13th international symposium on process systems engineering – PSE 2018. Elsevier B.V., San Diego, pp 127–132
29. Deb K, Beyer H (1995) Real-coded genetic algorithms with simulated binary crossover: studies on multi-modal and multi-objective problems. Complex Syst 9:431–454

Chapter 9
Enumeration-Based Properties of Processing Systems: Reliability and Resilience

Abstract Designers often consider several indicators for the selection of the process configuration in addition to cost. Some indicators, *e.g.*, sustainability, can be embedded into the mathematical model in a manner similar to cost. Other indicators, however, require additional tools for their incorporation. This is especially true for indicators such as reliability and resilience. In this chapter, it is shown that the enumeration-based algorithms of the P-graph framework can be applied in reliability and resilience analysis. Illustrative examples show the operation of the associated algorithms.

Keywords P-graph · Reliability · Resilience · Structural enumeration · Process synthesis · Combinatorial algorithms

Reliability in Process Network Synthesis

The reliability of a process, or processing system, generally relates to the probability that the system is able to perform its job. For processing systems, reliability is more specifically defined as the probability that a system is able to perform its job sufficiently, even if it experiences failures. Thus, reliability is one of the most important parameters of a processing system and its incorporation into process design is a desired extension to process network synthesis (PNS).

The combinatorial nature of the operability of a process gives rise to the main difficulty when reliability is considered in the design stage. The usual approach for improving the reliability of a process is to introduce redundant operating units, with which the process can be operational with several different subsets of the operating units, *i.e.*, structures. The complete enumeration of these structures can be done via combinatorial algorithms, or it can be estimated by stochastic optimization. The latter approach is implemented by Kumamoto et al. [1] wih Monte Carlo simulation and by Prowell and Poore [2] with Markov chains. The precision of these methods depends highly on the computational effort provided to them; usually a highly accurate result requires significant computational effort. On the side of combinatorial

F. Friedler et al., *P-graphs for Process Systems Engineering*,
https://doi.org/10.1007/978-3-030-92216-0_9

algorithms, Kovacs et al. [3] developed a P-graph based approach for the complete enumeration of the structures where the process is operational, *i.e.*, it is capable of fullfil the required production level. This extension of the P-graph framework is detailed in this chapter.

Several systems in practice can be structurally decomposed into specific subsystems such as parallel or sequential subsystems. Determining the reliability of such systems is performed on the basis of the reliabilities of these subsystems [4]. This simplistic approach, however, is not suitable for process design, as those systems often consist of a highly interconnected network of operating units involving several loops and parallel lines. In most cases, these systems cannot be decomposed into the predefined simple subsystems.

The current chapter shows that the P-graph framework can establish a direct connection between the fundamental areas of process design and reliability engineering. Namely, the connection between feasibility (for process design) and operability (for reliability engineering) is established. This recognition leads to a rigorously defined mathematical model and an appropriate algorithm.

The Extended PNS Model for Reliability Analysis

An additional parameter, reliability, is assigned to each unit, which is denoted by p_i $(0 \leq p_i \leq 1)$ for unit i. Depending on the type of systems and the design problem, the reliability of an operating unit can be constant, or a function of time and other parameters of the operating unit. For the current chapter, it is assumed that the failures of any two different units are independent of each other.

First, reliability of a process is determined on a structural basis. The formulation and the method given herein is general, it is independent of the mathematical model assigned to an operating unit. The proposed definitions, formulas, and algorithms are general; thus, they are applicable for any process with any structure. In parallel to the introduction of these formulas and algorithms, they will be illustrated through Example 9.1, shown in Fig. 9.1 [3]. The structural model of the example is given as $P = \{A\},\ R = \{G, K\},\ O = \{O1, O2, O3, O4\}$. Note that, since there is only one product in this system, the requirement for operability is the generation of this product.

The reliability parameters of the operating units are assumed to be constant for now. Unit i in a production process either performs its designated job on a sufficient level, *i.e.*, it is functional, or not. This is expressed by the state of the operating units, *i.e.*, a binary variable x_i, where

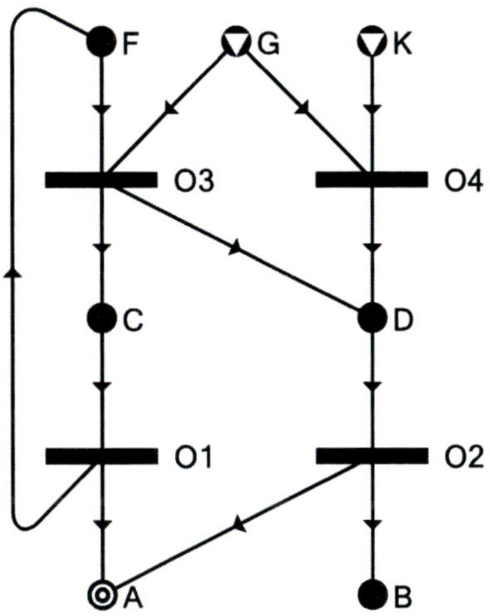

Fig. 9.1 Processing structure of Example 9.1

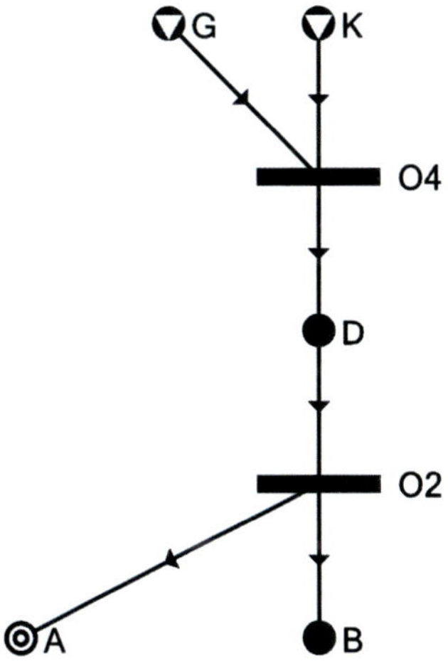

Fig. 9.2 Substructure of Example 9.1 represented by binary vector $X = (0, 1, 0, 1)$

$$x_i = \begin{cases} 1, & \text{if unit } i \text{ is functional} \\ 0, & \text{if unit } i \text{ is nonfunctional} \end{cases} \tag{9.1}$$

The state of the processing system containing n operating units is given by binary vector $X = (x_1, x_2, \ldots, x_n)$. Note that the set of functional units determines a structure, which is always a substructure of the structure of all units. For instance, $X = (0, 1, 0, 1)$ in Example 9.1 represents a structure with only operating units $O2$ and $O4$, see Fig. 9.2.

Depending on X, the structure determined by the functional units is either operational, *i.e.*, the selected system is capable of generating all products, or nonoperational. The structure represented in Fig. 9.2 is operational as it is capable of producing A, the only product of the system. As a result, the operability of a system depends solely on the states of its units; it is expressed by operability function Ψ for a system of n units as

$$\Psi(x_1, x_2, \ldots, x_n) = \begin{cases} 1, & \text{if the system is operational} \\ 0, & \text{if the system is nonoperational} \end{cases} \tag{9.2}$$

Let set Ω be defined as the domain of function Ψ, *i.e.*, the set of all binary n-vectors,

$$\Omega = \{0, 1\}^n = \{(0, 0, \ldots, 0), (0, 0, \ldots, 1), \ldots, (1, 1, \ldots, 1)\}. \tag{9.3}$$

As previously stated, each element of Ω represents a substructure of the original structure of the operating units. Naturally, the elements of Ω can be partitioned into two sets, the set of vectors representing an operational substructure and the set of vectors representing a nonoperational substructure. To be consistent with the parlance of probability theory, a state of the system can also be called as an event of the system, or system event. Let U be the set of operational elements of Ω (those vectors that represent an operational structure), *i.e.*,

$$U = \{(x_1, x_2, \ldots, x_n) \in \Omega : \Psi(x_1, x_2, \ldots, x_n) = 1\}. \tag{9.4}$$

In the case of Example 9.1, set U can be defined as

$$U = (0, 1, 0, 1), (0, 1, 1, 1), (1, 0, 1, 0), (1, 0, 1, 1), (1, 1, 0, 1), (1, 1, 1, 0), (1, 1, 1, 1).$$

All seven structures are given in Fig. 9.3. Note that some of the structures are not considered solution structures in terms of the P-graph axioms, since some of the included operating units would consume unproduced materials, *e.g.*, the structure with units *O*2, *O*3, and *O*4. These "unusable" operating units are the result of the failures of other units of the system. However, these structures are still operational since for each of them, there is at least a part of their structure that can perform the necessary production of *A*. Consequently, the set of operational structures is an extension of the set of solution structures. If reliability is only considered at the structural level, every solution-structure is also operational.

The reliability $\widehat{r}$ of the system is defined as the probability that the system is operational [3]. In the terms of probability theory, the reliability of the system is the probability of the set of events corresponding to set U, *i.e.*, $\widehat{r} = P(U)$. Since the reliability of unit i is p_i, the probability that $x_i = 1$ is p_i and that $x_i = 0$ is $1 - p_i$, *i.e.*,

$$p_i = P(x_i = 1) \text{ and } 1 - p_i = P(x_i = 0). \tag{9.5}$$

Since the failures of the units are independent, the stochastic variables assigned to their functionality are also independent, and thus, the probability of the system event $(b_1, b_2, \ldots, b_n)$ for $b_i \in \{0, 1\}$ and $i = 1, 2, \ldots, n$ is

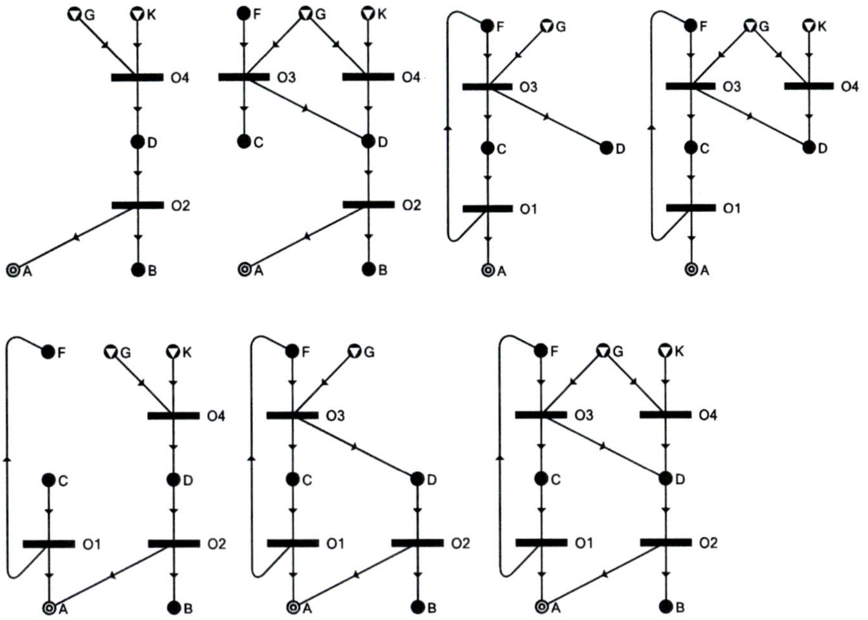

Fig. 9.3 Structurally operational substructures of process structure shown in Fig. 9.1

$$P((x_1, x_2, \ldots, x_n) = (b_1, b_2, \ldots, b_n)) = \prod_{i=1}^{n} p_i^{b_i}(1 - p_i)^{(1-b_i)} \tag{9.6}$$

Set Ω defines a complete event set with mutually exclusive events. As a result, the probability of set U of operational system events is the sum of the probabilities of the related events, therefore, the reliability $\widehat{r}$ of the system is calculated as

$$\widehat{r} = P(U) = \sum_{(x_1, x_2, \ldots, x_n) \in U} \left(\prod_{i=1}^{n} p_i^{x_i}(1 - p_i)^{(1-x_i)} \right) \tag{9.7}$$

Finally, following formula (9.7), the reliability $\widehat{r}$ of the system can easily be determined for any values of $p_1, p_2, \ldots, p_n$, $(0 \le p_i \le 1, i = 1, 2, \ldots, n)$ once the set of operational substructures is determined. For this reason, Eq. (9.7) is termed as the *system reliability formula* [3].

Going back to Example 9.1, the reliability of the system $\widehat{r}$ can be determined from Eq. (9.7) with $U = (0,1,0,1), (0,1,1,1), (1,0,1,0), (1,0,1,1), (1,1,0,1), (1,1,1,0), (1,1,1,1)$ as $\widehat{r} = (1 - p_1) * p_2 * (1 - p_3) * p_4 + (1 - p_1) * p_2 * p_3 * p_4 + p_1 * (1 - p_2) * p_3 * (1 - p_4) + p_1 * (1 - p_2) * p_3 * p_4 + p_1 * p_2 * (1 - p_3) * p_4 + p_1 * p_2 * p_3 * (1 - p_4) + p_1 * p_2 * p_3 * p_4$. If $p_1 = p_2 = 0.90$ and $p_3 = p_4 = 0.95$, the reliability of the process is 0.9789. For $p_1 = 0.95$, $p_2 = 0.98$, $p_3 = 0.96$, $p_4 = 0.99$, the system reliability is 0.9974.

The reliability of systems with simple structures, such as parallel and serial systems [4, 5], can be easily determined. These systems can be used for simple demonstrations of the system reliability formula. For example, a system of parallel units is operational if at least one unit is functional, i.e, an element of Ω $(x_1, x_2, \ldots, x_n)$, for which at least one of its entries (x_i) is equal to one, therefore,

$$U = \left\{ (x_1, x_2, \ldots, x_n) \in \Omega : \sum_{i=1}^{n} x_i \geq 1 \right\}$$

Clearly, the set of nonoperational structures is the difference between the entire set Ω and the set of operational structures U, which is denoted by $\Omega \backslash U$. Thus, the probability that a structure is operational is defined as

$$P(U) = 1 - P(\Omega \backslash U)$$

In the special case of a system of parallel units, only one nonoperational structure exists corresponding to that with no operational units, in other words,

$$\Omega \backslash U = \{(0, 0, \ldots, 0)\}$$

As mentioned earlier, the probability that a unit is operational $(x_i = 1)$ is p_i; hence, the probability that is not operational $(x_i = 0)$ is $(1 - p_i)$. The probability of this set of nonoperational structures can be calculated exactly the same way as the reliability formula (Eq. 9.7) indicates. The only difference is that the structures now are the nonoperational ones, and thus, in light of Eq. 9.7, we have

$$P(\Omega \backslash U) = \sum_{(x_1, x_2, \ldots, x_n) \in \Omega \backslash U} \left(\prod_{i=1}^{n} p_i^{x_i} (1 - p_i)^{(1 - x_i)} \right) =$$
$$= \prod_{i=1}^{n} p_i^{0} (1 - p_i)^{(1-0)} = \prod_{i=1}^{n} (1 - p_i)$$

Consequently, the reliability $\hat{r}$ of the parallel system, which is the probability of $P(U)$, is given by

$$\hat{r} = P(U) = 1 - \prod_{i=1}^{n} (1 - p_i) \tag{9.8}$$

Next, let us consider a system with no redundancy. A possible realization of such a system is one whose units are arranged in series; no redundancies are available. In this case, the system is operational if all units are functional, therefore, only one element of U exists, *i.e.*, (1, 1, . . ., 1), and then the operability function is

$$\Psi(x_1, x_2, \ldots, x_n) = x_1 \cdot x_2 \cdot \ldots \cdot x_n. \tag{9.9}$$

Consequently, since $U = \{(1, 1, \ldots, 1)\}$, the reliability of a redundancy-free system is exactly what is determined by the system reliability formula (Eq. 9.7); thus,

$$\begin{aligned}\widehat{r} = P(U) = \sum_{(x_1, x_2, \ldots, x_n) \in U} \left(\prod_{i=1}^{n} p_i^{x_i} (1 - p_i)^{(1-x_i)} \right) = \\ = \prod_{i=1}^{n} p_i^{1} (1 - p_i)^{(1-1)} = \prod_{i=1}^{n} p_i.\end{aligned} \tag{9.10}$$

With the use of Eq. (9.7), the reliability of any system can be calculated, as long as the set U can be determined. The question now is how the set U can be generated algorithmically; to do so, even for highly interconnected structure of operating units, a systematic procedure is required which will be presented next.

Synthesis Algorithms for Reliability Analysis

Now, the question is how to generate all operational substructures of a system algorithmically. In this regard, the P-graph framework provides convenient tools to perform this enumeration. For illustration, we look back to Chap. 3, and revisit Example 3.2 as another illustrative example. This process consists of seven units, and the structural model is repeated here: $P = \{A\}$, $R = \{E, J, G, K, L\}$, $O = \{O1, O2, O3, O4, O5, O6, O7\}$. In a substructure, each operating unit can be included or not, *i.e.*, it has two (2) possible existance states. Thus, the number of substructures is obtained as 2^n where n is the number of operating units. In our example, therefore, the system has $2^7 = 128$; only 19 of them satisfy the five axioms, these 19 are given in Chap. 3. The maximal structure is given in Fig. 9.4.

As we have seen before, there is a difference between solution structures and operational structures. If the operability is only considered from a structural aspect, then every solution structure is also operational. However, a structure can still be operational while not being a solution structure, *i.e.*, it violates some of the axioms, namely axioms (S2) and (S4). From now on, if a structure is operational from the strcutural aspect, it is called structurally operational.

Some more observations can be made for operational structures. First, any extension of an operational structure with additional units evidently remains operational. As a consequence, if a structure is nonoperational, removing any unit or units will still result in a nonoperational structure. Following from these observations, the final property can be declared: a structure of units is defined to be structurally operational, if it has a substructure that is a solution structure.

Set U of operational substructures can, therefore, be determined by testing whether or not each element of Ω is structurally operational. Since the maximal

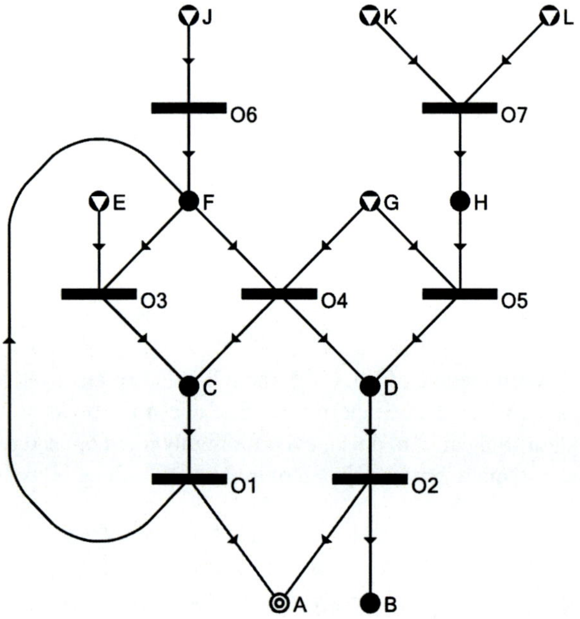

Fig. 9.4 Maximal structure of Example 3.2, revisited here for reliability analysis (Fig. 3.6)

structure is the union of all solution structures, a sufficient way of testing operability from a structural aspect is to determine the maximal structure with algorithm MSG [6, 7] for the structure of interest. If the maximal structure exists, then the structure has at least one substructure that is a solution structure, and so it is structurally operational, otherwise it is not. It is also worth mentioning that algorithm MSG is highly efficient and its complexity is polynomial on the number of units.

As a result, testing all elements of Ω by algorithm MSG provides the set of all structurally operational substructures. Despite its efficiency, note that the possibly large number of elements of Ω may render it difficult to test all elements with algorithm MSG. Additional tools and techniques can accelerate the algorithm, however, they are not the main focus here.

In real systems, that a particular structure be a solution structure is not enough for practical feasibility. Therefore, the described method also has to be able to consider more complex feasibility constraints, such as material balances in chemical processes or transportation capacities in logistic networks. There is no simple way to provide a general description of these constraints, but creating models for these constraints is not part of the P-graph methodology. However, as it will be shown here, the reliability determination algorithm presented is easily extended to consider additional constraints, provided that such constraints are appropriately modeled.

This can be achieved by introducing the function FEASIBLE, that receives a set of operating units and decides if the structure defined by those units is feasible with respect to the additional constraints. The exact details of this function are highly dependent on the process or system, but they are independent of the rest of the algorithm. If there are no additional constraints for feasibility, then the function FEASIBLE simply accepts every structure, and the algorithm determines the structural reliability.

Procedure for Determining the Reliability of a Processing System

The procedure for generating the reliability of a process is composed of previously available algorithms of the P-graph framework and additional algorithms introduced in this chapter. The first algorithm, algorithm SRP, given in Fig. 9.5, supervises the overall procedure for determining the reliability of systems.

Algorithm SRP invokes algorithms MSG, SON, and RBO; they are given in Figs. 9.6, 9.7, and 9.8, respectively. To facilitate the operation of algorithm SRP, Algorithm MSG is given in the convenient functional form presented in Fig. 9.6.

Two previously discussed examples will be revisited for illustrating the procedure of reliability analysis.

Example 9.1 (revisited) The input of the example represents the process with sets P, R, and O as $P=\{A\}$, $R=\{G,K\}$, and $O=\{O1, O2, O3, O4\}$ (see Fig. 9.1). The reliabilities of the units are given by vector $\boldsymbol{p}=(0.9, 0.9, 0.95, 0.95)$. The procedure calls algorithm MSG for sets P, R, and O, and returns the same set O since the structure given by O satisfies all five axioms.

After that, algorithm SON starts enumerating the structures contained in this maximal structure. Initially, set U is empty; subsequently, 16 iterations are executed as set Ω has 16 elements. Below are described in detail two of these iterations.

For $X = (0, 0, 1, 1)$; $O_X = \{O3, O4\}$ MSG gives $O = \emptyset$, *i.e.*, no maximal structure; this element is nonoperational.

SRP($P, R, O, \boldsymbol{p}, \hat{r}$)
Input:
  network of the process in the form of:
    set P of products: goals for the process to reach
    set R of raw materials: starting points for the process
    set O of operating units
  $\boldsymbol{p} = (p_1, p_2, \dots, p_n)$: p_i is the reliability of unit i
Output: $\hat{r}$: reliability of the process
Data: U: set of structurally operational subnetworks, given by algorithm SON

begin
MSG(P, R, O)
SON(P, R, O, U)
if $U \neq \emptyset$ **then begin**
  stop (no structurally operational subnetwork exists)
  end
RBO($O, U, \boldsymbol{p}, \hat{r}$)
write $\hat{r}$ to output
end

Fig. 9.5 Algorithm for determining the reliability of a process

Fig. 9.6 Algorithm for generating the maximal structure

```
MSG(P, R, O)
Input:
  network of the process or a subprocess in the form of:
    set P of products: goals for the process to reach
    set R of raw materials: starting points for the process
    set O of operating units
Output: set O: operating units in maximal structure
          (the algorithm reduces the original set)
```

Fig. 9.7 Algorithm for generating the set of operational substructures

```
SON(P, R, O, U)
Input: sets P, R, O
Output: set U: set of structurally operational subnetworks
Data:
  n: number of units (operations) (n = |O|)
  O_X: set (O_X ⊆ O)
  x = (x_1, x_2, ..., x_n): binary vector

begin
U = ∅
forall X ∈ {0,1}^n do begin
  O_X = {o_i: o_i ∈ O and x_i = 1 ∀ i = 1,2,...,n}
  MSG(P, R, O_X)
  if O_X ≠ ∅ and FEASIBLE(P, R, O_X) then begin
    U = U ∪ {X}
    end
  end
end
```

For $X = (0, 1, 0, 1)$; $O_X = \{O2, O4\}$, MSG gives $O = \{O2, O4\}$, *i.e.*, there exists a maximal structure; this element is operational, and thus, it is added to set U, *i.e.*, $U = \{(0, 1, 0, 1)\}$. Note that this is the first iteration which finds an operational structure.

At the end, set U={(0,1,0,1), (0,1,1,1), (1,0,1,0), (1,0,1,1), (1,1,0,1), (1,1,1,0), (1,1,1,1)}. These structures were already shown in Fig. 9.3.

Algorithm RBO determines $\widehat{r}$ according to formula (9.7); it is the reliability of the process. For this example $\widehat{r} = 0.9789$.

Example 3.2 (revisited) Let us examine the results of reliability analysis for Example 3.2. As the first step, algorithm MSG determines that the maximal structure contains all operating units of the problem, as shown in Fig. 9.4. Algorithm SON generates set U of structurally operational structures, which has 64 elements, see below.

Fig. 9.8 Algorithm for determining system reliability $\hat{r}$

RBO($O, U, \boldsymbol{p}, \hat{r}$)
Input:
  set O of operating units
  set U of structurally operational subnetworks
  $\boldsymbol{p} = (p_1, p_2, \ldots, p_n)$: p_i is the reliability of unit i
Output: $\hat{r}$: reliability
Data:
  n: number of units (operations) ($n = |O|$)
  $\boldsymbol{x} = (x_1, x_2, \ldots, x_n)$: binary vector

begin
$\hat{r} = 0$
foreach $x \in U$ **do begin**
  $\hat{r} = \hat{r} + \prod_{i=1}^{n} p_i^{x_i}(1 - p_i)^{(1-x_i)}$
  end
return $\hat{r}$
end

```
U={
 (0,1,0,0,1,0,1), (0,1,0,0,1,1,1), (0,1,0,1,0,1,0), (0,1,0,1,0,1,1),
(0,1,0,1,1,0,1),
 (0,1,0,1,1,1,0), (0,1,0,1,1,1,1), (0,1,1,0,1,0,1), (0,1,1,0,1,1,1),
(0,1,1,1,0,1,0),
 (0,1,1,1,0,1,1), (0,1,1,1,1,0,1), (0,1,1,1,1,1,0), (0,1,1,1,1,1,1),
(1,0,0,1,0,0,0),
 (1,0,0,1,0,0,1), (1,0,0,1,0,1,0), (1,0,0,1,0,1,1), (1,0,0,1,1,0,0),
(1,0,0,1,1,0,1),
 (1,0,0,1,1,1,0), (1,0,0,1,1,1,1), (1,0,1,0,0,0,0), (1,0,1,0,0,0,1),
(1,0,1,0,0,1,0),
 (1,0,1,0,0,1,1), (1,0,1,0,1,0,0), (1,0,1,0,1,0,1), (1,0,1,0,1,1,0),
(1,0,1,0,1,1,1),
 (1,0,1,1,0,0,0), (1,0,1,1,0,0,1), (1,0,1,1,0,1,0), (1,0,1,1,0,1,1),
(1,0,1,1,1,0,0),
 (1,0,1,1,1,0,1), (1,0,1,1,1,1,0), (1,0,1,1,1,1,1), (1,1,0,0,1,0,1),
(1,1,0,0,1,1,1),
 (1,1,0,1,0,0,0), (1,1,0,1,0,0,1), (1,1,0,1,0,1,0), (1,1,0,1,0,1,1),
(1,1,0,1,1,0,0),
 (1,1,0,1,1,0,1), (1,1,0,1,1,1,0), (1,1,0,1,1,1,1), (1,1,1,0,0,0,0),
(1,1,1,0,0,0,1),
 (1,1,1,0,0,1,0), (1,1,1,0,0,1,1), (1,1,1,0,1,0,0), (1,1,1,0,1,0,1),
(1,1,1,0,1,1,0),
 (1,1,1,0,1,1,1), (1,1,1,1,0,0,0), (1,1,1,1,0,0,1), (1,1,1,1,0,1,0),
(1,1,1,1,0,1,1),
 (1,1,1,1,1,0,0), (1,1,1,1,1,0,1), (1,1,1,1,1,1,0), (1,1,1,1,1,1,1)}
```

The system reliability formula is given based on set U, and with reliability vector p= (0.95, 0.999, 0.999, 0.95, 0.99, 0.99, 0.999), the structural reliability $\widehat{r}$ of the system is determined by algorithm RBO as 0.9999.

Redundancy Stategies for Processing Systems

If the failure of any of the equipment units of a processing system makes the system nonoperational, that system is called redundancy-free. Because of the high level of requirements on the availability, redundancy-free systems are not suitable for practical use, redundancy is required to ensure high availability. These redundancies in the designed process are supposed to be determined during its synthesis.

The general approach to increase a system's reliability is to implement unit-level redundancies, *i.e.*, to invest in redundant units performing the same task and have them serve as backup in case the original unit fails. However, this approach does not necessarily result in the optimal process for simple reasons. Most of the time, the process has no failures and operates as it is designed, *i.e.*, the process is in on-design operation. Compared to this, the off-design operations occur only during short periods of time. It is quite possible that the structure (flowsheet) of the off-design operation differs significantly from that of the on-design operation. Total site systems can be an example where the intermediate products can have alternatives, some of which have higher cost against reduced investment requirements.

Example 9.3 Figure 9.9 shows the maximal structure of a PNS problem where the goal is to produce material P1 [8] with at least 0.999 reliability. The structural model is represented as P={$P1$}, R={$R1$, $R2$, $R3$}, and O={$O1$, $O2$, $O3$, $O4$, $O5$, $O6$, $O7$, $O8$}. The reliabilities of the operating units are 0.997, 0.996, 0.998, 0.999, 0.970, 0.960, 0.960, and 0.999, respectively. Structure operability is considered here based on a linear programming model, however this is a result of the description of the PNS process and not the algorithm. Without considering reliability, the lowest cost design (with 9312 USD/year cost) can be determined with algorithm ABB, and it is shown in Fig. 9.10. However, this is a redundancy-free solution, as expected. The reliability of this process is only 0.989 which does not satisfy the requirement ($\geq$0.999). Moreover, none of the redundancy-free solutions reach the required reliability. Thus, redundancy must be considered in solving this PNS problem. Figure 9.11 shows a modified maximal structure where a redundant operating unit has been added for each original operating unit. These redundant units have equivalent parameters to the original ones. By generating all solution structures and determining their reliabilities, it is possible to generate all solutions to the PNS problem. From these solutions, the ones that satisfy the reliability requirement can be selected. Figure 9.12 shows the solution exhibiting the lowest cost among them. The on-design operating mode is indicated by black lines in the figure, while the grey lines represent the spare part of the system. Note that this solution contains both unit-level redundancy (for operating units $O4$ and $O8$), and process-level redundancy

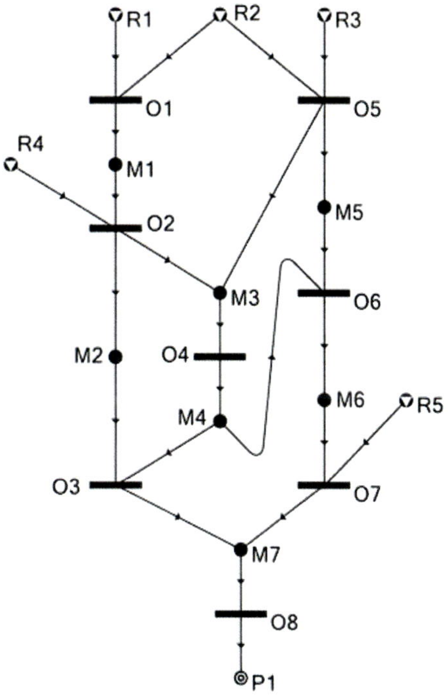

Fig. 9.9 Maximal structure of Example 9.3

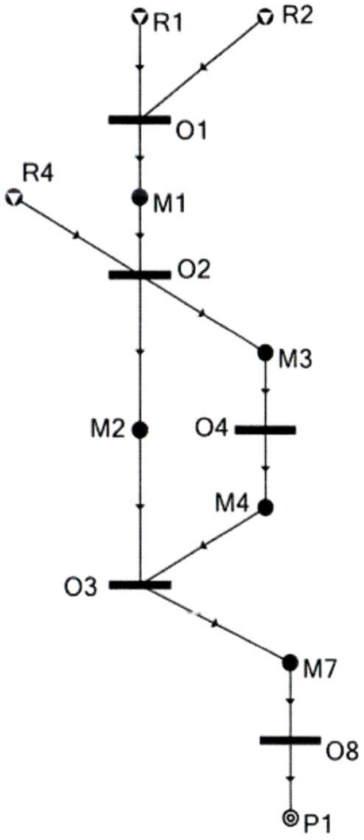

Fig. 9.10 Lowest-cost solution in Example 9.3 without redundancy, the reliability requirement is violated

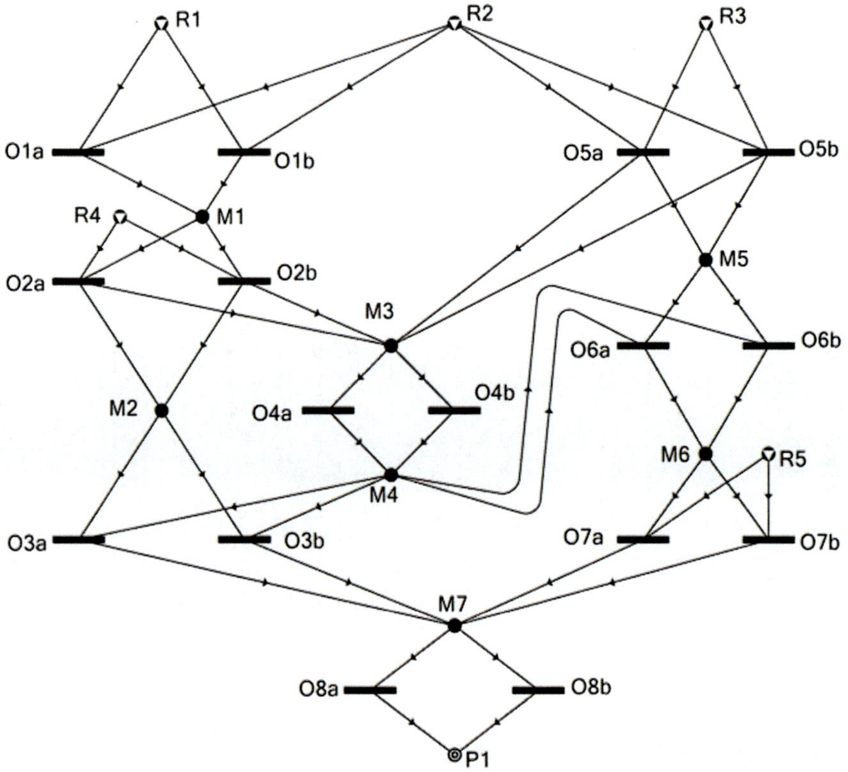

Fig. 9.11 Maximal structure of the PNS problem of Example 9.3 with redundant operating units

(both production pathways are included). Consequently, considering process-level as well as unit-level redundancies is essential to guarantee the optimality of a PNS problem with reliability constraints.

Resilience in PNS

Resilience is another property of systems, for which combinatorial enumeration is useful. The definition of resilience can greatly vary depending on the area of application. Generally speaking, resilience is the ability to resist change, and to recover from it. Since the original introduction by Holling [9], the definition has been adapted to several research areas, such as psychology, ecology, environmental sciences, engineering, computer science, etc. A thorough review of various application areas and their definitions of resilience was published by Hosseini et al. [10].

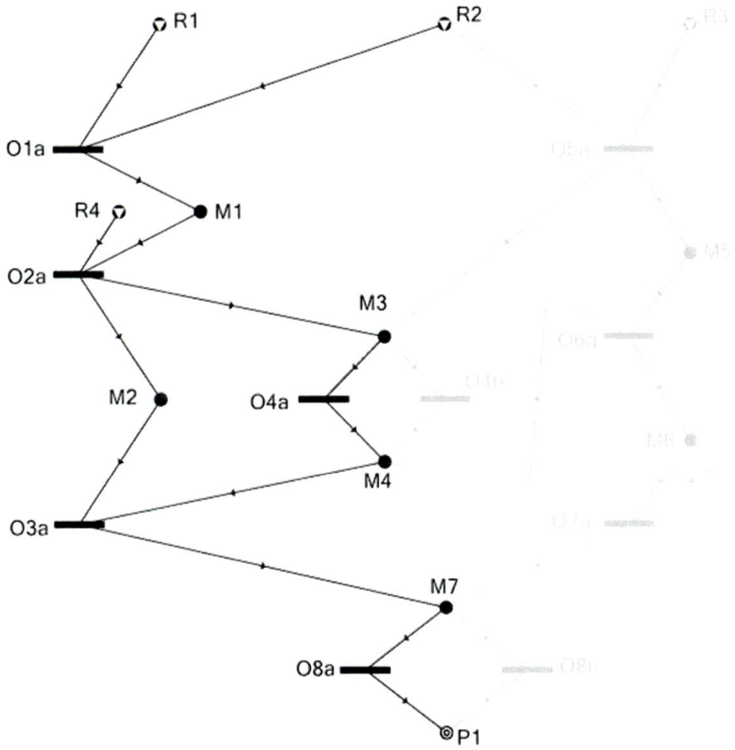

Fig. 9.12 The optimal solution of Example 9.3 considering the maximal structure given in Fig. 9.11 (black lines indicate the on-design operating mode)

Since we are considering resilience of processing systems, an appropriate definition is needed. In this case, a processing system is considered resilient if after any expected or unexpected failures, the system is (a) able to perform its designated functions in full or on a pre-specified partial level; and (b) able to return to its original state.

Resilience is similar to reliability in many ways. First, both are structural properties of the system, and thus, combinatorial enumeration-based algorithms are applicable. Second, the same principles of redundancy that can increase the reliability of the system can also increase the resilience. Finally, in some applications, the reliability of a system can be an actual variable in the resilience formula (in this book, however, resilience is determined independently of reliability).

To determine the resilience of a processing system, where the ability to return to the original state is also considered, it is necessary to be able to repair the failed units of the system. Hence, each operating unit will be assigned a repair time. At the same time, processing systems can be subject to failures that do not affect the whole system, only a part of it, thus resulting in partial operation. Partial operation can also occur if the redundant units are not as efficient as the main ones for the sake of reducing the cost of the system.

In a redundant system, there is a difference between the designed production rate of the system *i.e.*, the production during normal operation, and its maximal capacity *i.e.*, if the backup parts would also operate simultaneously. Formally, let the maximal production capacity of the system be α, and the designed production rate, β. In case of some failures, the production rate can decrease. However, in some systems some level of reduction is acceptable. Let γ denote the production rate that is still acceptable, and can be considered full production. If the production rate drops below γ, the system is considered to have partial production. Some systems also have a lower bound for the production, let us denote this with δ. If the production rate drops below δ, the system is turned off completely as the production below this low rate is either impossible, or unprofitable.

Due to the unpredictable nature of unexpected events, the failures in resilience estimation are only considered as events without probability. All failures are taken into account with the same weight. Furthermore, let us suppose that only the active units can experience failures. Multiple units can fail at the same time, *e.g.*, if the failures have a common cause, but it is supposed that there are no additional failures during the repair period (the repair period is short enough, thus another failure in the meantime is very unlikely).

If K denotes the number of active operating units in the on-design operation of the system, then theoretically there are 2^K different combinations of failures that could happen at any point, including the case when there is no failure. Let S_n^f denote the set of combinations with at most n failures ($0 \leq n \leq K$), and let N_n^f be the cardinality of set S_n^f. A time horizon h is defined during which the system's response to a given set of failures is analyzed. The basis for comparison is the case with no failures, since then the system is able to operate with full capacity.

Let s be an element of S_n^f, *i.e.*, a combination of failures. Function $F(s, t)$ denotes the system's production level at time $t \in [0, h]$ as a result of the failures in s. To be able to compare the results of different failures, the relative production of the systems $p(s, t)$ is defined as a normalized production given by Eq. 9.11

$$p(s,t) = \begin{cases} 1, & \gamma < F(s,t) \\ \dfrac{F(s,t)}{\gamma}, & \delta \leq F(s,t) \leq \gamma \\ 0, & F(s,t) < \delta \end{cases} \tag{9.11}$$

The resilience of the system to the combination of failures given by s is the production during the time horizon h relative to the maximum production. This is mathematically described as

$$q(s) = \frac{\int_0^h p(s,t)dt}{h} \tag{9.12}$$

Based on this, we can define the resilience of the system for any set of failures in S_n^f. The resilience to at most n failures (R_n) can be defined as the average of the

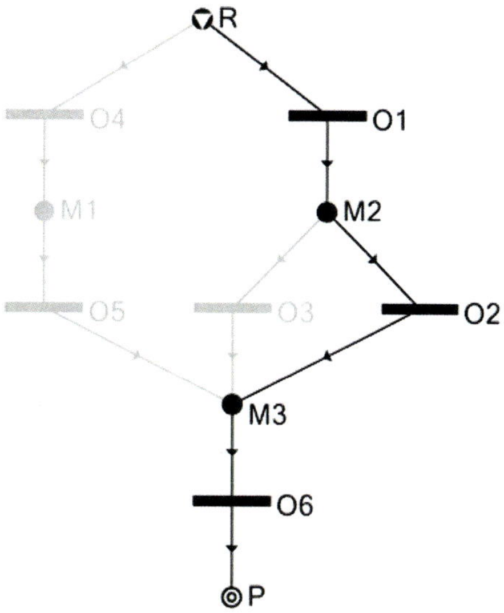

Fig. 9.13 Process and on-design operation of Example 9.4

relative production rates to the corresponding sets of failures (formula (9.13)). If the number of failures is limited to K (the number of units in the normal operation), than we get the resilience R of the system (formula (9.14)).

$$R_n = \frac{\sum_{s \in S_n^f} q(s)}{N_n^f} \tag{9.13}$$

$$R = R_K \tag{9.14}$$

With these formulas, the resilience of any structure can be determined by the combinatorial enumeration of all possible failure combinations.

Example 9.4 Let us consider the small illustrative example in Fig. 9.13. The system consists of six operating units, with the on-design operation shown with black lines, and the backup units shown in grey.

Since the process normally operates with 3 operating units ($K = 3$), the number of simultaneous failures can be 0, 1, 2, or 3. In case of some failures, the process remains operational through the redundant part, *e.g.*, if unit $O2$ fails, then unit $O3$ can be used instead. In other cases, *e.g.*, if unit 6 fails, then the system is not operational until the units are repaired. All eight failures combinations (including the case of no failure) are shown in Fig. 9.14. The units crossed over are the failed units, the black lines indicate the resulting mode of operation.

To determine the resilience, we need the repair times of the operating units. Let us assume that operating units $O1$, $O2$, and $O6$ can be repaired in 20, 15, and 9 days, respectively. It is possible to work on the repair of all units simultaneously. Also, let us define the time horizon h as 30 days. It is also important to define the capabilities

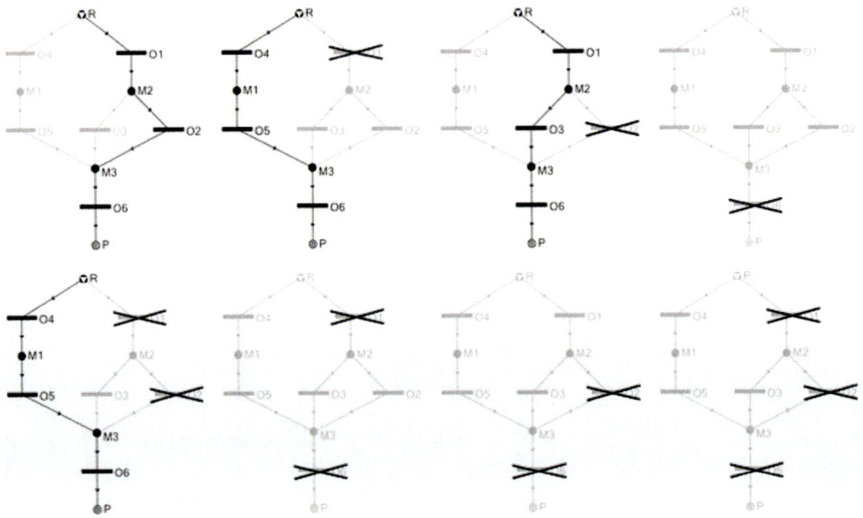

Fig. 9.14 All eight failure combinations of Example 9.4, and the resulting operating modes (if any)

of the redundant units. Operating unit $O3$ can perform the same job as operating unit $O2$ on the same level. However, if operating units $O4$ and $O5$ have to be used, the system is only capable of operating on 70% capacity.

Now, all data is given to determine the resilience for each set of failures. As an example, let us consider the case where operating units $O1$ and $O6$ fail. In this case, the system will be unable to operate, and the production halts. After 9 days, operating unit $O6$ is repared, and the process can restart at 70% production with operating units $O4$, $O5$, and $O6$. After another 11 days (20 days total since the failures), operating unit $O1$ is also repaired, thus the production level can return to 100%. Figure 9.15 shows the relative production level over time for the described case. The resilience to these set of failures is the relative production on the whole time horizon; thus,

$$q(s) = \frac{\int_0^h p(s,t)dt}{h} = \frac{(9-0)*0+(20-9)*0.7+(30-20)*1}{30}$$
$$= \frac{17.7}{30} = 0.59$$

Similarly, the resilience to all possible sets of failures can be determined. Table 9.1 shows the $q(s)$ values for all sets of failures.

The overall resilience of the system is computed as the average of its resiliences to the various sets of failures. In case of Example 9.4, the overall resilience is obtained as 0.7725.

This chapter has presented an insight into how the enumeration algorithms based on P-graphs can be deployed for analyzing specific properties of processing systems.

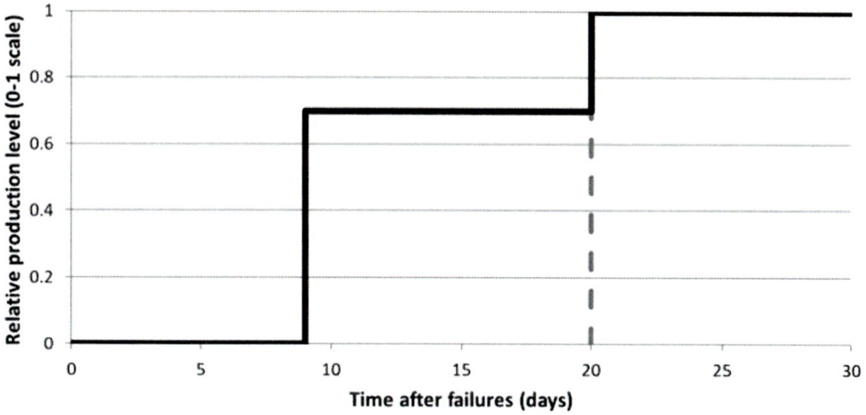

Fig. 9.15 Relative production level of Example 9.4 after failures of operating units $O1$ and $O6$

Table 9.1 Values of relative production levels of the system for different set of failures

Failures	Relative production level	Failures	Relative production level
∅	1	$\{O1,O2\}$	0.8
$\{O1\}$	0.8	$\{O1,O6\}$	0.59
$\{O2\}$	1	$\{O2,O6\}$	0.7
$\{O6\}$	0.7	$\{O1,O2, O6\}$	0.59

Herein, two of such properties have been addressed, reliability and resilience, whose mathematical description clearly reflects the highly combinatorial nature of processing systems. Naturally, proper incorporation of these properties in process design requires the use of combinatorial tools. This chapter has shown that the algorithms of the P-graph framework embody such tools.

References

1. Kumamoto H, Tanaka K, Inoue K (1997) Efficient evaluation of system reliability by Monte Carlo method. IEEE Trans Reliab 26(5):311–315
2. Prowell SJ, Poore JH (2004) Computing system reliability using Markov chain usage models. J Syst Softw 73(2):219–225
3. Kovacs Z, Orosz A, Friedler F (2018) Synthesis algorithms for the reliability analysis of processing systems. CEJOR 27(2):573–595
4. Birolini A (2007) Reliability engineering. Springer, Heidelberg
5. Rausand M, Arnljot HO (2004) System reliability theory: models, statistical methods, and applications. Wiley, New Jersey
6. Friedler F, Tarjan K, Huang YW, Fan LT (1992) Combinatorial algorithms for process synthesis. Comput Chem Eng 16:S313–S320

7. Friedler F, Tarjan K, Huang YW, Fan LT (1993) Graph-theoretic approach to process synthesis: polynomial algorithm for maximal structure generation. Comput Chem Eng 17(9):929–942
8. Orosz Á, Friedler F (2019) Synthesis technology for failure analysis and corrective actions in process systems engineering. Comput Aided Chem Eng 46:1405–1410
9. Holling CS (1973) Resilience and stability of ecological systems. Annu Rev Ecol Syst 4(1): 1–23
10. Hosseini S, Barker K, Ramirez-Marquez JE (2016) A review of definitions and measures of system resilience. Reliab Eng Syst Saf 145:47–61

Part II
Applications of the P-graph Framework

Chapters 1–6 of Part I introduced the P-graph framework along with its fundamental concepts, definitions, and core algorithms. Because the framework can be applied in a large variety of fields, Chap. 7 of Part II provides a review of literature concerning the P-graph framework, thus showing the versatility of the framework. Chapter 8 discusses the steps of the synthesis procedure of an industrial case study. The scope of the P-graph framework is certainly not restricted: It can be extended to various science and technology fields in new directions. In this regard, Chap. 9 presents the extension of the framework considering process reliability and resilience.

Chapter 10
Formal Proof of Algorithm MSG

Abstract The fundamentals of algorithm MSG, which rigorously constructs the maximal structure (superstructure) of a process synthesis problem, were given in Chap. 4. In this chapter, a formal proof of Algorithm MSG is developed on the basis of the axioms of combinatorially feasible process structures. Thus, it is guaranteed that algorithm MSG always generates the maximal structure of a process network synthesis (PNS) problem, if it exists. It is also shown that the algorithm's complexity is polynomial, thus rendering it highly convenient for process synthesis which usually involves numerous operating units.

Keywords P-graph · Maximal structure · Polynomial algorithm · Algorithm MSG · Process synthesis

In the current chapter, algorithm MSG [1] (presented in Chap. 4) will be validated. The algorithm is repeated here for convenience.

Algorithm for Maximal Structure Generation (Algorithm MSG)

Reduction

Initialization

Omit every operating unit from set O producing at least one raw material and denote the remaining set by O_0, *i.e.*, let $O_0 = O\{(\alpha, \beta) : (\alpha, \beta) \in O \,\&\, \beta \cap R = \varnothing\}$. Let M_0 be the set of input and output materials of operating units in O_0, *i.e.*, $M_0 = mat(O_0)$. If none of the remaining units in O_0 can produce the desired products, *i.e.*, $P \nsubseteq M_0$, then terminate indicating that no maximal structure exists for structural model $\mathbb{S}$. Otherwise, collect into set T_0 those materials from M_0 that are neither produced by the operating units in O_0 nor are raw materials.

F. Friedler et al., *P-graphs for Process Systems Engineering*,
https://doi.org/10.1007/978-3-030-92216-0_10

$$T_0 = \{X : X \in M_0 \backslash R \, \&((\alpha, \beta) \in O_0 \rightarrow X \notin \beta)\}.$$

As these materials are consumed but not produced by any operating unit and they are not defined as raw materials for the PNS problem, they must be excluded in subsequent steps of the algorithm. Finally, set $r := 0$.

Iteration

- *Step 1.1.* Proceed to the initialization of the composition phase if $T_r = \varnothing$. If not, choose one material from T_r; and denote it by X. As material X belongs to the group of materials to be excluded, the operating units that require it must also be removed. Therefore, collect those operating units from O_r, that consume X; denote it by O_X, *i.e.*, $O_X = \{(\alpha, \beta) : (\alpha, \beta) \in O_r \; \& \; X \in \alpha\}$. Let $O_{r+1} = O_r \backslash O_X$ and let $M_{r+1} = mat(O_{r+1})$. It is necessary to verify if the desired products can still be generated, *i.e.*, if $P \nsubseteq M_{r+1}$, then terminate the iteration since there is no maximal structure for $\mathbb{S}$. If $P \subseteq M_{r+1}$, collect into set $T_{r'}$ every material that is consumed by the remaining units (O_{r+1}) but is solely produced by the units in the set of excluded units O_X. In other words,

$$T_{r'} = \{Y : Y \in mat^{out}(O_X) \;\; \& \;\; Y \notin mat^{out}(O_{r+1}) \;\; \& \;\; Y \in mat^{in}(O_{r+1})\}.$$

These materials are to be excluded as their production becomes infeasible once the operating units in O_X are eliminated. Consequently, the new set of materials to be excluded, T_{r+1}, is formed by the union of the elements of T_r that are comprised in the remaining materials (M_{r+1}), and the new set of materials to be omitted, $T_{r'}$. Thus, let $T_{r+1} = (T_r \cap M_{r+1}) \cup T_{r'}$, $r := r+1$. Then, proceed to the following iteration for reduction.

Composition

Initialization

- This stage considers three sets, the set of required materials, W_s; the set of included materials, m_s; and the set of included operating units, o_s. For initialization, let $W_0 = P$; $m_0 = \varnothing$; $o_0 = \varnothing$; and set $s := 0$.

Iteration

- *Step 2.1.* Terminate if $W_s = \varnothing$. There exists at least one solution-structure for $\mathbb{S}$. In particular, $(\overline{m}, o_s)$ is the maximal structure of $\mathbb{S}$ where $\overline{m} = mat(o_s)$. Proceed to Step 2.2 if $W_s \neq \varnothing$.
- *Step 2.2.* Select one material from the set of required materials W_s; denote it by X; and let $m_{s+1} = m_s \cup \{X\}$. Then, form the set O_X^* of those operating units from O_r, that produce X, *i.e.*, let

$$O_X^* = \{(\alpha, \beta) : (\alpha, \beta) \in O_r \& X \in \beta\}.$$

- These operating units should be incorporated into the set of included units for the next iteration (o_{s+1}), *i.e.*, let $o_{s+1} = o_s \cup O_X^*$. Additionally, update the set W_s, incorporating the materials consumed by the units in O_X^* that are not raw materials and have not been already included in m_s. Let $W_{s+1} = \left(W_s \cup mat^{in}\left(O_X^*\right)\right) \backslash \left(R \cup m_{s+1}\right)$, set $s := s + 1$, and proceed to the succeeding iteration for composition.

Proof of Algorithm

Now, the proof of correctness follows. Since $\mathbb{S} = (P, R, O)$ is a structural model of PNS, the corresponding P-graph is (M, O), where $M = mat(O) \cup P \cup R$. In the initialization step of the reduction phase, set O_0 is given as the set of those operating units in O, that do not produce raw materials; moreover, set M_0 is determined as that of materials belonging to O_0. If $P \nsubseteq M_0$, then there is at least one product which cannot be generated by any of the remaining operating units; consequently, no maximal structure exists for this structural model. If $P \subseteq M_0$, then $O_0 \neq \varnothing$, and (M_0, O_0) is a P-graph where $(M_0, O_0) \subseteq (M, O)$; furthermore, $\mathbb{S}_0 = (P, R \cap M_0, O_0)$ is a PNS structural model whose P-graph is (M_0, O_0). Axiom (S2) prohibits any operating unit to generate a raw material in a solution-structure; therefore, P-graph (m, o) is a solution-structure for structural model $\mathbb{S}$ if and only if it is a solution-structure for structural model $\mathbb{S}_0$. Obviously, $O(|O| \cdot |R| \cdot k)$ time is required for this part of the reduction phase of the procedure, where $k = max\{|\beta| : (\alpha, \beta) \in O\}$ (note that O is a function called "big O" widely used in complexity theory, see Appendix I for details). It is worth noting that set T_0 contains all materials from $M_0 \backslash R$, which are sources in P-graph (M_0, O_0), *i.e.*, $T_0 = (M_0 \backslash R) \backslash (\cup\{\beta : (\alpha, \beta) \in O_0\})$; thus, it can be determined in $O(|O| \cdot |M|)$ time. Note that sources in a P-graph are those materials that are not produced by any operating unit in the P-graph, but are consumed by some.

In connection with the iteration part of the reduction phase of the algorithm, the following properties will be proved by induction on i provided that $P \subseteq M_i$.

(a) $\mathbb{S}_i = (P, R \cap M_i, O_i)$ is a structural model for PNS whose P-graph is (M_i, O_i) with $(M_i, O_i) \subseteq (M, O)$;
(b) P-graph $(m, o) \subseteq (M, O)$ is a solution-structure for $\mathbb{S}$ if and only if it is a solution-structure for $\mathbb{S}_i$;
(c) T_i is the set of all materials from set $M_i \backslash R$, which are sources in P-graph (M_i, O_i).

As previously verified, properties (a), (b), and (c) are valid for $i = 0$. At this point, let i be an arbitrary nonnegative integer, and let us suppose that $\mathbb{S}_i$ and T_i satisfy properties (a), (b), and (c) and that $T_i \neq \emptyset$ and $P \subseteq M_{i+1}$.

On the basis of the induction hypothesis, T_i is the set of all the materials from $M_i \backslash R$, which are sources in (M_i, O_i); hence, for $X \in T_i$, X is also an element of $M_i \backslash R$, and X is a source in (M_i, O_i). Since $M_i = mat(O_i)$, there is at least one operating unit in O_i that has X as an input material, *i.e.*, $O_X \neq \emptyset$, and thus, $O_{i+1} \subset O_i$; consequently $(M_{i+1}, O_{i+1}) \subseteq (M_i, O_i)$. Assumption $P \subseteq M_{i+1}$ implies that $O_{i+1} \neq \emptyset$; moreover, the induction hypothesis implies that $(M_i, O_i) \subseteq (M, O)$; therefore, $(M_{i+1}, O_{i+1}) \subseteq (M, O)$. It can be readily verified that $\mathbb{S}_{i+1} = (P, R \cap M_{i+1}, O_{i+1})$ is a structural model of PNS, thereby satisfying property (a).

In proving property (a), it has been ascertained that $\mathbb{S}_{i+1}$ is a structural model of PNS. It follows immediately from axiom (S2) that $(m, o) \subseteq (M, O)$ is a solution-structure for $\mathbb{S}_i$ if and only if it is a solution-structure for $\mathbb{S}_{i+1}$, thereby indicating that property (b) for $\mathbb{S}_{i+1}$ is validated by induction.

Finally, let us observe that $T_{i'}$ contains those materials from $M_{i+1} \backslash R$, which become sources in (M_{i+1}, O_{i+1}) as a result of deleting the operating units included in set O_X. On the other hand, $T_i \cap M_{i+1}$ contains those materials from $M_{i+1} \backslash R$, which are sources both in (M_i, O_i) and (M_{i+1}, O_{i+1}). Hence, T_{i+1} includes all the materials from set $M_{i+1} \backslash R$, which are sources in (M_{i+1}, O_{i+1}), *i.e.*, property (c) is valid.

Clearly, $O(|O_i| \cdot l)$ time is required to form set O_X, where $l = \max\{|\alpha| : (\alpha, \beta) \in O_i\}$; moreover, $T_{i'}$ and T_{i+1} can be determined in $O(|O_i| \cdot \max\{k, l\})$ time. Since, $O_0 \supset O_1 \supset O_2 \supset \ldots$, the number of iteration steps of the reduction phase is not greater than $|O_0| + 1$. As a consequence, the entire iteration part of the reduction phase can be carried out in $O(|O_0|^2 \cdot \max\{k, l\})$ time.

Now, let us suppose that $P \nsubseteq M_i$ for some positive integer i. Then, there is a product, for instance Y, such that $Y \notin M_i$. It will be proved that no solution-structure is available for $\mathbb{S}$. Contrary, let us suppose that $(m, o) \in S(\mathbb{S})$; then, according to property (b), $(m, o) \in S(\mathbb{S}_{i-1})$, thereby implying that $(m, o) \subseteq (M_{i-1}, O_{i-1})$, $P \subseteq m$, and thus, $Y \in m$. Based on Lemma 3.1, it will be proved by contradiction that there exists operating unit $u = (\alpha, \beta) \in o$ such that $Y \in \beta$. Suppose that u does not exist; then, assumption $P \cap R = \emptyset$ indicates that Y is a source in P-graph (m, o) while it is not raw material, thereby contradicting axiom (S2). Since $(m, o) \subseteq (M_{i-1}, O_{i-1})$, operating unit u is contained in O_{i-1}, also. Furthermore, $Y \notin M_i = mat(O_i)$ signifies that $u \notin O_i$; therefore, $u \in O_X$, where X is the material selected by the procedure. Then, $X \in mat(\{u\}) \subseteq mat(o)$; moreover, $mat(o) = m$ according to Lemma 3.1. As a result, $X \in m$. On the other hand, $X \in M_{i-1} \backslash R$, and X is a source in (M_{i-1}, O_{i-1}) because of property (c). Nevertheless, since $X \in m$ and $(m,$

$o) \subseteq (M_{i-1}, O_{i-1})$, $X \in m \backslash R$, *i.e.*, X is a source in (m, o), which contradicts axiom (S2); consequently, there is no solution-structure for $\mathbb{S}$.

Let us now assume that the reduction phase is concluded by $P \subseteq M_r$, and $T_r = \varnothing$ for some integer $r \geq 0$. Then, (M_r, O_r) satisfies properties (a), (b), and (c); in addition, it does not contain any material from $M_r \backslash R$, which is a source in (M_r, O_r).

Now, let us trace the building phase of the procedure starting with P-graph (M_r, O_r). We prove by induction on s that the following properties are satisfied:

(d) $m_s \subseteq M_r \backslash R$, $m_s \cap W_s = \varnothing$, moreover, there exists at least one operating unit $u = (\alpha, \beta) \in o_s$ with $X \in \beta$, for all $X \in m_s$,
(e) $o_s \subseteq O_r$,
(f) $\forall u \in o_s$, $\exists$ path $[u, Y]$ in $(mat(o_s), o_s)$ with $Y \in P$,
(g) $W_s \subseteq M_r \backslash R$.

Let $s = 0$; then, $P \neq \varnothing$ and $P \cap R = \varnothing$ since $\mathbb{S} = (P, R, O)$ is a structural model of PNS. Moreover, we have $\varnothing \neq W_0 \subseteq M_r \backslash R$ because $P \subseteq M_r$, and thus, property (g) is satisfied. The validity of properties (d), (e), (f) is obvious.

Now, let us suppose that the above properties are valid for an arbitrary integer $s \geq 0$. Furthermore, let us assume that $W_s \neq \varnothing$ and that X be an arbitrary element of W_s. Then, $X \in M_r \backslash R$ by the induction hypothesis on property (g). According to property (c) and the fact $T_r = \varnothing$, there is at least one operating unit, $u = (\alpha, \beta) \in O_r$, with $X \in \beta$; as a result, $O_X^* \neq \varnothing$. Then, from the definition of O_X^*, we have $O_X^* \subseteq O_r$. Again, by the induction hypothesis, we have $o_s \subseteq O_r$, thereby indicating that $o_{s+1} = o_s \cup O_X^* \subseteq O_r$, *i.e.*, property (e) is satisfied. The definitions of W_{s+1}, m_{s+1}, and o_{s+1} imply the validity of property (d) and the inclusion, $W_{s+1} \subseteq M_r \backslash R$, *i.e.*, the validity of property (g) as well. Finally, let $u = (\alpha, \beta) \in o_{s+1}$ be an arbitrary operating unit. If $u \in o_s$, then property (f) is satisfied on the basis of the induction hypothesis. If $u \notin o_s$, then $u \in O_X^*$ for some $X \in W_s$. On the other hand, the definitions of the sets, W_0, W_1, …, W_s, and the fact that $X \in W_s$, imply $X \in P \cup mat^{in}(o_s)$. Consequently, $X \in P$ or there is an operating unit $\overline{u} \in o_s$ with $X \in mat^{in}(\overline{u})$. Since $u \notin o_s$, $u \neq \overline{u}$. In the first case, property (f) is evidently satisfied. In the second case, based on the induction hypothesis, there is a path in $(mat(o_s), o_s)$ from $\overline{u}$ into a desired product. Then, it immediately follows that a path exists from u to the same product in $(mat(o_{s+1}), o_{s+1})$. Thus, property (f) is valid.

Based on property (d), $m_s \subseteq M_r \backslash R$ and $m_s \cap W_s = \varnothing$; moreover, $m_{s+1} = m_s \cup \{X\}$ where $X \in W_s$, and thus, $m_0 \subset m_1 \subset \ldots$, which indicates that the number of the iteration steps is at most $|M_r \backslash R| + 1$. Let us suppose that $W_j = \varnothing$ at the termination of the building phase for some nonnegative integer j. Since $P \neq \varnothing$ and $W_0 = P, j \neq 0$, consequently, $o_j \neq \varnothing$. First, by letting $\overline{m} = mat(o_j)$, it will be proven that $(\overline{m}, o_j)$ is a solution-structure for $\mathbb{S}$. Indeed, properties (a) and (e) and the fact that $o_j \neq \varnothing$ and $\overline{m} = mat(o_j)$ imply $(\overline{m}, o_j) \subseteq (M, O)$ and the validity of axiom (S5). From property (f), axiom (S4) is satisfied. Since $W_j = \varnothing$, axiom (S2) can be proven to be valid by contradiction. To accomplish this, let us suppose that $X \in \overline{m} \backslash R$ and that X is a source in $(\overline{m}, o_j)$. From property (d), $X \notin m_j$; on the other hand, $X \in \overline{m} = mat(o_j)$, and thus, operating unit $u = (\alpha, \beta) \in o_j$ with $X \in \alpha$ exists. Since $u \in o_j$, there exists a nonnegative integer $t < j$ such that $u \in O_Y^*$ for some $Y \in W_t$, thereby indicating that $X \in W_{t+1}$. Since $m_0 \subset m_1 \subset \ldots \subset m_j$ and $X \notin m_j$, $X \in W_{t+1}$ implies that $X \in W_j$ thereby giving rise to resulting in a contradiction. Finally, $W_0 = P$ and $W_j = \emptyset$ lead to $P \subseteq m_j$, and thus, axiom (S1) is also valid, *i.e.*, $(\overline{m}, o_j)$ is a solution-structure for $\mathbb{S}$.

To verify that $(\overline{m}, o_j)$ is the maxial structure of $\mathbb{S}$, let us suppose that (m, o) is a solution-structure for $\mathbb{S}$ and that $(\overline{m}, o_j) \subset (m, o)$. Based on propery (b), (m, o) is a solution-structure for $\mathbb{S}_r$; furthermore, $m = mat(o)$ from Lemma 3.1. Now, if $o_j = o$, then $\overline{m} = m$, which is a contradiction. Hence, $(\overline{m}, o_j) \subset (m, o)$ implies that $o_j \subset o$. Let $u \in O_r$ be an operating unit such that $u \notin o_j$ and $u \in o$. According to axiom (S4), path $[u, Y]$ exists in P-graph (m, o) with $Y \in P$. This implies that $[u, Y]$ is also a path in (M_r, O_r), and therefore, by resorting to the building phase of the procedure, we have $u \in o_j$ which is a contradiction; therefore, $(\overline{m}, o_j)$ is the maximal structure of $\mathbb{S}$.

With regards to the complexity of the building phase of the procedure, set O_X^* is determined in $O(|O_r| \cdot k)$ time; the actual W_s is determined in $O(|O_X| \cdot l)$ time. Considering the number of the iteration steps, the whole building phase takes $O(|M_r \backslash R| \cdot |O_r| \cdot max\{k, l\})$ time. Since $|O_r| \leq |O_0| \leq |O|$ and $2 \cdot |O| \cdot max\{k, l\} \geq |M_r \backslash R|$, the entire procedure can be carried out in $O((|O| \cdot max\{k, l\})^2)$ time. Thus, MSG is a polynomial algorithm, *i.e.*, qualified as efficient algorithm.

Reference

1. Friedler F, Tarján K, Huang YW, Fan LT (1993) Graph-theoretic approach to process synthesis: polynomial algorithm for maximal structure generation. Comput Chem Eng 17:929–942

Chapter 11
Simplification of the Maximal Structure

Abstract The maximal structure is the smallest structure that contains the structures of all feasible solutionsd of a process network synthesis (PNS) problem. However, sometimes the solution procedure of a PNS problem can be enhanced by solving a smaller, yet equivalent, problem. This chapter describes the method of merging reduction that is capable of generating a smaller, but identical, PNS problem.

Keywords P-graph · Maximal structure · Model transformation · Merging reduction

Usually, large number of decisions are to be made during process synthesis. Some of these decisions are, however, not independent of each other. For example, the involvement of an operating unit in a structure may imply the involvement of another operating unit even if they are not directly connected. For maximal structure shown on Fig. 11.1, whenever operating unit *O3* is involved in a solution-structure, this is the case also for operating unit *O6*.

A reduction procedure was presented in Chap. 4, where the main idea was to exclude the unnecessary operating units and materials from the investigation. Further reduction can be performed by a method called *merging reduction* that has originally been developed by Holló et al. [1]. The algorithm to be shown here may reduce the search space both for the generation of all solution-structures and for the accelerated branch-and bound algorithm (algorithms SSG and ABB). The merging algorithm transforms the maximal structure of a PNS problem to a reduced maximal structure. Algorithms SSG and ABB may solve the PNS problem related to the reduced maximal structure more effectively than the original problem.

The mergeable operating units are determined by an equivalence relation on the set of the operating units, and all of the operating units included in an equivalence class are merged into one new operating unit. The reduction mentioned above has the following property: any solution-structure of the original problem can be derived unambiguously from a solution-structure of the reduced problem and conversely.

If $u_1 = (\alpha_1, \beta_1)$ and $u_2 = (\alpha_2, \beta_2)$, then the merging of these two operating units result in operating unit $u = (\alpha_1 \cup \alpha_2, \beta_1 \cup \beta_2)$. The merging of operating units *O3* and *O6* in Fig. 11.1 is illustrated in Fig. 11.2.

F. Friedler et al., *P-graphs for Process Systems Engineering*,
https://doi.org/10.1007/978-3-030-92216-0_11

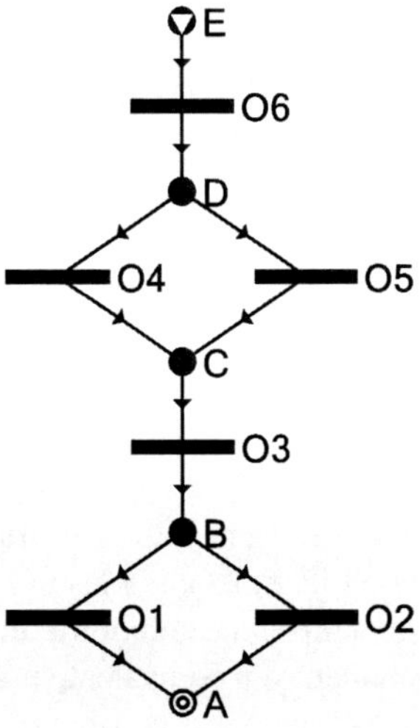

Fig. 11.1 Maximal structure with mergeable operating units

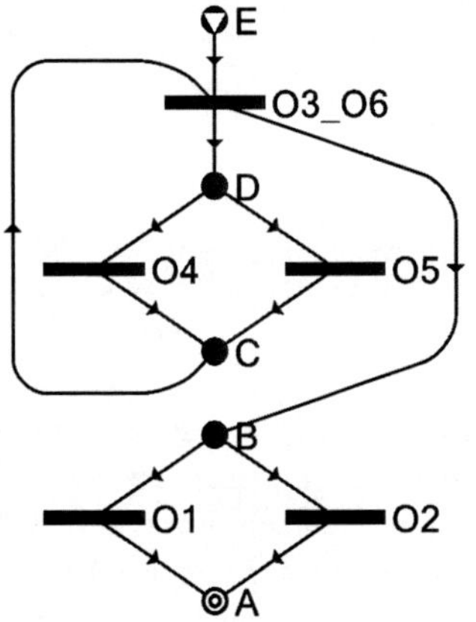

Fig. 11.2 Operating units *O3* and *O6* are merged for the combinatorial algorithms

After merging two or more operating units, we obtain a new structural model of PNS. If this new structural model is to be used for solving the original problem, a well defined relationship must be established between the solution-structures of the two problems. To establish this relationship, the first question is which operating units are mergeable.

Definition 11.1 Let $\mathbb{S} = (P, R, O)$ be a reduced structural model of PNS. Then, operating units u_1, $u_2 \in O$ are defined to be *mergeable* if operating unit u_1 is contained in a solution-structure if and only if u_2 is also contained. Formally, u_1 and u_2 are mergeable if $u_1 \in o$ implies $u_2 \in o$, and conversely, for every solution-structure $(m, o) \in S(\mathbb{S})$.

It can be easily shown that the relation of mergeability is reflexive, symmetric, and transitive, and thus, it is an equivalence relation on set O which is denoted by $\equiv$. Let structural model $\mathbb{S}/\equiv$ be defined as $\mathbb{S}/\equiv = (P, R, O^*)$ where

$$O^* = \{(\cup\{\alpha_t : u_t = (\alpha_t, \beta_t) \in C(u)\}, \cup\{\beta_t : u_t = (\alpha_t, \beta_t) \in C(u)\}) : u \in O\}$$

and $C(u)$ denotes the equivalence class containing u. The visual meaning of $\mathbb{S}/\equiv$ can be given as follows. For each equivalence class, all of the operating units belonging to this class are merged into a new operating unit. This new operating unit will substitute the original ones in $\mathbb{S}/\equiv$. Obviously, $\mathbb{S}/\equiv$ is a structural model of PNS and its maximal structure is (M, O^*). Now, we define a mapping φ of $M \cup O$ onto $M \cup O^*$. For every $X \in M$, let $\varphi(X) = X$, furthermore, for every $u_s \in C(u)$, let $\varphi(u_s) = (\cup\{\alpha_t : u_t \in C(u)\}, \cup\{\beta_t : u_t \in C(u)\})$. It will be shown as $\varphi(o) = \{\varphi(u) : u \in o\}$ and $\varphi(m) = \{\varphi(X) : X \in m\}$ for a subset o of O and for a subset m of M, respectively. Using this extension of φ, we can take the image of an arbitrary subgraph (m, o) of (M, O) under φ as $(\varphi(m), \varphi(o))$. This mapping is also denoted by φ.

The following theorem identifies the relationship between the sets of solution-structures of $S(\mathbb{S})$ and $S(\mathbb{S}/\equiv)$.

Theorem 11.1 *φ is a bijective mapping of $S(\mathbb{S})$ onto $S(\mathbb{S}/\equiv)$* [1].

Proof Let $(m, o) \in S(\mathbb{S})$ be an arbitrary solution-structure. First, it is shown that $(\varphi(m), \varphi(o))$ is a solution-structure of $\mathbb{S}/\equiv$. Obviously, $(\varphi(m), \varphi(o))$ is such a P-graph which is a subgraph of (M, O^*). Consequently, it is to be proved that $(\varphi(m), \varphi(o))$ satisfies axioms (S1) through (S5). Axiom (S1) is clearly valid, since $P \subseteq m = \varphi(m)$. To prove axiom (S2), let us observe that mapping φ preserves the sources. Axiom (S3) is clearly true, as $\varphi(o) \subseteq O^*$. As far as axiom (S4) is concerned, let $u \in \varphi(o)$ be an arbitrary operating unit. Then, there is at least one $u_j \in o$ such that $\varphi(u_j) = u$. On the other hand, $(m, o) \in S(\mathbb{S})$, then axiom (S4) verifies that there is a $[u_j, Y_n]$ path in (m, o) with $Y_n \in P$. Now, taking the images under φ of the nodes of this path, we obtain a path $[u, Y'_n]$ in $(\varphi(m), \varphi(o))$ where $Y'_n \in P$, thus axiom (S4) is satisfied. Finally, for axiom (S5), let $X \in \varphi(m)$ be arbitrary. Then, $X \in m$, and by axiom (S5), an operating unit $u_j = (\alpha_j, \beta_j)$ there exists such that $X \in \alpha_j \cup \beta_j$. Let $\varphi(u_j) = (\alpha, \beta)$. Then, by the definition of $\varphi(u_j)$, $X \in \alpha \cup \beta$, thereby validating axiom (S5).

Now, it is proven that φ is an injective mapping. For this purpose, let $(m, o) \neq (m', o') \in S(\mathbb{S})$. If $m \neq m'$, then $\varphi(m) \neq \varphi(m')$, and thus, the images are different. Otherwise, $o \neq o'$. To prove this case by contradiction, let us suppose that $(\varphi(m), \varphi(o)) = (\varphi(m'), \varphi(o'))$. Since $o \neq o'$, without loss of generality, we may assume that there exists a $u' \in o'$ with $u' \notin o$. Let $\varphi(u') = u$. Since $(\varphi(m), \varphi(o)) = (\varphi(m'), \varphi(o'))$, there exists a $\overline{u} \in o$ with $\varphi(\overline{u}) = u$. Then, by the definition of φ, $\overline{u} \equiv u'$, and thus, by the definition of the equivalence relation, $u' \subset o$ which is a contradiction. Consequently, φ is a one-to-one mapping.

Finally, we show that φ is a mapping of $S(\mathbb{S})$ onto $S(\mathbb{S}/\equiv)$. For this reason, let us consider an arbitrary solution-structure denoted by (m^*, o^*) of $S(\mathbb{S}/\equiv)$. Let $m = m^*$

and $o = \{u_j : u_j \in O \ \&\ \varphi(u_j) \in o^*\}$. Obviously, $\varphi(m, o) = (\varphi(m), \varphi(o)) = (m^*, o^*)$. Therefore, we have to prove that (m, o) is a solution-structure of $\mathbb{S}$. It can be easily seen that (m, o) is a P-graph and it is a subgraph of (M, O). Thus, it has to be proved that (m, o) satisfies all axioms.

Since $(m^*, o^*) \in S(\mathbb{S}/\equiv)$, axiom (S1) implies $P \subseteq m^*$. Since $m = m^*$, it indicates the validity of axiom (S1) for (m, o).

Since the ancestor of a source in (m^*, o^*) is a source in (m, o) under φ, and (m^*, o^*) satisfies axiom (S2), (m, o) satisfies axiom (S2) as well.

Axiom (S3) is obviously satisfied.

The validity of Axiom (S4) will be proved by contradiction. Suppose that it is not valid for (m, o). Let us denote by o_1 the set of operating units in o from which there is no path in (m, o) to any required product, *i.e.*, let

$$o_1 = \{u_j : u_j \in o \& no\ \ [u_j, Y]\ \ path\ \ exists\ \ with\ \ Y \in P\ \ in\ \ (m, o)\}.$$

By our assumption, $o_1 \neq \varnothing$. Now, let us consider P-graph (m', o') where $o' = o \backslash o_1$ and $m' = mat(o')$. Axiom (S5) follows from the definition of $\varphi(m, o)$. It has to be proven that (m', o') is a solution-structure of $\mathbb{S}$.

Since $(m^*, o^*) \in S(\mathbb{S}/\equiv)$, axiom (S1) implies that for any $X \in P$, there exists an operating unit u producing X directly. Taking an ancestor of u under φ, we obtain that there is an operating unit denoted by u' in o producing X directly, and thus, u' is not contained in o_1. Consequently, $u' \in o'$, thereby resulting in $P \subseteq m'$, *i.e.*, (m', o') satisfies axiom (S1).

To prove axiom (S2), let $X \in m'$ be arbitrary. If $X \in R$, then X is a source in (m^*, o^*), and since the ancestor of X under φ is a source in (m, o), X is a source of (m, o). Since $(m', o') \subseteq (m, o)$, X is a source in (m', o'). Conversely, let us suppose that X is a source in (m', o'); then, X is also a source in (m, o). In the opposite case, X would be an output material of at least one operating unit from o_1. Let u_1 denote such an operating unit. Then, there is a $[u_1, Y]$ path in (m, o) since X is a source in (m', o'), and thus, X is an input material for some operating unit in o'. This fact contradicts the definition of o_1. Hence, X is a source in (m, o). In this case, X is a source in (m^*, o^*), and since axiom (S2) is valid for (m^*, o^*), $X \in R$. Consequently, (m', o') satisfies axiom (S2).

The validity of axioms (S3), (S4), and (S5) follows from the definitions of o_1 and (m', o'), and thus, we obtain that (m', o') is a solution-structure of $\mathbb{S}$.

Now, let us observe that $\varphi(m', o') = (m^*, o^*) = \varphi(m, o)$, which implies $(m', o') = (m, o)$ since φ is injective. Hence, $o_1 = \varnothing$ which is a contradiction. Consequently, axiom (S4) is valid for (m, o).

In proceeding to prove the validity of axiom (S5), let $X \in m$ be an arbitrary material. Then, $X \in m^*$, and since (m^*, o^*) satisfies axiom (S5), there exists an operating unit $u = (\alpha, \beta) \in o^*$ such that $X \in \alpha \cup \beta$. This implies that there exists an operating unit $u_j = (\alpha_j, \beta_j) \in o$ such that $\varphi(u_j) = u$ and $X \in \alpha_j \cup \beta_j$. Indeed, in the opposite case, we would have that $X \notin \alpha \cup \beta$ which is a contradiction. This completes the proof of Theorem 11.1.

For illustration, let us suppose that the reduced structural model under consideration is equipped with a weight function denoted by w, *i.e.*, each operating unit has its "weight". Then, the corresponding PNS_w problem is

$$\min\left\{\sum\nolimits_{u\in o} w(u) : (m,o) \in S(\mathbb{S})\right\} \tag{11.1}$$

Let us now equip structural model $\mathbb{S}/\equiv$ by the weight function $\overline{w}$ defined as follows. For every $u \in O^*$, let $\overline{w}(u) = \sum\nolimits_{u_t \in C(u')} w(u_t)$ where $\varphi(u') = u$. Since the equivalent operating units have an identical image, function $\overline{w}$ is well-defined. The constructed new model is then

$$\min\left\{\sum\nolimits_{u\in o} \bar{w}(u) : (m,o) \in S(\mathbb{S}/\equiv)\right\} \tag{11.2}$$

Extend the weight functions for the solution-structures in the following way. For any $(m,o) \in S(\mathbb{S})$ and $(m^*,o^*) \in S(\mathbb{S}/\equiv)$, let $w(m,o) = \sum \{w(u) : u \in o\}$ and $\overline{w}(m^*,o^*) = \sum\{\overline{w}(u) : u \in o^*\}$. Then, $w(m,o) = \overline{w}(\varphi(m,o))$ is valid, for all solution-structures $(m,o) \in S(\mathbb{S})$. On the basis of this observation and Theorem 11.1, the validity of the following statement is obvious.

Theorem 11.2 *The image of an optimal solution-structure of problem* (Eq. 11.1) *under φ is an optimal solution-structure of* Eq. (11.2)*, and conversely, the image of an optimal solution-structure of problem* (Eq. 11.2) *under φ^{-1} is an optimal solution-structure of Eq.* (11.1).

To execute the merging reduction on an instance, we need to determine the equivalence relation of mergeability. For this reason, a further notation is introduced. Let $\mathbb{S} = (P,R,O)$ be a reduced structural model of PNS with $S(\mathbb{S}) \neq \varnothing$. Furthermore, let $u_j \in O$ be arbitrary. Then, a new structural model of PNS can be constructed by $\mathbb{S}(u_j) = (P,R,O\backslash\{u_j\})$. Let us denote the maximal structure of $\mathbb{S}(u_j)$ by (M_j, O_j) provided that it exists. If it does not exist, let $M_j = O_j = \varnothing$. Then, we have the following criterion.

Theorem 11.3 *For every u_i, $u_j \in O$, $u_i \equiv u_j$ if and only if $u_i \in O\backslash O_j$ and $u_j \in O\backslash O_i$ are simultaneously valid.*

Proof Let us suppose that $u_i \in O\backslash O_j$ and that $u_j \in O\backslash O_i$ for some $u_i \neq u_j \in O$. Let us consider an arbitrary solution-structure (m,o), then we have three cases.

Case 1. (m,o) does not contain u_i. Then, (m,o) is a subgraph of (M_i, O_i), and thus, by our assumption, (m,o) does not contain u_j.

Case 2. (m,o) does not contain u_j. In this case, (m,o) is a subgraph of (M_j, O_j), and hence, by our assumption, (m,o) does not contain u_i.

Case 3. (m,o) contains both u_i and u_j.

Since there is no further case, $u_i \equiv u_j$ has been proved.

In order to prove the necessity of the axiom, let us suppose that $u_i \equiv u_j$ for some $u_i \neq u_j \in O$. Let us consider the structural models $\mathbb{S}(u_i)$ and $\mathbb{S}(u_j)$. Then, (M_j, O_j) is the union of those solution-structures which do not contain u_j provided that there exist such solution-structures. Since $u_i \equiv u_j$, none of these solution-structures contains u_i. Consequently, their union does not contain u_i, *i.e.*, $u_i \in O \backslash O_j$. We can obtain by a similar argument that $u_j \in O \backslash O_i$. If every solution-structure contains u_j, *i.e.*, $O_j = \emptyset$, then from $u_i \equiv u_j$, it follows that every solution-structure contains u_i as well, and thus $O_i = \emptyset$, and the corresponding inclusions are obviously valid.

From Theorem 11.3, we immediately get the following corollary.

Corollary 11.1 *If $O_j = O\backslash\{u_j\}$, then u_j is not mergeable with any other operating unit.*

Now, by Theorem 11.3 and the algorithm for generating the maximal structure, we obtain the following procedure to determine the required equivalence relation, it is assumed that $O = \{u_1, u_2, \ldots, u_n\}$.

Procedure of Merging Reduction [1]

Initialization

- *Step 1.* Set $i := 1$, $k := 1$, $N = \{1, 2, \ldots, n\}$.
- *Step 2.* Determine the maximal structure of $\mathbb{S}(u_i)$ by the maximal structure generation algorithm. If $O_i = O\backslash\{u_i\}$, then let $V_k = \{u_i\}$, $N = N\backslash\{i\}$, $k = k + 1$. Proceed to Step 3.
- *Step 3.* If $i = n$, then proceed to Step 4. Otherwise, let $i = i + 1$, and proceed to Step 2.
- *Step 4.* Terminate if $N = \emptyset$. Otherwise, let i denote the smallest element of N. Let $J = \{t : t \in N \;\&\; u_t \in O\backslash O_i\}$. Let $V = \emptyset$, and proceed to Step 5.
- *Step 5.* If $J = \emptyset$, then let $N = N\backslash\{i\}$, $V_k = V \cup \{u_i\}$, $k = k + 1$, and proceed to Step 4. Otherwise, proceed to Step 6.
- *Step 6.* Choose an element j from J. Let $J = J\backslash\{j\}$. If $u_i \in O\backslash O_j$, then let $V = V \cup \{u_j\}$, $N = N\backslash\{j\}$. Proceed to Step 5.

As a result of this procedure, we obtain the equivalence classes belonging to the required equivalence relation as V_1, V_2, …, V_k.

For illustration, let us consider the following example.

Example 11.1 Let structural model $\mathbb{S} = (P, R, O)$ be given, where $P = \{A_1\}$; $R = \{A_9, A_{10}\}$; $O = \{u_1, u_2, \ldots u_8\}$. The operating units contained in O are listed in Table 11.1; the corresponding P-graph is shown in Fig. 11.3.

Determining the maximal structures, we obtain the following sets:

Table 11.1 Operating units.

Unit	Inputs	Outputs
u_1	A_2	A_1
u_2	A_3,A_4	A_2
u_3	A_5	A_2
u_4	A_6	A_3
u_5	A_7	A_4
u_6	A_8	A_5
u_7	A_9	A_6, A_7
u_8	A_{10}	A_7, A_8

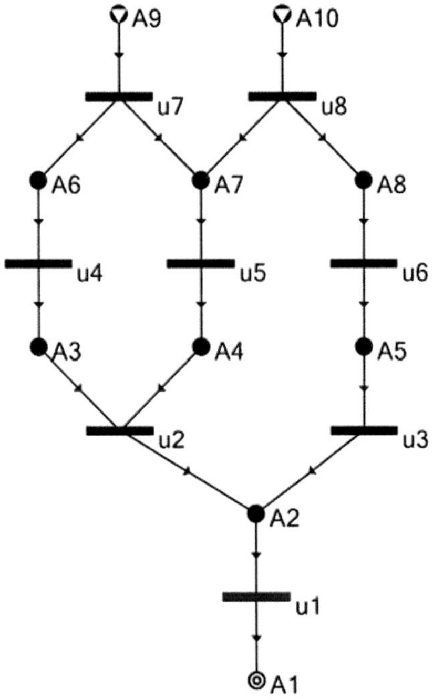

Fig. 11.3 P-graph belonging to Example 11.1

$$O_1 = \varnothing,$$

$$O_2 = O_4 = O_5 = O_7 = \{u_1, u_3, u_6, u_8\},$$

$$O_3 = O_6 = \{u_1, u_2, u_4, u_5, u_7, u_8\},$$

$$O_8 = \{u_1, u_2, u_4, u_5, u_7\}.$$

The equivalence relations, and thus the merged operating units are: $V_1 = \{u_1\}$, $V_2 = \{u_2, u_4, u_5, u_7\}$, $V_3 = \{u_3, u_6\}$, and $V_4 = \{u_8\}$. Now, after the merging of the equivalent operating units, we obtain the following four new operating units:

$$u'_1 = (\{A_2\}, \{A_1\}),$$
$$u'_2 = (\{A_3, A_4, A_6, A_7, A_9\}, \{A_2, A_3, A_4, A_6, A_7\}),$$
$$u'_3 = (\{A_5, A_8\}, \{A_2, A_5\}),$$
$$u'_4 = (\{A_{10}\}, \{A_7, A_8\}).$$

Reference

1. Holló C, Blázsik Z, Imreh C, Kovács K (1999) On a merging reduction of the process network synthesis problem. Acta Cybernet 14:251–261

Chapter 12
Formal Proof of Algorithm SSG

Abstract Chapter 5 introduced Algorithm SSG which is capable of identifying all combinatorially feasible process structures of any process network synthesis (PNS) problem. In this chapter, a formal proof is given for algorithm SSG.

Keywords P-graph · Solution-structure · Algorithm SSG · n-best · Process synthesis

Chapter 5 introduced algorithm SSG [1] to generate all solution structures of a PNS problem $\mathbb{S}(P, R, O)$. Now, the formal proof of this algorithm is given.

Further examinations of the decision-mapping is necessary for the formal proof of the algorithm. Suppose that we have made two decisions independently: One is to produce material *X* and the other is not to produce material *Y* in a single process by adopting an operating unit generating both materials *X* and *Y*. Obviously, these two independent decisions are contradictory. To circumvent such a contradiction, decision-mappings must be consistent. Hence, a special class of decision-mappings [2] need be introduced for this purpose, which is accomplished by conceiving the notion of consistency.

Definition 12.1 A decision-mapping $\delta[m]$ is said to be *consistent* if $\delta(X) \cap \Delta(Y) \subseteq \delta(Y)$ for all $X, Y \in m$.

The visual meaning of consistency can be given as follows: If material *X* is to be produced by operating units in $\delta(X)$, then the operating units in $\delta(X)$ that also produce material *Y*, $\delta(X) \cap \Delta(Y)$, must be considered for the production of material *Y*, *i.e.*, in $\delta(Y)$.

Definition 12.2 Let $\delta_1[m_1]$ and $\delta_2[m_2]$ be consistent decision-mappings. Decision-mapping $\delta_2[m_2]$ is termed to be an *extension* of $\delta_1[m_1]$ if $m_1 \subseteq m_2$ and $\delta_1(X) = \delta_2(X)$ for all $X \in m_1$; this is denoted by $\delta_1[m_1] \leq \delta_2[m_2]$. In particular, if $\delta_1[m_1] \leq \delta_2[m_2]$ and $m_1 \subset m_2$, we have a *proper extension of decision-mapping* $\delta_1[m_1]$; it is denoted by $\delta_1[m_1] \lessdot \delta_2[m_2]$.

Let us denote the set of all decision mappings with Ω. The relation *extension* is a binary relation on Ω. We see immediately that this relation is reflexive, antisymmetric and transitive; hence, it is a partial ordering on Ω. Let us denote the corresponding

F. Friedler et al., *P-graphs for Process Systems Engineering*,
https://doi.org/10.1007/978-3-030-92216-0_12

partially ordered set by pair $(\Omega, \leq)$. Obviously, $\delta[\emptyset]$ is the smallest element in $(\Omega, \leq)$. The following lemma facilitates the characterization of the maximal elements of $(\Omega, \leq)$.

Lemma 12.1 *If $\delta[m] \in \Omega$ and $m \neq M\backslash R$, then there exists a proper extension of $\delta[m]$.*

Proof Since $m \subseteq M\backslash R$ and $m \neq M\backslash R$, there is an $X \in M\backslash R$ with $X \notin m$. Let

$$Q = \{(\alpha, \beta) : (\alpha, \beta) \in \Delta(X) \ \& \ \exists \ Y \in m \ \textit{with} \ (\alpha, \beta) \in \delta(Y)\}.$$

Also let decision-mapping δ' on set $m' = m \cup \{X\}$ be defined as follows:

$$\delta'(Z) = \begin{cases} \delta(Z), & \textit{if } Z \in m \\ Q, & \textit{if } Z = X \end{cases}$$

for any $Z \in m'$.

It will be shown that $\delta'[m']$ is consistent, *i.e.*, $\delta'(V) \cap \Delta(W) \subseteq \delta'(W)$ for all $V, W \in m'$. For convenience, the following three cases are distinguished.

Case 1: Both V and W are included in m. Then, $\delta'(V) \cap \Delta(W) \subseteq \delta'(W)$ is satisfied since $\delta[m]$ is consistent.

Case 2: $V \in m$ and $W = X$. If $(\alpha, \beta) \in \delta'(V) \cap \Delta(X)$, then it is also an element of both $\delta(V) \cap \Delta(X)$ and Q. Thus, $(\alpha, \beta) \in \delta'(X)$; this results in the required inclusion, $\delta'(V) \cap \Delta(W) \subseteq \delta'(W)$.

Case 3: $V = X$ and $W \in m$. If $(\alpha, \beta) \in \delta'(V) \cap \Delta(W) = Q \cap \Delta(W)$, there exists a $Y \in m$ with $(\alpha, \beta) \in \delta(Y)$ since $(\alpha, \beta) \in Q$. On the other hand, $(\alpha, \beta) \in \Delta(W)$, and thus, $(\alpha, \beta) \in \delta(Y) \cap \Delta(W)$. But Y and W are in m and $\delta[m]$ is cosistent, thereby yielding $\delta(Y) \cap \Delta(W) \subseteq \delta(W)$. Consequently, $(\alpha, \beta) \in \delta(W) = \delta'(W)$; this implies the required inclusion.

In particular, if $V = W$, the inclusion holds obviously. This ends the proof of Lemma 12.1.

The above argument immediately leads to the following corollary.

Corollary 12.1 *$\delta[m] \in \Omega$ is a maximal element of partially ordered set $(\Omega, \leq)$ if and only if $m = M\backslash R$, or equivalently, the rank of $\delta[m]$ is equal to $|M\backslash R|$.*

Corollary 12.1 may allow us to simplify our notation. For any maximal element $\delta[M\backslash R]$ of Ω, the abbreviated notation, δ, can be used instead of $\delta[M\backslash R]$ whenever no misunderstanding arises.

We recall that (M, O) is the maximal structure, and hence, $\cup\{\Delta(X) : X \in M \backslash R\} = O$.

In what follows, a strong relationship is given between some maximal elements of Ω and the solution-structures of $\mathbb{S}$. For this purpose, we provide the following definition.

Definition 12.3 *Decision-mapping belonging to solution-structure* (m, o) of M is defined by

$$\delta(X) = \{(\alpha, \beta) : (\alpha, \beta) \in o \ \& \ X \in \beta\} \text{if } X \in m \backslash R$$
$$\text{and} \quad \delta(X) = \emptyset \quad \text{if } X \in M \backslash (R \cup m).$$

Note that the decision-mapping belonging to a given solution-structure (m, o) is uniquely determined and it is defined on the set $M \backslash R$; moreover, it is consistent as shown below.

Lemma 12.2 *If* (m, o) *is a solution-structure for* $\mathbb{S}$*, the decision-mapping belonging to* (m, o) *is consistent and is a maximal element of* $(\Omega, \leq)$.

Proof Since the rank of decision-mapping $\delta[M \backslash R]$ belonging to (m, o) is $|M \backslash R|$, it is sufficient to show that δ is consistent according to Corollary 12.1. For this purpose, let X, $Y \in M \backslash R$ be arbitrary materials and then prove that $\delta(X) \cap \Delta(Y) \subseteq \delta(Y)$. Indeed, if $\delta(X) \cap \Delta(Y) = \emptyset$, the inclusion required by the consistency is obviously satisfied. Now, let us suppose that $\delta(X) \cap \Delta(Y)$ is not empty and $(\alpha, \beta) \in \delta(X) \cap \Delta(Y)$; then, $X \in m \backslash R$ and $(\alpha, \beta) \in o$ from the definition of δ. On the other hand, $(\alpha, \beta) \in \Delta(Y)$ implying that $Y \in \beta$. Since (m, o) is a solution-structure, $m = mat(o) \supseteq \beta \ni Y$ according to Lemma 3.1, and thus, $Y \in m$. The assumption that $Y \in M \backslash R$ implies that $Y \notin R$, and hence, $Y \in m \backslash R$. This leads to $(\alpha, \beta) \in \delta(Y)$, thereby resulting in the required inclusion.

The following lemma indicates that different decision-mappings belong to different solution-structures.

Lemma 12.3 *Let* (m, o) *and* (m', o') *be different solution-structures, i.e.,* $(m, o), (m', o') \in S(\mathbb{S})$ *and* $(m, o) \neq (m', o')$*. Then, the decision-mappings,* δ *and* δ'*, belonging to* (m, o) *and* (m', o')*, respectively, are different.*

Proof We recall that P-graph (m, o) is uniquely determined by set o if (m, o) is a solution-structure of $\mathbb{S}$. From our assumptions, therefore, it follows that $o \neq o'$. Then, without loss of generality, we may assume that there exists an operating unit $(\alpha, \beta) \in O$ with $(\alpha, \beta) \in o$ and $(\alpha, \beta) \notin o'$. Since $(m, o) \in S(\mathbb{S})$, $m = mat(o)$ from Lemma 3.1, and thus, $\beta \subseteq m$. From axiom (S4), $|\beta| \geq 1$, *i.e.*, there exists at least one material, X, such that $X \in \beta$; then, $X \in m$. Since (M, O) is a solution-structure and $X \in \beta$, furthermore, $(\alpha, \beta) \in O$, we have $X \notin R$ from axiom (S2), and thus, $X \in m \backslash R$. Moreover, $(\alpha, \beta) \in \delta(X)$ from the definition of δ. Now let us distinguish the following two cases.

If $X \notin m'$, then $\delta'(X) = \emptyset$ from the definition of δ', and consequently, $\delta(X) \neq \delta'(X)$.
If $X \in m'$, then $(\alpha, \beta) \notin \delta'(X)$ from $(\alpha, \beta) \notin$ o$'$ and the definition of δ', and therefore, $\delta(X) \neq \delta'(X)$.

This completes the proof of Lemma 12.3.

Let Ω_{max} denote the set of all maximal elements of $(\Omega, \leq)$. Moreover, let us define the fuction, ρ, on $S(\mathbb{S})$ as follows: For any solution-structure $(m, o) \in S(\mathbb{S})$, let $\rho((m, o))$ be the decision-mapping belonging to (m, o). By Lemmas 12.2 and 12.3, function ρ is injective and maps $S(\mathbb{S})$ into Ω_{max}. On the other hand, it is easy to see that ρ does not map $S(\mathbb{S})$ onto Ω_{max}. For example, decision-mapping δ_0, defined by $\delta_0(X) = \emptyset$ for all $X \in M \backslash R$, is contained in Ω_{max}, but there is no solution-structure $(m, o) \in S(\mathbb{S})$ with $\rho((m, o)) = \delta_0$. The following example presents further elements of Ω_{max}, which are also not the image of any solution-structure under ρ; moreover, it also demonstrates the notions introduced.

Example 12.1 Let $M = \{A, B, C, P_1, R_1, R_2, R_3\}$, $R = \{R_1, R_2, R_3\}$, and $P = \{P_1\}$, and also let O contain the operating units listed below.

$$u_1 = (\{A, B\}, \{A, P_1\}), \qquad u_2 = (\{B, C\}, \{P_1\}),$$
$$u_3 = (\{R_1, R_2\}, \{A, B\}), \qquad u_4 = (\{R_3\}, \{B, C\}).$$

Now, $S(\mathbb{S})$ contains the maximal structure, $s = (M, O)$, which itself is a solution-structure, and six additional solution-structures, (m_i, o_i), $i = 1, 2, \ldots, 6$. They are given in Figs. 12.1 and 12.2.

Regarding the elements of Ω_{max}, let us observe that for any decision-mapping $\delta \in \Omega_{max}$ in this example,

$$\delta(P_1) \subseteq \Delta(P_1) = \{u_1, u_2\}, \quad \delta(A) \subseteq \Delta(A) = \{u_1, u_3\},$$
$$\delta(B) \subseteq \Delta(B) = \{u_3, u_4\}, \text{and} \quad \delta(C) \subseteq \Delta(C) = \{u_4\}.$$

On the basis of this observation, the elements of Ω_{max} can be determined as given in Table 12.1, where the rows correspond to the materials and the columns correspond to the elements of Ω_{max}. Under the mappings in the last row, we present each solution-structure in the column of the decision-mapping belonging to the solution-

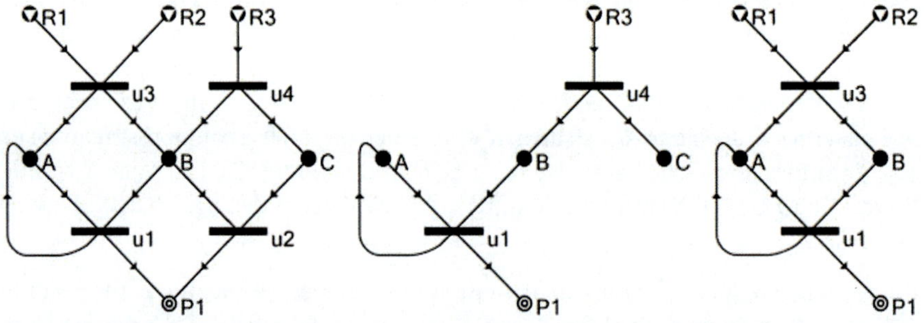

Fig. 12.1 Solution-structures, s, s_1, s_2, of the structural model of Example 12.1

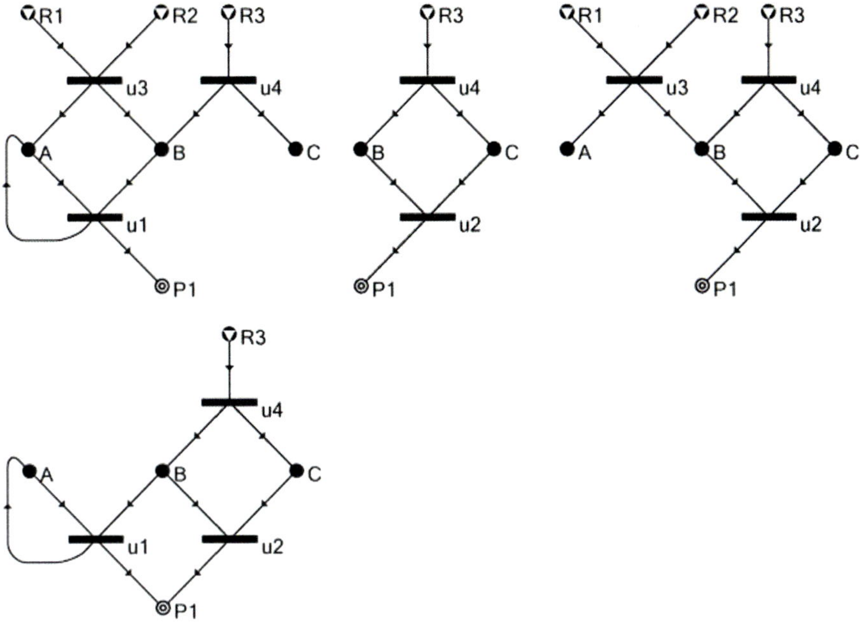

Fig. 12.2 Additional solution-structures, s_3, s_4, s_5, s_6, of the structural model of Example 12.1

structure considered. Specifically, $\rho(s) = \delta_{15}$, $\rho(s_1) = \delta_5$, $\rho(s_2) = \delta_6$, $\rho(s_3) = \delta_7$, $\rho(s_4) = \delta_9$, $\rho(s_5) = \delta_{11}$, and $\rho(s_6) = \delta_{13}$.

Since the solution-structures will be obtained through consistent decision-mappings, it is of utmost importance to resolve the following question. How can the maximal consistent decision-mappings belonging to the solution-structures of $\mathbb{S}$ be characterized, *i.e.*, how can set $\rho(S(\mathbb{S}))$ be described?

As a prelude to further elaboration we define a function *op* on Ω. This function selects the set of those operating units which are to be involved in the production of materials in set *m* based on the decisions given by consistent decision-mapping $\delta[m]$. Formally, for any $\delta[m] \in \Omega$, let

$$op(\delta[m]) = \bigcup\{\delta(X) : X \in m\}.$$

For example, $op(\delta_0) = \varnothing$, $op(\delta_1) = \{u_4\}$, $op(\delta_2) = \{u_3\}$, and $op(\delta_3) = \{u_3, u_4\}$ in Example 12.1.

Now, let $\delta \in \Omega_{max}$ be arbitrary; moreover, let us define the sequences, m_i, $i = 0$, 1, ..., and o_i, $i = 0$, 1, ..., as follows:

$$m_0 = P, o_0 = \cup\{\delta(X) : X \in P\}, \text{and for all } i \geq 0,$$

Table 12.1 Elements of Ω_{max}

	δ_0	δ_1	δ_2	δ_3	δ_4	δ_5	δ_6	δ_7
P_1	$\varnothing$	$\varnothing$	$\varnothing$	$\varnothing$	$\{u_1\}$	$\{u_1\}$	$\{u_1\}$	$\{u_1\}$
A	$\varnothing$	$\varnothing$	$\{u_3\}$	$\{u_3\}$	$\{u_1\}$	$\{u_1\}$	$\{u_1, u_3\}$	$\{u_1, u_3\}$
B	$\varnothing$	$\{u_4\}$	$\{u_3\}$	$\{u_3, u_4\}$	$\varnothing$	$\{u_4\}$	$\{u_3\}$	$\{u_3, u_4\}$
C	$\varnothing$	$\{u_4\}$	$\varnothing$	$\{u_4\}$	$\varnothing$	$\{u_4\}$	$\varnothing$	$\{u_4\}$
	–	–	–	–	–	s_1	s_2	s_3
	δ_8	δ_9	δ_{10}	δ_{11}	δ_{12}	δ_{13}	δ_{14}	δ_{15}
P_1	$\{u_2\}$	$\{u_2\}$	$\{u_2\}$	$\{u_2\}$	$\{u_1, u_2\}$	$\{u_1, u_2\}$	$\{u_1, u_2\}$	$\{u_1, u_2\}$
A	$\varnothing$	$\varnothing$	$\{u_3\}$	$\{u_3\}$	$\{u_1\}$	$\{u_1\}$	$\{u_1, u_3\}$	$\{u_1, u_3\}$
B	$\varnothing$	$\{u_4\}$	$\{u_3\}$	$\{u_3, u_4\}$	$\varnothing$	$\{u_4\}$	$\{u_3\}$	$\{u_3, u_4\}$
C	$\varnothing$	$\{u_4\}$	$\varnothing$	$\{u_4\}$	$\varnothing$	$\{u_4\}$	$\varnothing$	$\{u_4\}$
	–	s_4	–	s_5	–	s_6	–	s

$$m_{i+1} = m_i \cup mat^{in}(o_i), o_{i+1} = \cup\{\delta(X) : X \in m_{i+1} \backslash R\}.$$

From these definitions, it follows that

$$m_i \subseteq m_{i+1} \subseteq M \text{ and } o_i \subseteq o_{i+1} \subseteq O \text{ for all } i \geq 0,$$

furthermore, if $m_i = m_{i+1}$ for some nonnegative integer i, then $m_i = m_j$ and $o_i = o_j$, for all $j \geq i$. Let k denote the least nonnegative integer, for which $m_k = m_{k+1}$. Obviously, $k \leq |M \backslash R|$.

Now, let us introduce the new function, *unit*(δ), on set Ω_{max} in the following way. For any $\delta \in \Omega_{max}$, let

$$unit(\delta) = o_k$$

where o_k denotes the set of operating units defined above. The visual meaning of *unit*(δ) is the set of those operating units which are necessary in the production of the desired products directly or indirectly under the decision system given by δ. To compare the two functions defined, *i.e.*, *op* and *unit*, let us observe that *unit*(δ) $\subseteq$ *op*(δ) for all $\delta \in \Omega_{max}$. In general, the equality does not hold as demonstrated in Example 12.1, where *unit*(δ_3) = $\varnothing \subset \{u_3, u_4\}$ = *op*(δ_3).

Now, we are ready to characterize the decision-mappings which are maps of the solution-structures under ρ.

Theorem 12.1 For a decision-mapping $\delta \in \Omega_{max}$, there exists a solution-structure (m, o) of $\mathbb{S}$ with $\rho((m, o)) = \delta$, if and only if

1. *unit*(δ) = *op*(δ),
2. $\delta(X) \neq \varnothing$, for all $X \in$ *mat*(*op*(δ))$\backslash R$,
   and
3. $P \subseteq$ *mat*(*op*(δ)).

Proof To prove the necessity, let us suppose that $\delta = \rho((m, o))$ for some $(m, o) \in S(\mathbb{S})$. Then,

$$\delta(X) = \begin{cases} \{(\alpha,\beta) : (\alpha,\beta) \in o \quad \& \ X \in \beta\}, & \textit{if } X \in m \backslash R \\ \emptyset, & \textit{if } X \in M \backslash (R \cup m) \end{cases}$$

and therefore, $op(\delta) \subseteq o$. On the other hand, if $(\alpha, \beta) \in o$, then $\beta \neq \emptyset$ because of axiom (S4). Let $X \in \beta$; then, $X \in m \backslash R$ by $m = mat(o)$ and axiom (S2). Moreover, from the definition of δ, we have $(\alpha, \beta) \in \delta(X)$; this implies that $o \subseteq op(\delta)$, and hence, $op(\delta) = o$ and $mat(op(\delta)) = mat(o) = m$. In the light of the last equality, (2) and (3) are validated, respectively, because of axioms (S2) and (S1). Finally, (1) also holds because of axiom (S4) and $op(\delta) = o$.

Now, the sufficiency of conditions (1), (2), and (3) is to be proved. Let us suppose that conditions (1), (2), and (3) hold for $\delta \in \Omega_{max}$. Let $o = op(\delta)$ and $m = mat(o)$. Then, it need be proved that (m, o) is an element of $S(\mathbb{S})$.

Axiom (S3) holds for (m, o), since $o \subseteq O$ is naturally true. Since $\delta \in \Omega_{max}$, we have $m \subseteq M$ and $o \subseteq O$. On the other hand, $P \neq \emptyset$ and property (3) imply that $o \neq \emptyset$. Therefore, (m, o) is a P-graph and $(m, o) \subseteq (M, O)$. Equality $m = mat(o)$ yields axiom (S5). To verify axiom (S2), let $X \in m$ be arbitrary. If there is no (Y, X) arc in (m, o), then $X \notin m \backslash R$ from (2), and thus, $X \in R$. Conversely, let us suppose that $X \in R$. Then, there is no (Y, X) arc in (M, O) satisfying axiom (S2). This implies that there is no (Y, X) arc in (m, o) since $(m, o) \subseteq (M, O)$. Therefore, axiom (S2) holds for (m, o). Axiom (S1) follows from (3). Finally, the validity of axiom (S4) can be proved by induction on i for o_i where o_i, $i = 1, 2, \ldots, k$, denote the sets of the operating units in the definition of $unit(\delta)$. Moreover, $o_k = unit(\delta) = op(\delta) = o$ from (1), and therefore, we see that axiom (S4) also holds. This completes the proof of our theorem.

Remark 12.1 Let $S'(\subset \Omega_{max})$ denote the set of all maximal consistent decision-mappings satisfying conditions (1), (2), and (3). Then, on the basis of Theorem 12.1, we can decide if a given maximal consistent decision-mapping is included in S'. It can be completed in polynomial time, especially in $O(|M| \cdot |O|)$.

Remark 12.2 Note that in proving the sufficiency, we assign to δ a solution-structure (m, o) defined by $o = op(\delta)$ and $m = mat(o)$. As such, we have defined a function of S' into $S(\mathbb{S})$. It is easy to see that this function is the inverse function of ρ, and thus, ρ is bijective. This very strong connection between S' and $S(\mathbb{S})$ will be extensively exploited; specifically, we shall substitute each solution-structure with the corresponding decision-mapping.

For the formal proofs of the previously introduced algorithms, such as algorithm SSG, a different formalism is applied. With this formalism, we shall introduce various branch-and-bound algorithms for solving different classes of PNS problems, each of which will be based on a finite sequence, $\Theta_0 \geq \Theta_1 \geq \ldots \geq \Theta_t$, of partitions of Ω_{max} where Θ_0 is the trivial partition, $\{\Omega_{max}\}$. The classes of the r-th partition correspond to the leaves of the branch-and-bound tree associated with the r-th iteration step. To represent the corresponding branching rule, let us consider an arbitrary element of $\Omega \backslash \Omega_{max}$, *i.e.*, $\delta[m]$, with the rank less than $|M \backslash R|$. Let us define set $\omega(\delta[m])$ as follows:

$$\omega(\delta[m]) = \{\delta : \delta \in \Omega_{max} \ \& \ \delta[m] \leq \delta\},$$

i.e., $\omega(\delta[m])$ is the set of all consistent maximal extensions of $\delta[m]$. From Lemma 12.1, $\omega(\delta[m])$ is nonempty; It is worth noting that $\omega(\delta[\varnothing]) = \Omega_{max}$.

The meaning of $\omega(\delta[m])$ can be visualized as follows: We specify some decisions presented by $\delta[m]$. The set defined, *i.e.*, $\omega(\delta[m])$, leads to all the successful completions of the partial decision system under consideration, each of which is complete in the sense that it specifies all the operating units to be involved directly in the production of every material in $M \backslash R$.

The sets assigned to different decision-mappings from $\Omega \backslash \Omega_{max}$ by ω are disjoint, provided that they differ on at least one common material, as shown below.

Lemma 12.4 If $\delta[m]$, $\delta'[m'] \in \Omega \backslash \Omega_{max}$ and $\delta(X) \neq \delta'(X)$ for some $X \in m \cap m'$, then $\omega(\delta[m]) \cap \omega(\delta'[m']) = \varnothing$.

Proof By definition, for any extension $\overline{\delta}[\overline{m}]$ of $\delta[m]$, and any other extension $\overline{\delta'}[\overline{m'}]$ of $\delta'[m']$,

$$\overline{\delta}(X) = \delta(X) \neq \delta'(X) = \overline{\delta'}(X) \ \text{ if } \ X \in m \cap m'.$$

This implies $\omega(\delta[m]) \cap \omega(\delta'[m']) = \varnothing$.

In the light of the above property of ω, we can define a partition of $\omega(\delta[m])$ for any $\delta[m] \in \Omega \backslash \Omega_{max}$ as follows: Since $\delta[m] \notin |\Omega_{max}|$, the rank of $\delta[m]$ less than $|M \backslash R|$ from Corollary 12.1. Moreover, $m \subseteq M \backslash R$ implies that there exists an element, X, in set $M \backslash (R \cup m)$. Since (M, O) is the maximal structure, $\Delta(X) \neq \varnothing$. For each subset Q_i of $\Delta(X)$, $i = 1, 2, \ldots, 2^{|\Delta(X)|}$, the decision-mapping $\delta_i[m \cup \{X\}]$ can be defined by

$$\delta_i[m \cup \{X\}] = \delta[m] \cup \{(X, Q_i)\}.$$

The proof of Lemma 12.1 indicates that, at least one consistent decision-mapping exists among $\delta_i[m \cup \{X\}]$, $i = 1, 2, \ldots, 2^{|\Delta(X)|}$. Let us denote the consistent members of the decision-mappings by $\delta_{i_t}[m \cup \{X\}]$, $t = 1, 2, \ldots, k$. Without loss of generality, i_t can be replaced by t for $t = 1, 2, \ldots, k$; then, the following theorem can be stated on the partitioning of $\omega(\delta[m])$.

Theorem 12.2 The sets, $\omega(\delta_t[m \cup \{X\}])$, $t = 1, 2, \ldots, k$, constitute a (not necessarily nontrivial) partition of $\omega(\delta[m])$.

Proof Since $\delta_i(X) \neq \delta_j(X)$ if $1 \leq i \neq j \leq k$, we obtain from Lemma 12.4, $\omega(\delta_i[m \cup \{X\}]) \cap \omega(\delta_j[m \cup \{X\}]) = \varnothing$ for all $1 \leq i \neq j \leq k$; in other words, the sets $\omega(\delta_i[m \cup \{X\}])$, $i = 1, 2, \ldots, k$, are pairwise disjoint. It is, therefore, sufficient to prove that

$$\omega(\delta[m]) = \bigcup\{\omega(\delta_i[m \cup \{X\}]) : 1 \leq i \leq k\}.$$

Let $1 \leq i \leq k$ be an arbitrary integer and $\overline{\delta} \in \omega(\delta_i[m \cup \{X\}])$; then, $\delta_i[m \cup \{X\}] \leq \overline{\delta}$. On the other hand, $\delta[m] < \delta_i[m \cup \{X\}]$, and thus, $\delta[m] \leq \overline{\delta}$. Consequently, $\omega(\delta_i[m \cup \{X\}]) \subseteq \omega(\delta[m])$. Conversely, suppose that $\overline{\delta} \in \omega(\delta[m])$. Then, $\delta[m] < \overline{\delta}$, and there is an integer $1 \leq j \leq 2^{|\Delta(X)|}$ with $\overline{\delta}(X) = Q_j \subseteq \Delta(X)$. Since $m \subseteq M \backslash R$ and $X \in M \backslash (R \cup m)$,

$$\delta^*[m \cup \{X\}] = \delta[m] \cup \{(X, Q_j)\}$$

is a decision-mapping. Furthermore, since $\overline{\delta}$ is consistent, $\overline{\delta}(V) \cup \Delta(W) \subseteq \overline{\delta}(W)$ for all $V, W \in M \backslash R$; in addition, $\delta^*(V) \cap \Delta(W) \subseteq \delta^*(W)$ for all $V, W \in m \cup \{X\}$, thus yielding the consistency of $\delta^*[m \cup \{X\}]$. There exists, therefore, an integer i $(1 \leq i \leq k)$ with $\delta_i[m \cup \{X\}] = \delta^*[m \cup \{X\}]$. From the definition of δ^* and $\overline{\delta}$, we have $\delta[m] < \delta^*[m \cup \{X\}] \leq \overline{\delta}$: hence, $\overline{\delta} \in \omega(\delta_i[m \cup \{X\}])$ which, in turn, results in $\omega(\delta[m]) \subseteq \cup \{\omega(\delta_i[m \cup \{X\}]) : 1 \leq i \leq k\}$. This completes the proof.

Theorem 12.2 is illustrated here with Example 12.1. Let $m = \{A\}$ and $\delta(A) = \{u_1, u_3\}$; then, $\delta[m]$ is consistent, and $\omega(\delta[m]) = \{\delta_6, \delta_7, \delta_{14}, \delta_{15}\}$. Let us select P_1 as X; then, $\Delta(P_1) = \{u_1, u_2\}$ and the corresponding subsets, Q_i, $i = 1, 2, 3, 4$, are $\emptyset$, $\{u_1\}$, $\{u_2\}$, and $\{u_1, u_2\}$, respectively. Now, consistent decision-mappings can be obtained in two cases, *i.e.*, the cases involving $\{u_1\}$ and $\{u_1, u_2\}$. Thus, the extensions are

$$\delta_{i_1}[\{P_1, A\}] = \{(P_1, \{u_1\}), (A, \{u_1, u_3\})\},$$
$$\delta_{i_2}[\{P_1, A\}] = \{(P_1, \{u_1, u_2\}), (A, \{u_1, u_3\})\}.$$

The corresponding partition is $\{\{\delta_6, \delta_7\}, \{\delta_{14}, \delta_{15}\}\}$ where

$$\omega(\delta_{i_1}[\{P_1, A\}]) = \{\delta_6, \delta_7\} \text{ and } \omega(\delta_{i_2}[\{P_1, A\}]) = \{\delta_{14}, \delta_{15}\}.$$

As noted in Theorem 12.2, the sets, $\omega(\delta_t[m \cup \{X\}])$, $t = 1, 2, \ldots, k$, constitute a (not necessarily nontrivial) partition of $\omega(\delta[m])$, *i.e.*, if $k = 1$, the same set as the original is recovered. There is only one significant difference: the rank of the new decision-mapping is greater than that of the original one. For illustration, let us consider Example 12.1. If $m = \{A, B\}$, $\delta(A) = \{u_1\}$, and $\delta(B) = \{u_4\}$, then $\delta[m]$ is consistent, and $\omega(\delta[m]) = \{\delta_5, \delta_{13}\}$. If C is chosen to be X, the corresponding subsets are $Q_1 = \emptyset$ and $Q_2 = \{u_4\}$ since $\Delta(C) = \{u_4\}$. Only the second case, for which $Q_2 = \{u_4\}$, generates a consistent decision-mapping; therefore, $k = 1$ and the same set, *i.e.*, $\{\delta_5, \delta_{13}\}$, is obtained for both $\omega(\delta[m \cup \{C\}])$ and $\omega(\delta[m])$.

Theorem 12.2 together with an additional rule on material selection establishes a branching rule of an enumeration algorithm to generate a partition of $\omega(\delta[m])$ for all $\delta[m] \in \Omega \backslash \Omega_{max}$. By resorting to this branching rule, a search or enumeration tree can be built for set Ω_{max} according to the enumeration algorithm which proceeds as follows. Note that this algorithm essentially performs the same steps as algorithm SSG, but it is written according to the formalism introduced in this chapter, to help with the proof. Instead of recursive function calls, the algorithm presented here utilizes the open node list method to keep track of all the nodes of the branching tree it has yet to examine. This is just a detail of implementation for a more flexible operation.

Structure Generation Algorithm

Initialization

- Let $\omega(\delta[\varnothing]) = \Omega_{max}$ be the root of the tree and set $r := 0$.

  Iteration (r-th iteration)

- *Step 1.* Terminate if the rank of the decision-mapping of each leaf is $|M\backslash R|$; othervise, proceed to Step 2.
- *Step 2.* Choose a leaf, $\omega(\delta[m])$, of the actual tree whose rank is less than $|M\backslash R|$ and apply the branching rule to $\omega(\delta[m])$ to obtain the successor nodes of the leaf under consideration. Set $r := r + 1$ and proceed to the next iteration.

This procedure is correct since each node of the enumeration tree is given in the form of $\omega(\delta[m])$ where $\delta[m] \in \Omega$. On the other hand, Corollary 12.1 implies that the rank of $\delta[m]$ is equal to $|M\backslash R|$ if and only if $\delta[m] \in \Omega_{max}$. Finally, the branching rule can be applied to any decision-mapping $\delta[m] \in \Omega\backslash\Omega_{max}$.

With respect to the finiteness, let us observe that each node has at most *max* $\{2^{|\Delta(X)|} : X \in M\backslash R\}$ successors, and the depth of the tree is at most $|M\backslash R|$. Furthermore, it is easy to see that at the outset of each iteration, the leaves of the actual tree constitute a partition of Ω_{max} which is equal the trivial partition induced by the equality relation when the algorithm terminates.

To illustrate the procedure, Example 12.1 is revisited. Two iteration steps are given where the material selection rule is based on the alphabetic ordering, *i.e.*, a letter is selected from set $M\backslash(R \cup m)$ if it is the first in the alphabetic ordering.

If $r = 0$, we have a tree of one node, *i.e.*, $\omega(\delta[\varnothing]) = \Omega_{max}$. In this case, $M \backslash(R \cup m) = \{A, B, C, P_1\}$, and the material selection rule determines material A. Since $\Delta(A) = \{u_1, u_3\}$, the corresponding subsets are $\varnothing$, $\{u_1\}$, $\{u_3\}$, and $\{u_1, u_3\}$, all of which yield consistent decision-mappings denoted by $\delta_i^{(1)}[\{A\}]$, $i = 1, 2, 3, 4$, respectively. Then, the tree given in Fig. 12.3 exists at the outset of the first iteration step, where

$$\Omega_1 = \omega\left(\delta_1^{(1)}[\{A\}]\right) = \{\delta_0, \delta_1, \delta_8, \delta_9\},$$

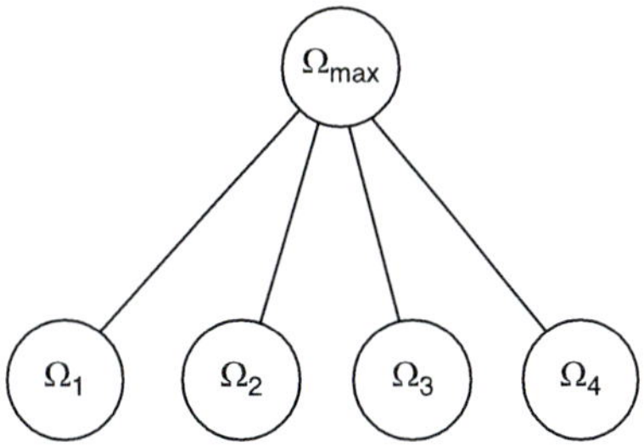

Fig. 12.3 Tree of the first iteration step

$$\Omega_2 = \omega\left(\delta_2^{(1)}[\{A\}]\right) = \{\delta_4, \delta_5, \delta_{12}, \delta_{13}\},$$

$$\Omega_3 = \omega\left(\delta_3^{(1)}[\{A\}]\right) = \{\delta_2, \delta_3, \delta_{10}, \delta_{11}\},$$

and

$$\Omega_4 = \omega\left(\delta_4^{(1)}[\{A\}]\right) = \{\delta_6, \delta_7, \delta_{14}, \delta_{15}\}.$$

Partition Θ_1 of Ω_{max} determined by the leaves of this tree is

$$\Theta_1 = \{\{\delta_0, \delta_1, \delta_8, \delta_9\}, \{\delta_4, \delta_5, \delta_{12}, \delta_{13}\}, \{\delta_2, \delta_3, \delta_{10}, \delta_{11}\}, \{\delta_6, \delta_7, \delta_{14}, \delta_{15}\}\}.$$

By choosing the leaf, $\Omega_3 = \omega\left(\delta_3^{(1)}[\{A\}]\right)$, the selection rule determines material B. Since $\Delta(B) = \{u_3, u_4\}$, the corresponding subsets are $\varnothing$, $\{u_3\}$, $\{u_4\}$, and $\{u_3, u_4\}$. Consistent decision-mappings are obtained only in cases $\{u_3\}$ and $\{u_3, u_4\}$, denoted by $\delta_1^{(2)}[\{A, B\}]$ and $\delta_2^{(2)}[\{A, B\}]$, respectively; moreover,

$$\Omega_5 = \omega(\delta_1^{(2)}[\{A, B\}]) \quad \text{and} \quad \Omega_6 = \omega(\delta_2^{(2)}[\{A, B\}]).$$

The actual tree at the outset of the second iteration step is given in Fig. 12.4. Partition Θ_2 of Ω_{max} determined by the leaves of this tree is

$$\Theta_2 = \{\{\delta_0, \delta_1, \delta_8, \delta_9\}, \{\delta_4, \delta_5, \delta_{12}, \delta_{13}\}, \{\delta_2, \delta_{10}\}, \{\delta_3, \delta_{11}\}, \{\delta_6, \delta_7, \delta_{14}, \delta_{15}\}\}.$$

Note that this partition is a proper refinement of partition Θ_1.

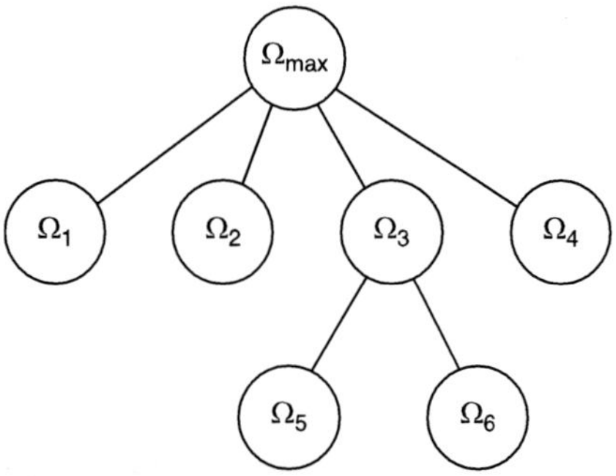

Fig. 12.4 Tree of the second iteration step

References

1. Friedler F, Tarján K, Huang YW, Fan LT (1992) Combinatorial algorithms for process synthesis. Comput Chem Eng 16(suppl):S313–S320
2. Friedler F, Varga JB, Fan LT (1995) Decision-mapping: a tool for consistent and complete decisions in process synthesis. Chem Eng Sci 50:1755–1768

Chapter 13
Formal Proof of Algorithm ABB

Abstract Chapter 6 discussed the accelerated branch-and-bound algorithm that was created to exploit the combinatorial and structural properties of the P-graph framework for accelerating the solution search. This chapter contains the formal proof of algorithm ABB.

Keywords P-graph · Branch-and-bound · Algorithm ABB · Process synthesis · Combinatorial acceleration

As far as it has been stated in Chap. 6, algorithm ABB is capable of synthesizing processes independently of the mathematical model and constraints of PNS (*i.e.*, the model and constraints can be linear, nonlinear, or implicit). Because of that, a simplified cost function is considered for the proof of the algorithm. Let $\mathbb{S} = (P, R, O)$ be a reduced structural model of process network synthesis (PNS) with the maximal structure, (M, O); moreover, let $z : S(\mathbb{S}) \to R$ be a given real-valued function, the *objective function*. The basic model is then

$$\min \{z(s) : s \in S(\mathbb{S})\} \tag{13.1}$$

Since $S(\mathbb{S})$ is a nonempty finite set, there exists at least one optimal solution of problem (13.1). Using the strong connection between $S(\mathbb{S})$ and S' given by bijective function ρ in Chap. 12, problem (13.2) can be considered instead of Eq. (13.1)

$$\min \{z'(\delta) : \delta \in S'\} \tag{13.2}$$

where $z'(\delta) = z(\rho^{-1}(\delta))$ for all $\delta \in S'$. Since ρ is a bijective mapping, it is easy to verify that if s^* is an optimal solution of problem (13.1), then $\delta^* = \rho(s^*)$ is an optimal solution of problem (13.2), and conversely, if δ^* is an optimal solution of problem (13.2), then $s^* = \rho^{-1}(\delta^*)$ is an optimal solution of problem (13.1).

Consequently, we may restrict ourselves to the investigation of problem (13.2). To simplify our discussion, the elements of S' are called *feasible solutions*.

F. Friedler et al., *P-graphs for Process Systems Engineering*,
https://doi.org/10.1007/978-3-030-92216-0_13

In order to provide validation of the algorithm, the next parts of the chapter will explain the main concepts of the Accelerated Branch-and-Bound algorithm through a different formalism than what was presented in Chap. 6. Similarly to algorithm SSG, this is done for presenting a proper proof of the algorithm. The following description of the algorithm builds on the definitions of decision mapping. Also, instead of a recursive approach, the set of open nodes will be considered.

The Formal Descriptions of Branch-and-Bound Algorithms

One of the main component of a branch-and-bound algorithm is the branching rule which divides some subsets of S' into their partitions. Here, S' is given in an implicit way and this takes difficult to give an appropriate branching rule. This difficulty can be eliminated by a suitable including set which can be treated easier than S'. We shall use set Ω_{max} for this purpose. In general, the *branching rule* or *branching operation* can be defined as a mapping assigning to some nonempty subsets of Ω_{max} their not necessarily nontrivial partitions. Here, a more general function is applied. Chapter 12 introduced a branching rule based on Theorem 12.2 that defines how the partitions $(\omega(\delta_t\ [m \cup \{X\}]),\ t=1,2,\ldots,k)$ of the set of consistent maximal extensions $(\omega(\delta[m]))$ is determined for decision-mapping $\delta[m]$. Let the function that generates the partitions for any $\delta[m]$ be denoted by φ. Its visual meaning can be given as follows. For any consistent decision-mapping $\delta[m] \in \Omega\backslash\Omega_{max}$, function φ classifies its maximal extensions depending on the operating units which are participating directly in the production of a fixed material $X \in M\backslash(R \cup m)$.

It is worth noting that $\omega(\delta[m]) = \omega(\delta'[m'])$ can occur for some $\delta[m] \neq \delta'[m']$ and $\delta[m], \delta'[m'] \in \Omega_{max}$. To illustrate this fact, let us consider Example 12.1. Let $m = \{A, C\}$, $\delta(A) = \varnothing$, and $\delta(C) = \{u_4\}$, furthermore, $m' = \{A, B\}$, $\delta'(A) = \varnothing$, and $\delta'(B) = \{u_4\}$. Then, both $\delta[m]$ and $\delta'[m']$ are contained in $\Omega\backslash\Omega_{max}$, and $\omega(\delta[m]) = \omega(\delta'[m']) = \{\delta_1, \delta_9\}$. Now, choosing B and P_1 from $M\backslash(R \cup m)$ and $M\backslash(R \cup m')$, respectively, we obtain the partitions $\{\delta_1, \delta_9\}$ and $\{\{\delta_1\}, \{\delta_9\}\}$ which are different. This is the reason why we have to define a more general function φ on set $\Omega\backslash\Omega_{max}$.

Another main component of a B&B procedure is the *bounding rule* or *bounding operation*. It is defined as a real-valued function on set Ω which gives a lower bound for values $z'(\delta')$, $\delta' \in S' \cap \omega(\delta[m])$, for all $\delta[m] \in \Omega$, provided that $\delta[m] \in \Omega\backslash\Omega_{max}$, while it provides value $z'(\delta)$ if $\delta[m] \in S'$, and ∞ if $\delta[m] \in \Omega_{max}\backslash S'$.

More precisely, let g be a real-valued function on Ω having the following properties:

$$g(\delta[m]) = \begin{cases} z'(\delta), & \delta[m] \in S' \\ \infty, & \delta[m] \in \Omega_{max}\backslash S' \end{cases}$$

$$g(\delta[m]) \leq \min\{z'(\delta') \ : \delta' \in S' \cap \omega(\delta[m])\} \text{if } \delta[m] \in \Omega \backslash \Omega_{max} \text{ and } S' \cap \omega(\delta[m]) \neq \emptyset,$$

$g(\delta[m])$ is an abitrary real number in all other cases.

The meaning of the definition above can be represented as follows. The bounding operation is defined on each element $\delta[m] \in \Omega$ depending on the set of the maximal extensions of $\delta[m]$. This set can be visualized as the set of the completions of the considered decision system given by $\delta[m]$. With regard to $\delta[m]$, two cases are distinguished.

If the decision system is complete, *i.e.*, $\delta[m] \in \Omega_{max}$, then it has only itself as its completion, and thus, $\omega(\delta[m]) = \{\delta\}$. If this complete decision system given by δ is also a feasible solution, then the bounding operation provides the value $z'(\delta)$, while g gives ∞ when δ is not a feasible solution.

If the decision system considered is not complete, *i.e.*, $\delta[m] \in \Omega \backslash \Omega_{max}$, then $\omega(\delta[m])$ is the set of its proper maximal completions. If this set contains feasible solutions, *i.e.*, $\delta[m]$ has *feasible solution extensions*, then g provides a lower bound for the values of the objective fuction, z', on these feasible solutions, while g gives an arbitrary real value when there is no feasible solution in the set of the completions considered.

By *node selection rule*, we mean a search strategy to select a leaf from an actual tree.

A branch-and-bound algorithm can be considered as a special intelligent enumerating procedure based on an enumeration tree. Its intelligence means that using the bounding operation, the algorithm does not build up the whole enumeration tree necessarily, but only a part of this tree. Specifically, it does not branch the classes of the partition determined by the leaves of the actual tree for which it is known that they do not contain feasible solutions or they do not contain better feasible solutions than a known one. A class satisfying one of the last two conditions is called *fathomed* or *dead leaf*, and otherwise, it is referred to as a *live leaf*. Now, we are ready to build up our branch-and-bound algorithm for solving problem (13.2).

Branch-and-Bound Algorithm

Initialization

- Let $L := \{\omega(\delta[\emptyset])\}$, $z^* := \infty$, $s := \emptyset$, and set $r := 0$. Calculate the value $g(\delta[\emptyset])$.

Iteration (r-th iteration)

- *Step 1. (Termination)* Terminate if $L = \emptyset$; s contains the optimal solution, z^* provides the optimal value. Otherwise, proceed to Step 2.
- *Step 2. (Selection)* Choose an element denoted by $\omega(\delta[m])$ from L by the node selection rule and proceed to Step 3.

- *Step 3. (Branching)* Form the partition $\varphi(\delta[m])$ of $\omega(\delta[m])$; denote this partition by $\{\omega(\delta_i[m_i]) : i = 1, 2, \ldots, k\}$. Proceed to Step 4.
- *Step 4. (Bounding)* For every i, $i = 1, 2, \ldots, k$, calculate lower bound $g(\delta_i[m_i])$, furthermore, if the rank of $\delta_i[m_i]$ is $|M\backslash R|$; moreover, $\delta_i \in S'$ and $g(\delta_i) < z^*$, then let $z^* = g(\delta_i)$, $s := \{\delta_i\}$. Proceed to Step 5.
- *Step 5. (Fathoming)* Redefine set L in the following way:

$$L := \{\omega(\delta'[m']) : \omega(\delta'[m']) \in (L\backslash\{\omega(\delta[m])\}) \cup \varphi(\delta[m]), g(\delta'[m']) < z^*\}.$$

Set $r := r + 1$ and proceed to the following iteration.

Now, we provide proof that the algorithm is correct, *i.e.,* it determines the best solution of the PNS problem. First, let us observe that it makes possible to build up a subtree of the enumeration tree given in Chap. 5. This subtree is called the *branch-and-bound tree*. Its root is $\omega(\delta[\varnothing]) = \Omega_{max}$. In each iteration step, the leaves of the actual tree constitute a partition of Ω_{max} and each of them is given in the form $\omega(\delta[m])$ where $\delta[m] \in \Omega$; moreover, L is a subset of the leaves of the actual tree. By Steps 4 and 5, L consists of those leaves which may contain better solution than the best known one stored in s, *i.e.*, L is the set of the live leaves of the actual tree. By Steps 4 and 5, if $\omega(\delta[m])$ is a leaf of the actual tree with $|\delta[m]| = |M\backslash R|$, then $z^* \leq g(\delta)$, and thus, $\omega(\delta[m])$ is not contained in L. Therefore, $\omega(\delta[m]) \in L$ implies $\delta[m] \in \Omega\backslash\Omega_{max}$ in each iteration step, and therefore, function φ is defined on any leaf selected by Step 2 from L. Hence, we obtain a larger subtree of the enumeration tree than the previous one in each iteration step. This yields the finiteness of the algorithm.

Now, assume that $L = \varnothing$ in some iteration step. Let us denote the leaves of the actual tree associated with the given iteration step by $\omega(\delta_i[m_i])$, $i = 1, 2, \ldots, t$. Then, $z^* \leq g(\delta_i[m_i])$, $i = 1, 2, \ldots, t$, according to Steps 4 and 5. By the definition of the bounding operation,

$$g(\delta_i[m_i]) \leq \min\{z'(\delta) : \delta \in S' \cap \omega(\delta_i[m_i])\}$$

is valid for all $i = 1, 2, \ldots, t$, provided that $S' \cap \omega(\delta_i[m_i]) \neq \varnothing$. Since $\{\omega(\delta_i[m_i]): i = 1, 2, \ldots, t\}$ is a partition of Ω_{max} and $S' \subseteq \Omega_{max}$, we obtain that $z^* \leq \min\{z'(\delta) : \delta \in S'\}$. On the other hand, if $s \neq \varnothing$, then there exists a $\delta^* \in s$ with $z^* = z'(\delta^*)$, *i.e.*, δ^* is an optimal solution.

Now, it is proven that $s = \varnothing$ is impossible under our assumption. Indeed, if $s = \varnothing$, then $z^* = \infty$ during the procedure, and the whole enumeration tree is built up. Moreover, we have that $\delta \notin S'$, for all $\delta \in \Omega_{max}$, which contradicts the fact that $\mathbb{S}$ is a reduced structural model of PNS. Thus, this case can not occur.

As far as the introduced procedure is concerned, it is important to note that this algorithm is not complete in the following sense.

1. The objective function is not specified.
2. The node selection rule is not fixed; there is only one assumption, namely, we have to select a leaf from L containing the live leaves of the actual tree associated with the corresponding iteration step.
3. In Step 3, the material selection rule (to choose a material from set $M \setminus (R \cup m)$) is not fixed.
4. The bounding operation is not given, only some properties are prescribed.

Therefore, to present a complete procedure, we have to give (1) the objective function; (2) the node selection rule; (3) the material selection rule; (4) a suitable bounding operation.

Illustration by PNS_w Problem

Here, we shall apply very simple rules to demonstrate this procedure on Example 12.1. Afterwards we take some observations which allow us to build up an acceleration. This new version will also be illustrated on another example.

First of all, we consider the PNS_w problem, which is a subclass of PNS problems. Let $\mathbb{S} = (P, R, O)$ be a reduced structural model and w is a positive real valued function defined on set O. By assigning a positive real number (weight), $w(u)$, to each operating unit u, the weight of an arbitrary solution-structure, (m, o), is defined as $\sum_{u \in o} w(u)$, *i.e.*, the sum of the weights of its operating units. The PNS_w problem is then

$$\min\left\{\sum\nolimits_{u \in o} w(u) : (m, o) \in S(\mathbb{S})\right\} \tag{13.3}$$

By the proof of Theorem 12.1, $o = op(\delta)$ if $\delta = \rho((m, o))$ for each $(m, o) \in S(\mathbb{S})$, and consequently, the equivalent problem is

$$\min\left\{\sum\nolimits_{u \in op(\delta)} w(u) : \delta \in S'\right\} \tag{13.4}$$

Let us define the node selection rule in the following standard way. If $L = \{\omega(\delta_i[m_i]) \colon i = 1, 2, \ldots, k\}$, then choose such a leaf, $\omega(\delta_i[m_i])$, for which the corresponding bound, $g(\delta_i[m_i])$, is minimal. If there are more possible candidates, then choose one of them in a random way.

For defining the material selection rule, let us suppose that the available materials are denoted by the letters of the alphabet. Now, let us apply the same material selection rule as in the previous chapter, *i.e.*, let us choose the first letter from $M \setminus (R \cup m)$ with respect to the alphabetic ordering.

Finally, for any $\delta[m] \in \Omega$, let

$$g(\delta[m]) = \sum\nolimits_{u \in op(\delta[m])} w(u) \text{ if } \delta[m] \notin \Omega_{max}$$

and

$$g(\delta) = \begin{cases} \sum\nolimits_{u \in op(\delta)} w(u), & \delta[m] \in S', \\ \infty, & otherwise \end{cases}$$

It is easy to check that g satisfies the required properties given in (a), (b), (c). Furthermore, by Corollary 12.1, we can decide if $\delta[m] \in \Omega_{max}$; moreover, if $\delta[m] \in \Omega_{max}$, then by Theorem 12.1, we can decide if $\delta \in S'$. Therefore, the values $g(\delta[m])$, $g(\delta)$ can be computed.

Completing the given branch-and-bound algorithm with the aboves, we obtain a procedure for solving problem (13.4). This procedure is illustrated on the following example.

Example 13.1 Let us consider the reduced structural model given by Example 12.1. Let us assign to the operating units, u_1, u_2, u_3, and u_4, the weights 8, 2, 5, 2, respectively.

The root of the tree is $\omega(\delta[\varnothing]) = \Omega_{max}$; the bound belonging to it is 0, $z^* = \infty$, and $s = \{\varnothing\}$. In Step 3, material A is chosen from $\{A, B, C, P_1\}$. Then, $\Delta(A) = \{u_1, u_3\}$, the corresponding subsets are $\varnothing$, $\{u_1\}$, $\{u_3\}$, and $\{u_1, u_3\}$, furthermore, we obtain consistent decision-mappings in all cases which are denoted by $\delta_i[\{A\}]$, $i = 1$, 2, 3, 4. Furthermore, let $\Omega_i = \omega(\delta_i[\{A\}])$, $i = 1$, 2, 3, 4. They are the successors of Ω_{max}, and the corresponding bounds are 0, 8, 5, 13, respectively.

In the following iteration step, the node selection rule provides the leaf, $\Omega_1 = \omega(\delta[\{A\}])$. In Step 3, material B is selected. $\Delta(B) = \{u_3, u_4\}$, and we obtain two consistent decision-mappings, $\delta_5[\{A, B\}]$ and $\delta_6[\{A, B\}]$ for the subsets, $\varnothing$ and $\{u_4\}$. Therefore, Ω_1 has two successors, $\Omega_5 = \omega(\delta_5[\{A, B\}])$ and $\Omega_6 = \omega(\delta_6[\{A, B\}])$; the corresponding bounds are 0 and 2.

In the second iteration step, leaf Ω_5 is branched. The material selected is C, and we get only one consistent decision-mapping, $\delta_7[\{A, B, C\}]$ for subset $\varnothing$, and thus, Ω_5 has one successor $\Omega_7 = \omega(\delta_7[\{A, B, C\}])$ whose bound is 0.

In the following iteration step, this leaf is selected. For material P_1, we obtain two extensions. Both of them are maximal elements, specifically, δ_0 and δ_8, and none of them is contained in S'. Therefore, the bound is ∞ for these leaves.

Proceeding to the fourth iteration step, leaf Ω_6 is selected. For material C, we get only one consistent decision-mapping, $\delta_8[\{A, B, C\}]$ for subset $\{u_4\}$. Let us denote $\omega(\delta_8[\{A, B, C\}])$ by Ω_8. The bound belonging to Ω_8 is 2.

In the following iteration step, this leaf is chosen. For material P_1, we obtain two extensions for subsets $\varnothing$ and $\{u_2\}$. Both of them are maximal elements. The first one is $\delta_1 \notin S'$ and thus, the corresponding bound is ∞. The second one is a feasible solution extension of the decision-mapping considered, namely, it is δ_9. The corresponding bound is 4, and we have that $z^* = 4$, $s = \{\delta_9\}$. By Step 5, we obtain

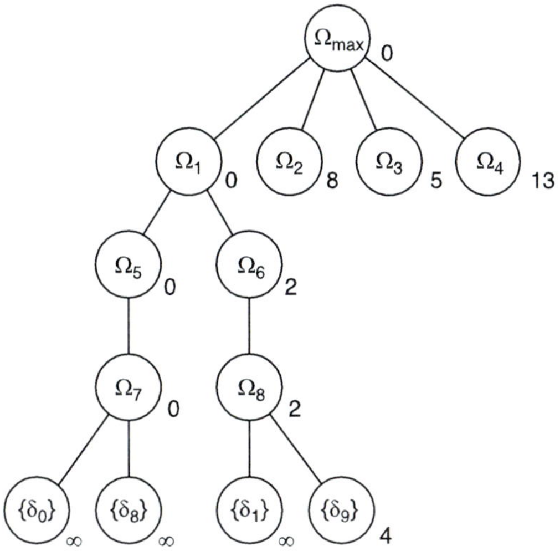

Fig. 13.1 Branch-and-bound tree built up for Example 13.1

that the leaves of the actual tree do not contain better feasible solution than δ_9, and therefore, the procedure terminates in the sixth iteration step. The optimal solution-structure is $\rho^{-1}(\delta_9) = s_4$.

The tree built up by our procedure is shown in Fig. 13.1.

With regard to Example 13.1, one can observe that the empty set is assigned to the selected materials in too many cases, and in general, these cases do not yield feasible solutions. For eliminating this situation, a new material selection rule is presented which makes our procedure more effective.

Branching rules for the general PNS

Now the general PNS problem is examined that covers all types of mathematical models and objective functions.

Definition 13.1 Let $\delta[m] \in \Omega \backslash \Omega_{max}$ be an arbitrary decision-mapping. Let us choose a material X from $(mat^{in}(op(\delta[m])) \cup P) \backslash (R \cup m)$ provided that such an X there exists. Then, the consistent decision-mappings of the form $\delta_t[m \cup \{X\}]$ where $\delta_t(X) = Q_t \subseteq \Delta(X)$ and $Q_t \neq \emptyset$, $t = 1, 2, \ldots k$, are called *regular extensions* of $\delta[m]$ and it is said that X is chosen by the *input material selection rule*. This fact is denoted by $\delta[m] \prec \delta_t[m \cup \{X\}]$, $t = 1, 2, \ldots, k$.

Concerning Definition 13.1, it is an interesting question if there exists a regular extension for any $\delta[m] \in \Omega \backslash \Omega_{max}$. The answer is no. For example, if we consider Example 12.1 and take the consistent decision-mapping, $\delta[m]$, where $m = \{A, B, P_1\}$, furthermore, $\delta(A) = \emptyset$, $\delta(B) = \emptyset$, and $\delta(P_1) = \emptyset$, then $\delta[m]$ has only one

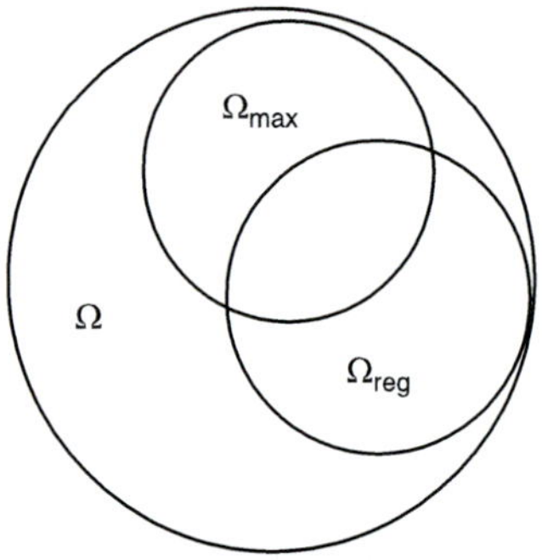

Fig. 13.2 Comparison of the sets Ω, Ω_{max}, Ω_{reg}

extension which is not regular. To treat this difficulty, a new subset of Ω is introduced. Let us define the sets $\Omega_i \subseteq \Omega$, $i = 0, 1, \ldots$, in the following way:

$$\Omega_0 = \{\delta[\emptyset]\}, \text{and for all integer } i \geq 0,$$

$$\Omega_{i+1} = \Omega_i \cup \{\delta'[m'] : \delta'[m'] \in \Omega \ \& \ \exists \ \delta[m] \in \Omega_i \text{ with } \delta[m] \prec \delta'[m']\}.$$

Then, $\Omega_0 \subseteq \Omega_1 \subseteq \ldots$, and it is easy to see that there is an integer $j \leq |M \backslash R|$ for which $\Omega_j = \Omega_i$ if $i \geq j$. Let us denote this Ω_j by Ω_{reg}, and for the sake of simplicity, we call *regular decision-mappings* the elements of Ω_{reg}.

The visual meaning of a regular decision-mapping can be given in the following way. A consistent decision-mapping is regular if and only if it can be obtained by such a sequence of decision-mappings which is started from $\delta[\emptyset]$ and each further element of this sequence is a regular extension of the previous one. The comparison of the sets, Ω, Ω_{max}, and Ω_{reg}, is shown in Fig. 13.2.

The following statement can be immediately obtained from the definition of the regular decision-mapping.

Lemma 13.1 *If* $\delta[m]$ *is a regular decision-mapping, then* $\delta(X) \neq \emptyset$ *for all* $X \in m$, *furthermore,* $m \subseteq mat(op(\delta[m]))$.

Now, we are going to prove some properties of the regular decision-mappings. For this purpose, we have to extend function *unit* defined on set Ω_{max} in Chap. 12. Let $\delta[m] \in \Omega$ be arbitrary. Let us define the sequences, m_i, $i = 0, 1\ldots$, and o_i, $i = 0, 1, \ldots$, in the following way:

$$m_0 = m \cap P, o_0 = \cup\{\delta(X) : X \in m_0\}, \text{and for all } i \geq 0,$$

$$m_{i+1} = m_i \cup mat^{in}(o_i), o_{i+1} = \cup\{\delta(X) : X \in m \cap (m_{i+1} \backslash R)\}.$$

Obviously, $m_i \subseteq m_{i+1} \subseteq M$ and $o_i \subseteq o_{i+1} \subseteq O$ for all $i \geq 0$, and if $m_i = m_{i+1}$ for some integer i, then $m_i = m_j$ and $o_i = o_j$ for all $j \geq i$. Let k denote the smallest integer with $m_k = m_{k+1}$; such an integer there exists and it is less than $|M \backslash R|$. Let us define function *unit* on $\delta[m]$ by $unit(\delta[m]) = o_k$. It is worth noting that $unit(\delta[m]) \subseteq op(\delta[m])$ is valid from the definitions.

The meaning of *unit*($\delta[m]$) is the set of all operating units from the fixed set, *op*($\delta[m]$), which are participating directly or indirectly in the production of those desired products that are contained in m.

The next lemma immediately follows from the definitions.

Lemma 13.2 *If $\delta[m]$ and $\delta'[m']$ are consistent decision-mappings and $\delta[m] \leq \delta'[m']$, then unit($\delta[m]$) $\subseteq$ unit($\delta'[m']$).*

Now, we are ready to prove the following statement.

Lemma 13.3 *If $\delta[m]$ is a regular decision-mapping, then unit($\delta[m]$) = op($\delta[m]$).*

Proof Since *unit*($\delta[m]$) $\subseteq$ *op*($\delta[m]$) for all $\delta[m] \in \Omega$, it is sufficient to prove that *op*($\delta[m]$) $\subseteq$ *unit*($\delta[m]$) for any regular decision-mapping, $\delta[m]$. Let $\delta[m] \in \Omega_{reg}$ be arbitrary. We prove that *op*($\delta[m]$) $\subseteq$ *unit*($\delta[m]$) by induction on the rank of $\delta[m]$ which is denoted by i.

If $i = 0$, *i.e.*, the rank of $\delta[m]$ is equal to 0, then $\delta[m] = \delta[\emptyset]$, and the statement obviously holds.

Now, let us suppose that $i \geq 0$ is an arbitrary integer and that the statement is valid for i. Furthermore, let us assume that $\delta[m] \in \Omega_{reg}$ and that the rank of $\delta[m]$ is equal to i + 1. Since $\delta[m] \in \Omega_{reg}$ and its rank is i+1, there exists a $\bar{\delta}[\bar{m}] \in \Omega_{reg}$ having rank i with $\bar{\delta}[\overline{m}] \prec \delta[m]$. Let $\{X\} = m \backslash \overline{m}$. Then, $X \in \left(mat^{in}\left(op\left(\bar{\delta}[\overline{m}]\right)\right) \cup P\right) \backslash (R \cup \overline{m})$ since X is chosen by the input material selection rule. On the other hand, by $\bar{\delta}[\overline{m}] \prec \delta[m]$, we have that $\bar{\delta}(V) = \delta(V)$ for all $V \in \overline{m}$. Thus,

$$op(\delta[m]) = \cup_{V \in m}\delta(V) = (\cup_{V \in \overline{m}}\delta(V)) \cup \delta(X) = op\left(\bar{\delta}[\overline{m}]\right) \cup \delta(X).$$

Let $u \in$ *op*($\delta[m]$) be an arbitrary operating unit. By the above equality, $u \in op\left(\bar{\delta}[\overline{m}]\right)$ or $u \in \delta(X)$. Let us consider the case $u \in op\left(\bar{\delta}[\overline{m}]\right)$ firstly. From our induction hypothesis, $op\left(\bar{\delta}[\overline{m}]\right) \subseteq unit\left(\bar{\delta}[\overline{m}]\right)$, and by Lemma 13.2, $unit\left(\bar{\delta}[\overline{m}]\right) \subseteq unit(\delta[m])$. Therefore, $u \in$ *unit*($\delta[m]$), yielding the required inclusion.

In order to prove the second case, let us suppose that $u \in \delta(X)$. We distinguish the following two cases depending on X.

If $X \in P$, then $X \in m \cap P = m_0$ and $u \in \delta(X) \subseteq o_0$. But $o_0 \subseteq$ *unit*($\delta[m]$), and thus, $u \in$ *unit*($\delta[m]$), implying the required inclusion.

Now, let us suppose that $X \notin P$. Then, $X \in mat^{in}\left(op\left(\bar{\delta}[\overline{m}]\right)\right) \backslash (R \cup \overline{m})$. Therefore, there exists at least one operating unit $\bar{u} \in op\left(\bar{\delta}[\overline{m}]\right)$ for which $X \in mat^{in}(\{\bar{u}\})$. By our induction hypothesis, $op\left(\bar{\delta}[\overline{m}]\right) = unit\left(\bar{\delta}[\overline{m}]\right)$, and $unit\left(\bar{\delta}[\overline{m}]\right) \subseteq unit(\delta[m])$ from Lemma 13.2. Thus, $\bar{u} \in unit(\delta[m])$. But then, by the definition of function *unit*, there is a nonnegative integer j with $\bar{u} \in o_j$, furthermore, $X \in mat^{in}(o_j)$. From the definition of m_{j+1}, we have $X \in m_{j+1}$. But $X \in m$ and $X \notin R$, resulting in $X \in m \cap (m_{j+1} \backslash R)$. Then, $u \in \delta(X) \subseteq o_{j+1} \subseteq$ *unit*($\delta[m]$), yielding the required inclusion. This completes the proof of Lemma 13.3.

The following theorem presents the usefulness of the regular decision-mappings.

Theorem 13.1 *If* $\delta[m] \in \Omega_{reg} \backslash \Omega_{max}$, *and* $\delta'[m'] = \delta[m] \cup \{(X, \emptyset)\}$ *is an extension of* $\delta[m]$ *with* $X \in (mat^{in}(op(\delta[m])) \cup P) \backslash (R \cup m)$, *then there is no feasible solution extension of* $\delta'[m']$, *i.e.*, $\omega(\delta'[m']) \cap S' = \emptyset$.

Proof According to the assumptions of the statement, $m' = m \cup \{X\}$ and $X \in (mat^{in}(op(\delta[m])) \cup P) \backslash (R \cup m)$. Then, $X \in mat^{in}(op(\delta[m])) \backslash R$ or $X \in P$. We distinguish two cases depending on the set containing X.

Let us suppose that $X \in P$. Let $\delta^* \in \omega(\delta'[m'])$ be arbitrary. Then, $\delta'[m'] \leq \delta^*$, and thus, $\delta^*(X) = \emptyset$. If $\delta^* \in S'$, then property (3) of Theorem 12.1 yields $P \subseteq mat(op(\delta^*))$, and therefore, $X \in mat(op(\delta^*)) \backslash R$. But then, $\delta^*(X) \neq \emptyset$ by (2) of Theorem 12.1, which is a contradiction.

Now let us assume that $X \in mat^{in}(op(\delta[m])) \backslash R$. Then, we have that $X \in mat(op(\delta[m])) \backslash R$, and thus, $X \in mat(unit(\delta[m])) \backslash R$ from Lemma 13.3. Let $\delta^* \in \omega(\delta'[m'])$ be arbitrary. Then, $\delta[m] \leq \delta'[m'] \leq \delta^*$, furthermore, $unit(\delta[m]) \subseteq unit(\delta^*)$ from Lemma 13.2. On the other hand, $unit(\delta^*) \subseteq op(\delta^*)$. Therefore, $X \in mat(op(\delta^*)) \backslash R$. Since $\delta'[m'] \leq \delta^*$ and $\delta'(X) = \emptyset$, we have $\delta^*(X) = \emptyset$. Then, δ^* does not satisfy condition (2) of Theorem 12.1, and thus, $\delta^* \notin S'$ which ends the proof of our statement.

In the light of Theorem 13.1, we obtain the following observations.

Remark 13.1 Let us use the input material selection rule instead of the alphabetic material selection rule in our procedure. Then, in the branching step of the procedure, when we are branching set $\omega(\delta[m])$ into its subsets $\omega(\delta_i[m \cup \{X\}])$, $i = 1, 2, \ldots, k$, we can desregard from the investigation of that class $\omega(\delta_j[m \cup \{X\}])$ for which $\delta_j(X) = \emptyset$ is valid provided that such a class there exists. Indeed, we do not lose any feasible solution since $\omega(\delta[m] \cup \{(X, \emptyset)\}) \cap S' = \emptyset$ according to Theorem 13.1. Therefore, this leaf would get ∞ as a bound and it would be fathomed in the branch-and-bound tree. Applying the input material selection rule, we do not establish this unnecessary leaf.

Remark 13.2 Let us consider such a situation when $\delta[m] \in \Omega_{reg} \backslash \Omega_{max}$, $X \in (mat^{in}(op(\delta[m])) \cup P) \backslash (R \cup m)$, and there is no regular extension of $\delta[m]$ with respect to X. In this case, by Lemma 12.1, $\delta[m]$ has only one consistent extension, $\delta'[m']$, where $m' = m \cup \{X\}$ and $\delta'(X) = \emptyset$. Then, $\omega(\delta[m]) = \omega(\delta'[m'])$, and $\omega(\delta'[m']) \cap S' = \emptyset$ from Theorem 13.1. Therefore, $\omega(\delta[m]) \cap S' = \emptyset$. This provides that if there is no regular extension of $\delta[m] \in \Omega_{reg} \backslash \Omega_{max}$, then there is no feasible solution extension of $\delta[m]$. Thus, the corresponding leaf in the branch-and-bound tree may get bound ∞ and this leaf can be fathomed.

If we perform such a modified version of our procedure on Example 13.1, then it is interrupted in the 3th iteration step at the leaf $\omega(\delta[m])$ where $m = \{P_1, B, C\}$, $\delta(P_1) = \{u_2\}$, $\delta(B) = \{u_4\}$, and $\delta(C) = \{u_4\}$. The reason of the interruption is that $(mat^{in}(op(\delta[m])) \cup P) \backslash (R \cup m) = \emptyset$. Fortunately, this situation does not exclude the application of the input material selection rule, but improve the efficiency of our procedure. Namely, in this case, we prove that there is exactly one feasible solution extension of $\delta[m]$, *i.e.*, $\omega(\delta[m]) \cap S'$ is singleton, and this feasible solution extension of $\delta[m]$ can be constructed from $\delta[m]$ uniquely. For this purpose, we need some preparation.

Definition 13.2 Let $\delta[m]$ be an arbitrary regular decision-mapping with $(mat^{in}(op(\delta[m])) \cup P)\backslash(R \cup m) = \emptyset$. Let us define decision-mapping δ' by $\delta'(X) = \{(\alpha,\beta) : (\alpha,\beta) \in op(\delta[m]) \;\&\; X \in \beta\}$ for all $X \in M\backslash R$. Then, δ' is called the *closure* of $\delta[m]$.

We note that the closure of a regular decision-mapping is determined uniquely. An important property of the closure is given by the followig statement.

Lemma 13.4 *Let $\delta[m]$ be an arbitrary regular decision-mapping. If $(mat^{in}(op(\delta[m])) \cup P)\backslash(R \cup m) = \emptyset$, then closure δ' of $\delta[m]$ is a maximal extension of $\delta[m]$, i.e., $\delta' \in \omega(\delta[m])$.*

Proof If $\delta[m] \in \Omega_{max}$, then its closure is equal to itself, and thus, the statement obviously holds. Now, let us suppose that $\delta[m] \in \Omega_{reg}\backslash\Omega_{max}$. First, we prove that closure δ' is a consistent decision-mapping, *i.e.*, $\delta'(X) \cap \Delta(Y) \subseteq \delta'(Y)$ holds for all $X, Y \in M\backslash R$. Indeed, let $X, Y \in M\backslash R$ be arbitrarily fixed materials and let us suppose that $u = (\alpha,\beta) \in \delta'(X) \cap \Delta(Y)$. Then, $Y \in \beta$. Since $u \in \delta'(X)$, we have $u \in op(\delta[m])$ from the definition of δ'. Therefore, $(\alpha,\beta) \in op(\delta[m])$ and $Y \in \beta$, implying $u = (\alpha,\beta) \in \delta'(Y)$. Thus, closure δ' is consistent.

Now, we prove that δ' is an extension of $\delta[m]$. Since δ' is defined on $M\backslash R$ and $m \subset M\backslash R$, it is sufficient to prove that $\delta(X) = \delta'(X)$ for all $X \in m$. Let $X \in m$ be arbitrary. If $u = (\alpha,\beta) \in \delta(X)$, then $u \in op(\delta[m])$, and thus, $u \in \delta'(X)$ by the definition of δ'. Conversely, let $u = (\alpha,\beta) \in \delta'(X)$. Then, $u \in op(\delta')$ and $X \in \beta$, furthermore, $u \in \Delta(X)$. Since $u \in op(\delta')$, there exists a $Y \in m$ with $u \in \delta(Y)$. But $\delta[m]$ is consistent, and thus, $\delta(Y) \cap \Delta(X) \subseteq \delta(X)$, yielding $u \in \delta(X)$.

Regarding the maximality of δ', since the rank of δ' is $|M\backslash R|$, we obtain that $\delta' \in \Omega_{max}$ by Corollary 12.1.

The closure of a regular decision-mapping has further properties. Another one is presented by the following statement.

Lemma 13.5 *Let $\delta[m]$ be an arbitrary regular decision-mapping. If $(mat^{in}(op(\delta[m])) \cup P)\backslash(R \cup m) = \emptyset$, then closure δ' of $\delta[m]$ is a feasible solution, i.e., $\delta' \in S'$.*

Proof Closure δ' is contained in $\omega(\delta[m])$ from Lemma 13.4. Thus, in the light of Theorem 12.1, it is sufficient to prove that conditions (1), (2), and (3) are satisfied by δ'. Since $unit(\delta') \subseteq op(\delta')$, to show the equality $unit(\delta') = op(\delta')$, we have to prove that $u \in unit(\delta')$ holds for all $u \in op(\delta')$. Indeed, let $u \in op(\delta')$ be an arbitrary operating unit. By the definition of δ', $op(\delta') = op(\delta[m])$. On the other hand, $\delta[m] \in \Omega_{reg}$, and thus, by Lemma 13.3, $op(\delta[m]) = unit(\delta[m])$. From Lemma 13.4, δ' is an extension of $\delta[m]$, and thus, $unit(\delta[m]) \subseteq unit(\delta')$ by Lemma 13.2. Therefore, $u \in unit(\delta')$, resulting in the required equality.

In order to prove the validity of condition (2), let $X \in mat(op(\delta'))\backslash R$ be an arbitrary material. If $X \in m$, then, by Lemma 13.1, $\delta(X) \neq \emptyset$, but then $\delta'(X) \neq \emptyset$ since $\delta[m] \leq \delta'$.

Now, let us suppose that $X \notin m$. Then, $X \in mat(op(\delta'))\backslash(R \cup m)$. By the definition of δ', $op(\delta') = op(\delta[m])$. Thus,

$$X \in \left(mat^{in}(op(\delta[m]))\backslash(R \cup m)\right) \cup \left(mat^{out}(op(\delta[m]))\backslash(R \cup m)\right).$$

Now, let us observe that the first set of the above union is empty from our assumption. Therefore, $X \in mat^{out}(op(\delta[m]))\backslash(R \cup m)$. Then, there exists at least one operating unit $u = (\alpha, \beta) \in op(\delta[m])$ with $X \in \beta$. Thus, by the definition of δ', we obtain that $\delta'(X) \neq \emptyset$, *i.e.*, condition (2) is satisfied by δ'.

By our assumption, $(mat^{in}(op(\delta[m])) \cup P)\backslash(R \cup m) = \emptyset$. Thus, $P \subseteq m$. From Lemma 13.1, $m \subseteq mat(op(\delta[m]))$. But $op(\delta[m]) = op(\delta')$, and thus, $m \subseteq mat(op(\delta'))$. Therefore, $P \subseteq mat(op(\delta'))$ resulting in (3). This completes the proof of Lemma 13.5.

Theorem 13.2 *Let $\delta[m]$ be an arbitrary regular decision-mapping. If $(mat^{in}(op(\delta[m])) \cup P)\backslash(R \cup m)) = \emptyset$, then $\delta[m]$ has exactly one feasible solution extension, which is closure δ' of $\delta[m]$, i.e., $\omega(\delta[m]) \cap S' = \{\delta'\}$.*

Proof Let us denote the closure of $\delta[m]$ by δ', and let us suppose that $\delta[m] \leq \delta^*$ for some $\delta^* \in \Omega_{max}$. Consider the sequences contained in the definitions of $unit(\delta')$ and $unit(\delta^*)$; denote them by $m'_0, m'_1, \ldots; o'_0, o'_1, \ldots$ and $m^*_0, m^*_1, \ldots; o^*_0, o^*_1, \ldots$, respectively. It will be proved that $m'_i = m^*_i$, $m'_i\backslash R \subseteq m$, and $o'_i = o^*_i$ for all nonnegative integer i. We proceed by induction on i.

Let $i = 0$. From $(mat^{in}(op(\delta[m])) \cup P)\backslash(R \cup m)) = \emptyset$, we have that $P \subseteq m$. Then, $P = m'_0 = m^*_0 \subseteq m$. Relations $\delta[m] \leq \delta'$ and $\delta[m] \leq \delta^*$ imply that $\delta'(X) = \delta^*(X)$ for all $X \in m$. From this it follows that $o'_0 = o^*_0$.

Now, let i be an arbitrary nonnegative integer, and let us suppose that the statement is valid for i, *i.e.*, $m'_i = m^*_i$, $m'_i\backslash R \subseteq m$, and $o'_i = o^*_i$. Then, by the induction hypothesis and the definition, $m'_{i+1} = m^*_{i+1}$ and $m'_{i+1} = m'_i \cup mat^{in}(o'_i)$. Moreover, by the induction hypothesis, $m'_i\backslash R \subseteq m$. On the other hand, $o'_i \subseteq unit(\delta')$. By Theorem 12.1, we have $unit(\delta') = op(\delta')$. From the definition of the closure, $op(\delta') = op(\delta[m])$. Therefore, $o'_i \subseteq op(\delta[m])$. But then $mat^{in}(o'_i) \subseteq mat^{in}(op(\delta[m]))$. According to our assumption, $mat^{in}(op(\delta[m])) \cup P \subseteq R \cup m$, and thus, $mat^{in}(o'_i) \subseteq R \cup m$ yielding the validity of $mat^{in}(o'_i)\backslash R \subseteq m$. Consequently,

$$m'_{i+1}\backslash R = (m'_i\backslash R) \cup (mat^{in}(o'_i)\backslash R) \subseteq m.$$

Then, $m'_{i+1}\backslash R = m^*_{i+1}\backslash R \subseteq m$, and we obtain that $o'_{i+1} = o^*_{i+1}$ by the definitions of o'_{i+1} and o^*_{i+1}.

Since $o'_i = o^*_i$, $i = 0, 1, \ldots$, we have $unit(\delta') = unit(\delta^*)$ from the definition of the *unit* function. Now, let us distinguish the following two cases.

Case 1. If $unit(\delta^*) \subset op(\delta^*)$, then we get that $\delta^* \notin S'$ by Theorem 12.1.

Case 2. Let us suppose that $unit(\delta^*) = op(\delta^*)$. By Lemma 13.5, $\delta' \in S'$, and thus, $unit(\delta') = op(\delta')$ from Theorem 12.1. Therefore, $op(\delta') = op(\delta^*)$. Using this equality, we prove that $\delta'(X) = \delta^*(X)$ holds for all $X \in M \backslash R$. Indeed, let $u = (\alpha, \beta) \in \delta'(X)$ be arbitrary. Then, $u \in op(\delta')$ and $u \in \Delta(X)$. But $op(\delta') = op(\delta^*)$, and thus, $u \in op(\delta^*)$. Therefore, there exists a $Y \in M \backslash R$ with $u \in \delta^*(Y)$. Since δ^* is a consistent decision-mapping, $\delta^*(Y) \cap \Delta(X) \subseteq \delta^*(X)$, implying $u \in \delta^*(X)$. Consequently, $\delta'(X) \subseteq \delta^*(X)$. The converse inclusion, $\delta^*(X) \subseteq \delta'(X)$, can be seen in a similar way. This ends the proof of Theorem 13.2.

As a representation of the above theorem, let us consider Example 13.1 with $\delta[m] \in \Omega_{reg}$ where $m = \{P_1, B, C\}$, furthermore, $\delta(P_1) = \{u_2\}$, $\delta(B) = \{u_4\}$, and $\delta(C) = \{u_4\}$. Then, $(mat^{in}(op(\delta[m])) \cup P) \backslash (R \cup m) = \emptyset$, closure δ' of $\delta[m]$ is δ_9; moreover, $\omega(\delta[m]) = \{\delta_9\} = \{\rho^{-1}(s_4)\}$. We note that $\omega(\delta[m])$ is not singleton in general under the assumptions of Theorem 13.2.

Now, we are ready to present a more efficient version of our branch-and-bound algorithm. The idea of this version was introduced in [1] and can be found in it as *accerelated branch-and-bound method* for solving PNS problems. It is based on the branch-and-bound algorithm given at the beginning of this chapter with the following specifications and modifications.

1. It is assumed that $M = \{A_1, \ldots, A_n\}$, $P = \{A_1, \ldots, A_v\}$ and $R = \{A_w, \ldots, A_n\}$ for some integers v, w, n with $1 \leq v < w \leq n$.
2. The following node selection rule will be applied. Leaf $\omega(\delta[m])$ is selected from L if $g(\delta[m])/\mid m\mid$ is minimal provided that $m \neq \emptyset$. If there are more candidates with a same ratio, then we choose one of them in a random way.
3. For choosing a material, the input material selection rule is used. For providing a deterministic selection, we choose the candidate having the smallest index.
4. We complete the procedure with a testing step to treat that case separately when $(mat^{in}(op(\delta[m])) \cup P) \backslash (R \cup m) = \emptyset$. Then, in the light of Theorem 13.2, $\omega(\delta[m]) \cap S' = \{\delta'\}$ where δ' is the closure of $\delta[m]$.
5. In the branching step, we separate that case when there is no regular extension to $\delta[m]$ with respect to the chosen material X. Then, $\omega(\delta[m]) \cap S' = \emptyset$ from Theorem 13.1.

By the goal of completeness, we present here algorithm ABB in completely detailed form.

Accelerated Branch-and-Bound Algorithm

Initialization

- Let $L := \{\omega(\delta[\varnothing])\}$, $z^* := \infty$, $s := \varnothing$, and set $r := 0$. Calculate the value $g(\delta[\varnothing])$.

Iteration (r-th iteration)

- *Step 1. (Termination)* Terminate if $L = \varnothing$; s contains the optimal solution, z^* provides the optimal value. Otherwise, proceed to Step 2 if $r > 0$ and to Step 3 if $r = 0$.
- *Step 2. (Selection)* Choose an element $\omega(\delta[m])$ from L for which the ratio, $g(\delta[m])/ \mid m|$, is minimal. (If there are more candidates with the same ratio, then choose one of them in a random way.) Denote it by $\omega(\delta[m])$, and proceed to Step 3.
- *Step 3. (Testing)* If $(mat^{in}(op(\delta[m])) \cup P)\backslash(R \cup m) = \varnothing$, then form closure δ' of $\delta[m]$. Redefine the values of z^* and s by $z^* := g(\delta')$ and $s := \{\delta'\}$ if $g(\delta') < z^*$, otherwise, z^* and s do not change. Furthermore, set $\Phi := \varnothing$ and proceed to Step 6. If $(mat^{in}(op(\delta[m])) \cup P)\backslash(R \cup m) \neq \varnothing$, then proceed to Step 4.
- *Step 4. (Branching)* Choose the material with the smallest index from set $(mat^{in}(op(\delta[m])) \cup P)\backslash(R \cup m)$; denote it by X. Form the regular extensions of $\delta[m]$ with respect to X. If there is no such regular extension of $\delta[m]$, then let $L := L \backslash\{\omega(\delta[m])\}$, and proceed to Step 1. Otherwise, denote the regular extensions of $\delta[m]$ with respect to X by $\delta_i[m_i]$, $i = 1, 2, \ldots, k$; let $\Phi = \{\omega(\delta_i[m_i]) : 1 \leq i \leq k\}$, and proceed to Step 5.
- *Step 5. (Bounding)* Calculate the value, $g(\delta_i[m_i])$, for all i, $i = 1, \ldots, k$. If for some index i, $1 \leq i \leq k$, the rank of $\delta_i[m_i]$ is equal to $|M\backslash R|$, and $\delta_i \in S'$, furthermore, $g(\delta_i) < z^*$, then redefine z^* and s by $z^* = g(\delta_i)$ and $s := \{\delta_i\}$. Proceed to Step 6.
- *Step 6. (Fathoming.)* Redefine set L in the following way:

$$L{:=}\{\omega(\delta'[m']) : \omega(\delta'[m']) \in (L\backslash\{\omega(\delta[m])\}) \cup \Phi, g(\delta'[m']) < z^*\}.$$

- Set $r := r + 1$ and continue by the next iteration.

To illustrate our procedure, let us consider the following PNS_w example.

Example 13.2 Let $\mathbb{S} = (P, R, O)$ be a reduced structural model of PNS_w where $M = \{A_1, A_2, \ldots, A_{16}\}$, $P = \{A_1, A_2\}$, $R = \{A_{14}, A_{15}, A_{16}\}$, $O = \{u_1, u_2, \ldots, u_{12}\}$, and the descriptions of the operating units given by Table 13.1. The corresponding maximal structure with the weights is shown in Fig. 13.3.

Our procedure is started by $L = \{\omega(\delta[\varnothing])\} = \{\Omega_{max}\}$, $s := \varnothing$, $z^* = \infty$, $g(\omega(\delta[\varnothing]) = 0$. In Step 4, material A_1 is selected. $\Delta(A_1) = \{u_8, u_{11}\}$, and we obtain three regular extensions, $\delta_i[\{A_1\}]$, $i = 1, 2, 3$, where $\delta_1(A_1) = \{u_8\}$, $\delta_2(A_1) = \{u_{11}\}$, and $\delta_3(A_1) = \{u_8, u_{11}\}$. Let $\Omega_i = \omega(\delta_i[\{A_1\}])$, $i = 1, 2, 3$. The corresponding bounds are 2, 7, 9, respectively, and $L = \{\Omega_1, \Omega_2, \Omega_3\}$.

Table 13.1 Operating units of Example 13.2

Unit	Inputs	Outputs	Weight
u_1	A_{14}	A_5, A_6, A_9	5
u_2	A_{15}	A_7	3
u_3	A_{16}	A_4, A_5, A_9	7
u_4	A_5, A_{15}	A_8	1
u_5	A_5, A_7	A_{11}	2
u_6	A_6, A_8, A_9	A_3, A_4, A_{12}	3
u_7	A_4, A_5	A_{10}, A_{11}	5
u_8	A_8, A_9	A_1, A_2, A_6	2
u_9	A_{10}, A_{11}	A_2, A_{13}	4
u_{10}	A_3, A_4	A_2	6
u_{11}	A_3, A_{12}	A_1	7
u_{12}	A_7, A_{10}	A_2	5

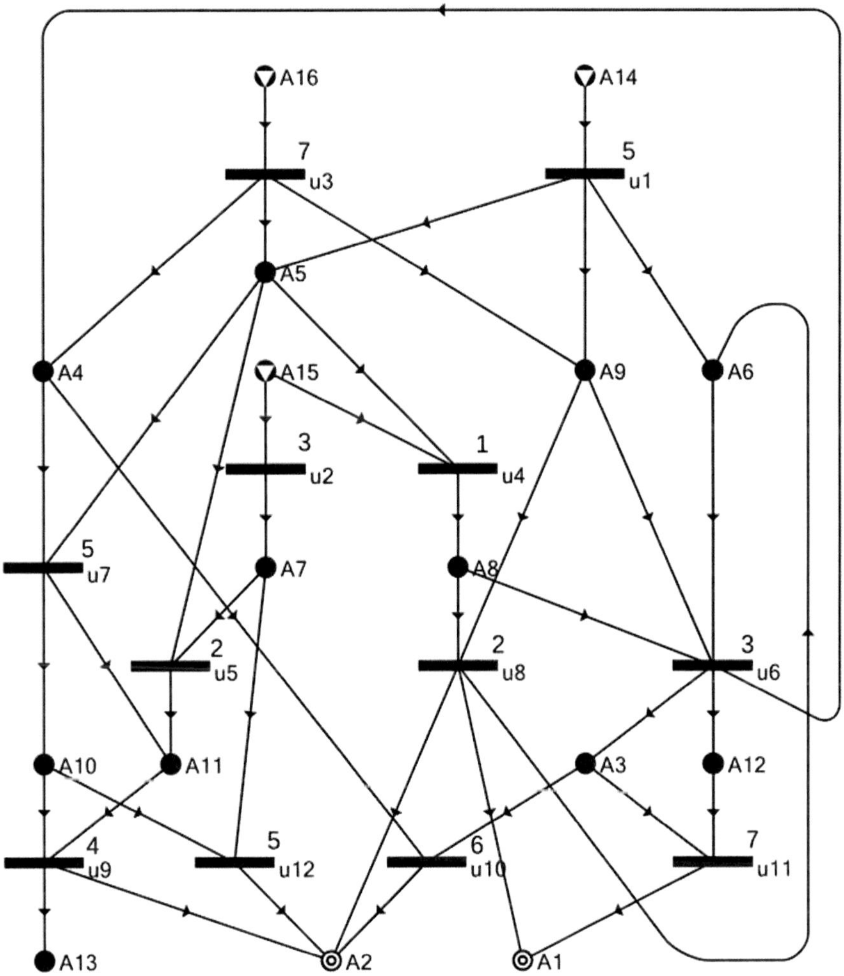

Fig. 13.3 Maximal structure determined by $\mathbb{S}$ in Example 13.2

In the next iteration step, node Ω_1 is selected. Then, we have to choose a material from $(mat^{in}(op(\delta_1[\{A_1\}])) \cup P)\backslash(R \cup m) = \{A_2, A_8, A_9\}$. By our material selection rule, material A_2 is chosen in Step 4. $\Delta(A_2) = \{u_8, u_9, u_{10}, u_{12}\}$, and we obtain eight regular extensions, $\delta_i[\{A_1, A_2\}]$, $i = 4, 5, \ldots, 11$, where $\delta_4(A_2) = \{u_8\}$, $\delta_5(A_2) = \{u_8, u_9\}$, $\delta_6(A_2) = \{u_8, u_{10}\}$, $\delta_7(A_2) = \{u_8, u_{12}\}$, $\delta_8(A_2) = \{u_8, u_9, u_{10}\}$, $\delta_9(A_2) = \{u_8, u_9, u_{12}\}$, $\delta_{10}(A_2) = \{u_8, u_{10}, u_{12}\}$, and $\delta_{11}(A_2) = \{u_8, u_9, u_{10}, u_{12}\}$. Let $\Omega_i = \omega(\delta_i[\{A_1, A_2\}]$, $i = 4, 5, \ldots, 11$. The corresponding bounds are 2, 6, 8, 7, 12,11,13,17. For the set of the live leaves, we obtain that $L = \{\Omega_2, \Omega_3, \ldots, \Omega_{11}\}$.

In the following step, node Ω_4 is selected. Now, the available materials in Step 4 are $\{A_8, A_9\}$, and material A_8 is chosen. Since $\Delta(A_8) = \{u_4\}$, $\delta_4[\{A_1, A_2\}]$ has only one regular extension denoted by $\delta_{12}[\{A_1, A_2, A_8\}]$ where $\delta_{12}(A_8) = \{u_4\}$. Let $\Omega_{12} = \omega(\delta_{12}[\{A_1, A_2, A_8\}])$. The corresponding bound is 3, and the set of the live leaves is $\{\Omega_2, \Omega_3, \Omega_5, \Omega_6, \ldots, \Omega_{12}\}$.

Now, node Ω_{12} is selected. Here, $(mat^{in}(op(\delta_{12}[\{A_1, A_2, A_8\}])) \cup P)\backslash(R \cup m) = \{A_5, A_9\}$. For material A_5 chosen in Step 4, $\Delta(A_5) = \{u_1, u_3\}$, and we obtain three regular extensions, $\delta_i[\{A_1, A_2, A_5, A_8\}]$, $i = 13,14,15$, where $\delta_{13}(A_5) = \{u_1\}$, $\delta_{14}(A_5) = \{u_3\}$, and $\delta_{15}(A_5) = \{u_1, u_3\}$. Let $\Omega_i = \delta_i[\{A_1, A_2, A_5, A_8\}]$, $i = 13,14,15$. The corresponding bounds are 8,10,15, and

$$L = \{\Omega_2, \Omega_3, \Omega_5, \Omega_6, \ldots, \Omega_{11}, \Omega_{13}, \Omega_{14}, \Omega_{15}\}.$$

Proceeding to the following iteration step, node Ω_{13} is selected. There is only one available material in Step 4, specifically, we have to choose material A_9. Since $\Delta(A_9) = \{u_1, u_3\}$, there exists exactly one regular extension, $\delta_{16}[\{A_1, A_2, A_5, A_8, A_9\}]$ where $\delta_{16}(A_9) = \{u_1\}$. Let Ω_{16} denote set $\omega(\delta_{16}[\{A_1, A_2, A_5, A_8, A_9\}])$. The corresponding bound is 8 and

$$L = \{\Omega_2, \Omega_3, \Omega_5, \Omega_6, \ldots, \Omega_{11}, \Omega_{14}, \Omega_{15}, \Omega_{16}\}.$$

Here, Ω_{16} is selected. The set of the available materials for Step 4 is empty, *i.e.*, $(mat^{in}(op(\delta_{16}[\{A_1, A_2, A_5, A_8, A_9\}])) \cup P)\backslash(R \cup m) = \emptyset$. Thus, the closure of $\delta_{16}[\{A_1, A_2, A_5, A_8, A_9\}]$ is a feasible solution. Let us denote this feasible solution by δ'. According to Step 3, variables z^* and s obtain new values, specifically, $z^* = 8$ and $s = \{\delta'\}$. Using the object value of the obtained feasible solution, we can reduce the set of the live leaves in Step 6. The redefined set is $\{\Omega_2, \Omega_5, \Omega_7\}$. The corresponding actual tree is given in Fig. 13.4 where the fathomed leaves are marked by star.

In the next iteration step, node Ω_5 is selected. The available materials are A_8, A_9, A_{10}, A_{11}, and A_8 is selected in Step 4. $\Delta(A_8) = \{u_4\}$, and thus, we get exactly one regular extension which is denoted by $\delta_{17}[\{A_1, A_2, A_8\}]$. Let $\Omega_{17} = \omega(\delta_{17}[\{A_1, A_2, A_8\}])$. The corresponding bound is 7, and $L = \{\Omega_2, \Omega_7, \Omega_{17}\}$.

Now, node Ω_{17} is selected from L. The available materials in Step 4 are A_5, A_9, A_{10}, A_{11}, and A_5 is selected. $\Delta(A_5) = \{u_1, u_3\}$, and thus, we obtain three regular extensions, $\delta_i[\{A_1, A_2, A_5, A_8\}]$, $i = 18,19,20$, where $\delta_{18}(A_5) = \{u_1\}$, $\delta_{19}(A_5) = \{u_3\}$, and $\delta_{20}(A_5) = \{u_1, u_3\}$. Let us denote set $\omega(\delta_i[\{A_1, A_2, A_5, A_8\}])$ by Ω_i for

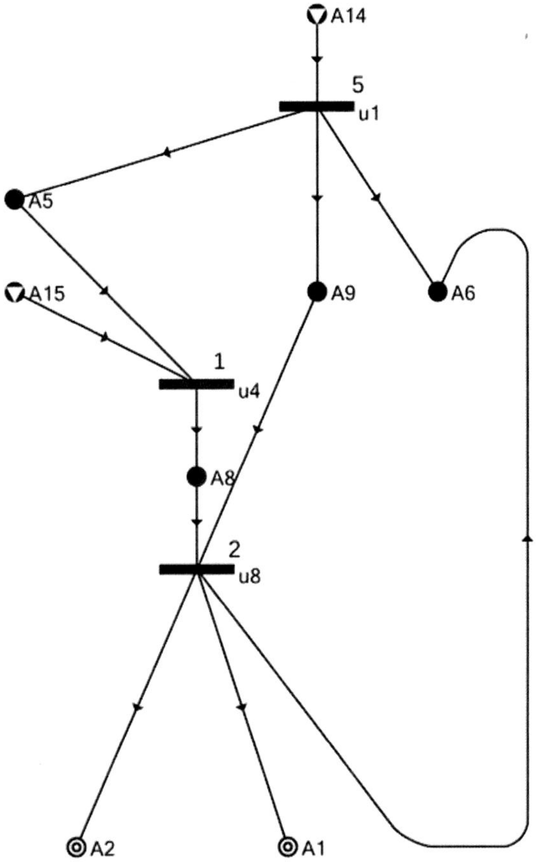

Fig. 13.5 Optimal solution-structure in Example 13.2

i = 18,19,20. The corresponding bounds are 12,14,19. The set of the live leaves is $\{\Omega_2, \Omega_7\}$.

In the following iteration step, Ω_7 is selected. The available materials are A_7, A_8, A_9, A_{10}, and A_7 is chosen in Step 4. $\Delta(A_7) = \{u_2\}$, and thus, we obtain only one regular extension, $\delta_{21}[\{A_1, A_2, A_7\}]$ where $\delta_{21}(A_7) = \{u_2\}$. Let $\Omega_{21} = \omega(\delta_{21}[\{A_1, A_2, A_7\}])$. The corresponding bound is 10 and $L = \{\Omega_2\}$.

Choosing Ω_2 from L, the available materials are A_2, A_3, A_{12}, and A_2 is selected in Step 4. Then, $\delta_2[\{A_1\}]$ has seven regular extensions, $\delta_i[\{A_1, A_2\}]$, i = 22, 23, ..., 28, where the corresponding subsets are $\{u_9\}$, $\{u_{10}\}$, $\{u_{12}\}$, $\{u_9, u_{10}\}$, $\{u_9, u_{12}\}$, $\{u_{10}, u_{12}\}$, and $\{u_9, u_{10}, u_{12}\}$. Let $\Omega_i = \omega(\delta_i[\{A_1, A_2\}])$, i = 22, 23, ..., 28. The corresponding bounds are 11,13,12,17,16,18,22, and $L = \varnothing$. Therefore, the procedure terminates; δ' is an optimal solution. P graph belonging to δ' is shown in Fig. 13.5.

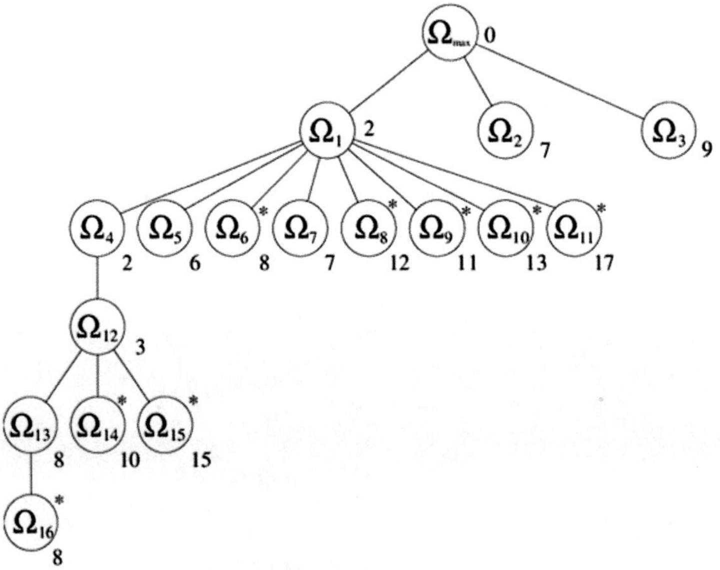

Fig. 13.4 The branch-and-bound tree belonging to Example 13.2

Reference

1. Friedler F, Varga JB, Fehér E, Fan LT (1996) Combinatorially accerelated branch-and -bound method for solving the MIP model of process network synthesis. In: State of the art in global optimization. Springer, Boston, pp 609–626

Chapter 14
Refined B&B Procedure

Abstract Algorithm ABB has been presented in Chap. 6. The algorithm guarantees the optimal solution with regard to the model implemented. The efficiency of algorithm ABB can be enhanced by improving the bounding procedure. This chapter introduces algorithm ABB-RB, which incorporates a refined bounding procedure to determine a sharper bound for the partial problems.

Keywords P-graph · Branch-and-bound · Algorithm ABB · Search tree · Sharpened bound

Regarding the different branch-and-bound procedures, their efficiencies highly depend on the applied bounding functions. In general, model reduction together with a sharp bounding function provides a good efficacy of the considered procedure. The current chapter presents a method to reduce the structural model of a partial problem during the branch-and-bound procedure [1]. The method reduces the sizes of the partial problems while also providing a basis for sharper bounding functions. Inherently, this leads also to a reduction in the number of partial problems, increasing the efficiency of the algorithm. At the end of the chapter, an example of sharper bounding is presented for PNS_w problems. Without additional structure-based improvements, when a relatively small number of operating units are selected in a partial problem defined by $\delta[m]$, a large gap exists between the bound, $g(\delta[m])$, and the minimum of $w(\overline{\delta})$ for all $\overline{\delta}$ feasible solution extensions of $\delta[m]$, *i.e.*, $\overline{\delta} \in \omega(\delta[m]) \cap S'$.

The deficiency of the bounding function is attributable to the following. Suppose that $\delta[m] \in \Omega_{reg} \backslash \Omega_{max}$ and that $\omega(\delta[m])$ is a leaf of the current branch-and-bound tree. This implies that the operating units included in $op(\delta[m])$ are fixed and that the bounding function, g, assigns to this leaf the sum of weights of the operating units in $op(\delta[m])$. If $\overline{\delta} \in \omega(\delta[m]) \cap S'$, then inclusion

$$(mat(op(\delta[m])), op(\delta[m])) \subseteq \left(mat\left(op\left(\overline{\delta}\right)\right), op\left(\overline{\delta}\right)\right)$$

is valid for the P-graphs belonging to $\delta[m]$ and $\overline{\delta}$. Moreover, the operating units appearing in the extension of $\delta[m]$, *i.e.*, in set $op\left(\overline{\delta}\right) \backslash op(\delta[m])$, are disregarded in

F. Friedler et al., *P-graphs for Process Systems Engineering*,
https://doi.org/10.1007/978-3-030-92216-0_14

determining the bound, $g(\delta[m])$. If the bounding function is to be improved, these operating units also need be taken into account. As we shall see, to establish this idea needs a detailed deep investigation.

To achieve the goal sketched above, some preparation is needed. For this purpose, let a reduced structural model of process network synthesis (PNS_w) be given by $\mathbb{S} = (P, R, O)$, and also let (M, O) denote its maximal structure. Let us extend the weight function, w, to Ω as follows. For every $\delta[m] \in \Omega$, let

$$w(\delta[m]) = \sum_{u \in op(\delta[m])} w(u).$$

Obviously, if $\delta' \in S'$, then $w(\delta')$ presents the value of the object function of problem (13.4) which belongs to δ'.

We recall that in a special case, we are able to determine the exact value, min $\{w(\delta') : \delta' \in \omega(\delta[m]) \cap S'\}$, for a given decision-mapping, $\delta[m]$. Namely, by Theorem 13.2, if $\delta[m] \in \Omega_{reg}$ and $(mat^{in}(op(\delta[m])) \cup P)\backslash(R \cup m) = \emptyset$, then $\omega(\delta[m]) \cap S' = \{\overline{\delta}\}$ where $\overline{\delta}$ denotes the closure of $\delta[m]$. Therefore, the above minimum is equal to $w(\overline{\delta})$. The following statements present another particular case which allows to determine the exact minimal value again.

Theorem 14.1 *Let $\delta[m]$ be an arbitrary regular decision-mapping. If $F = (mat^{in}(op(\delta[m])) \cup P)\backslash(mat^{out}(op(\delta[m])) \cup R) = \emptyset$ is valid, then $\delta^* \in \omega(\delta[m]) \cap S'$ where $\delta^*(X) = \{(\alpha, \beta) : (\alpha, \beta) \in op(\delta[m])\ \&\ X \in \beta\}$ for all $X \in M\backslash R$.*

Proof It is easy to see that δ^* is a consistent decision-mapping. On the other hand, from the definition of δ^* it follows that $\delta(X) \subseteq \delta^*(X)$ for all $X \in m$. Now, let us suppose that $\delta(X) \subset \delta^*(X)$ is valid for some $X \in m$. Then, there exists an operating unit $(\alpha, \beta) = u \in op(\delta[m])$ with $X \in \beta$ and $u \notin \delta(X)$. Thus, $u \in \Delta(X)$, and since $u \in op(\delta[m])$, there is a material $Y \in m$ with $u \in \delta(Y)$. Therefore, $u \in \delta(Y) \cap \Delta(X)$ where $X, Y \in m$. By the consistency of $\delta[m]$, $\delta(Y) \cap \Delta(X) \subseteq \delta(X)$, but then, $u \in \delta(X)$ which is a contradiction. Consequently, $\delta(X) = \delta^*(X)$ for all $X \in m$, and thus, δ^* is an extension of $\delta[m]$. Since the rank of δ^* is $|M\backslash R|$, δ^* is a maximal extension of $\delta[m]$. Therefore, $\delta^* \in \omega(\delta[m])$.

Now, we shall show that $\delta^* \in S'$. In the light of Theorem 12.1, it is sufficient to prove that δ^* satisfies the following conditions:

1. $unit(\delta^*) = op(\delta^*)$,
2. $\delta^*(X) \neq \emptyset$, for all $X \in mat(op(\delta^*))\backslash R$,
3. $P \subseteq mat(op(\delta^*))$.

Since $\delta[m] \in \Omega_{reg}$, we have that $unit(\delta[m]) = op(\delta[m])$ from Lemma 13.3. On the other hand, $op(\delta[m]) = op(\delta^*)$ by the definition of δ^*. Since δ^* is an extension of $\delta[m]$, $unit(\delta[m]) \subseteq unit(\delta^*)$ from Lemma 13.2. Thus, $op(\delta^*) \subseteq unit(\delta^*)$. Finally, $unit(\delta^*) \subseteq op(\delta^*)$ by the definition of the *unit* function. Therefore, $unit(\delta^*) = op(\delta^*)$, *i.e.*, condition (1) is satisfied.

Obviously, $op(\delta[m]) = op(\delta^*)$ implies $mat(op(\delta[m])) = mat(op(\delta^*))$, furthermore, $mat(op(\delta[m])) \subseteq mat^{out}(op(\delta[m])) \cup R$ by our assumption on F. Thus, $X \in mat^{out}(op(\delta[m]))\backslash R$ is valid for any $X \in mat(op(\delta^*))\backslash R$. Therefore, $\delta^*(X) \neq \emptyset$ if $X \in mat(op(\delta^*))\backslash R$, and subsequently, δ^* satisfies condition (2).

By $F = \emptyset$ and $P \cap R = \emptyset$, we have that $P \subseteq mat^{out}(op(\delta[m]))$. On the other hand, $mat^{out}(op(\delta[m])) \subseteq mat(op(\delta^*))$ since δ^* is an extension of $\delta[m]$. These two inclusions imply (3).

Corollary 14.1 *If the conditions of Theorem 14.1 are satisfied for a decision-mapping* $\delta[m]$, *then* $w(\delta^*) = min\ \{w(\delta') : \delta' \in \omega(\delta[m]) \cap S'\}$ *where* δ^* *denotes the decision-mapping defined in Theorem 14.1.*

Indeed, $\delta^* \in \omega(\delta[m]) \cap S'$ from Theorem 14.1, and $op(\delta^*) = op(\delta[m])$. On the other hand, $op(\delta[m]) \subseteq op(\overline{\delta})$ for any feasible solution extension $\overline{\delta}$ of $\delta[m]$. This yields that $w(\delta^*) \leq w(\overline{\delta})$ for all $\overline{\delta} \in \omega(\delta[m]) \cap S'$.

Using Corollary 14.1, we can take the following important observation.

Remark 14.1 Regarding the bound determination, it is sufficient to deal with such type of decision-mappings of $\Omega_{reg}\backslash\Omega_{max}$ for which the corresponding set F is nonempty.

For demonstrating this situation, the following example is presented.

Example 14.1 Let us consider the reduced structural model of PNS, $\mathbb{S} = (P, R, O)$, where $P = \{A_1, A_2\}$, $R = \{A_{16}, A_{17}, A_{18}\}$, and $O = \{u_1, u_2, \ldots, u_{12}\}$. The descriptions of the operating units contained in set O are presented by Table 14.1; the corresponding P-graph is given in Fig. 14.1.

Investigate decision-mapping $\delta[m]$ defined by $m = \{A_1, A_5, A_8, A_{13}\}$, $\delta(A_1) = \{u_1\}$, $\delta(A_5) = \{u_4\}$, $\delta(A_8) = \{u_8\}$, and $\delta(A_{13}) = \{u_{11}\}$.

Table 14.1 Operating units of Example 14.1

Unit	Inputs	Outputs	Weight
u_1	A_3, A_4, A_5	A_1, A_2	3
u_2	A_5, A_6	A_1, A_2, A_{11}	6
u_3	A_6, A_7	A_2, A_{11}	5
u_4	A_8, A_9	A_3, A_4, A_5	2
u_5	A_9, A_{10}, A_{11}	A_4, A_5, A_{13}	4
u_6	A_{10}, A_{11}	A_4, A_5, A_6	6
u_7	A_{11}, A_{14}	A_7	5
u_8	A_{13}, A_{14}, A_{16}	A_8, A_9, A_{10}	7
u_9	A_{14}, A_{15}	A_9, A_{10}, A_{11}	9
u_{10}	A_{16}	A_{12}, A_{13}	6
u_{11}	A_{17}	A_{13}, A_{14}	4
u_{12}	A_{11}, A_{18}	A_{14}, A_{15}	3

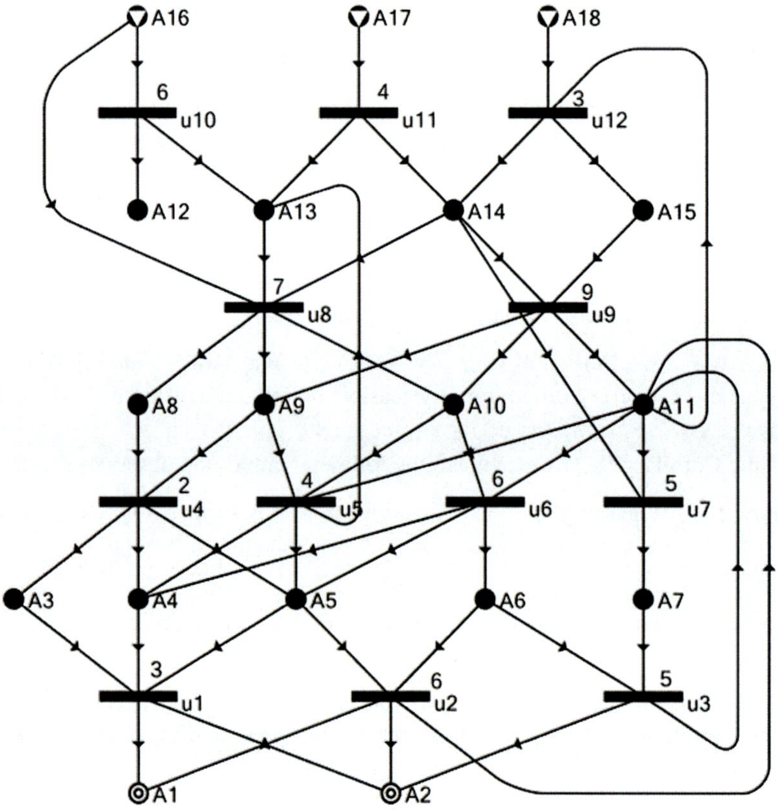

Fig. 14.1 P-graph belonging to $\mathbb{S}$ in Example 14.1

Obviously, $\delta[m] \in \Omega_{reg}$; moreover, we have the following equalities:

$$mat^{in}(op(\delta[m])) = \{A_3, A_4, A_5, A_8, A_9, A_{13}, A_{14}, A_{16}, A_{17}\} \text{ and}$$
$$mat^{out}(op(\delta[m])) = \{A_1, A_2, A_3, A_4, A_5, A_8, A_9, A_{10}, A_{13}, A_{14}\}.$$

Thus,

$$F = \left(mat^{in}(op(\delta[m])) \cup P\right) \setminus (mat^{out}(op(\delta[m])) \cup R) = \emptyset.$$

Taking the corresponding decision-mapping, δ^*, it is a feasible solution extension of $\delta[m]$. P-graph belonging to δ^* is given in Fig. 14.2. It is easy to see that $\omega(\delta[m]) \cap S' = \{\delta^*, \delta'\}$ now, where δ' is determined by the set of operating units, $\{u_1, u_4, u_8, u_9, u_{11}, u_{12}\}$. Regarding the weights, $16 = w(\delta^*) < w(\delta') = 28$.

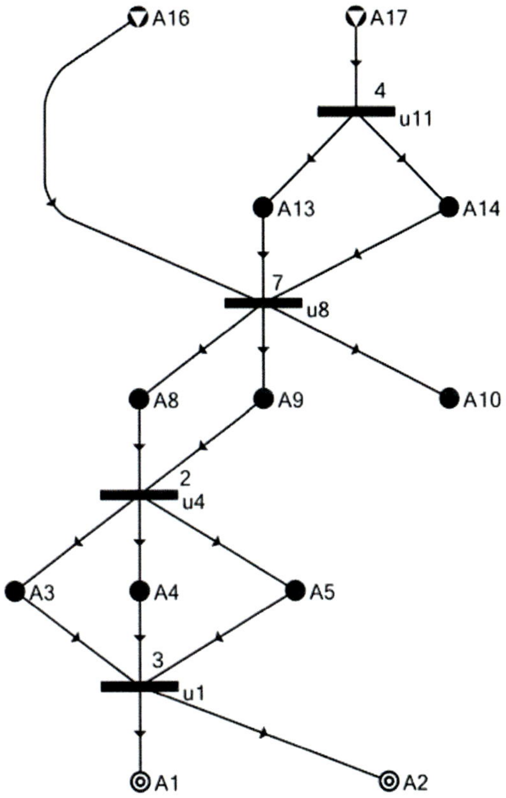

Fig. 14.2 P-graph belonging to δ^* in Example 14.1

Now, we are going to establish our goal of refinement. The main idea of the refinement of the bounding function comprises of two main parts: a structural reduction of the partial problems, and an enhanced bounding based on the materials.

Firstly, regarding the structural reduction, the *greatest feasible solution extension* of $\delta[m]$ is determined, provided that it exists. Here, the ordering is the inclusion relation of the corresponding P-graphs. We shall show that if $\delta[m] \in \Omega_{reg}$ has at least one feasible solution extension, its greatest feasible solution extension exists as well.

Non-existence of a feasible solution extension of $\delta[m]$ means $\omega(\delta[m]) = \emptyset$; thus, $\omega(\delta[m]))$ can be disregarded from further assessment. If $\omega(\delta[m]) \neq \emptyset$, P-graph belonging to the greatest feasible solution extension of $\delta[m]$ is denoted by (M', O'). Thus, each operating unit incorporated into one of the feasible solution extensions of $\delta[m]$ is included in O'.

Then, for the enhancement of the bounding function, a nonnegative value called *material bound* is assigned to all materials represented on P-graph (M', O'), *i.e.*, to the elements of M'; zero is assigned to the raw materials, as well as to each material which is an output from at least one operating unit involved in set $op(\delta[m])$. Since the weights of the operating units included in $op(\delta[m])$ are already considered in the bound, $g(\delta[m])$, the production of their output materials does not increase the bound.

On the basis of the weights of the operating units included in $O'\backslash op(\delta[m])$ and the bounds of materials already determined, the remaining materials can be assigned with a material bound each of which is minimal in the following sense. For any feasible solution extension of $\delta[m]$, $\overline{\delta} \in \omega(\delta[m]) \cap S'$, and material $X \in mat(op(\overline{\delta}))\backslash R$, the material bound of X does not exceed the sum of the weights of those operating units included in $op(\overline{\delta})\backslash op(\delta[m])$ participating directly or indirectly in the production of X in the process given by $\overline{\delta}$. Suppose that these material bounds are determined for those materials of the P-graph of $\delta[m]$, which neither are produced by any of the operating units given by $\delta[m]$ nor are raw materials; moreover, determined for desired products not produced by any operating unit given by $\delta[m]$. Thus, this bound is supposedly determined for the materials included in the following set,

$$(mat^{in}(op(\delta[m])) \cup P)\backslash(mat^{out}(op(\delta[m])) \cup R).$$

Observe that the elements of this set include both those sources of the P-graph belonging to $\delta[m]$ which are not raw materials and those desired products which are not produced by any operating unit in $op(\delta[m])$. Take the maximum of their bounds and denote it by h^*. Then, for any $\overline{\delta} \in \omega(\delta[m]) \cap S'$, h^* is not greater than the sum of the weights of the operating units necessary to produce set $(mat^{in}(op(\delta[m])) \cup P)$ $\backslash(mat^{out}(op(\delta[m])) \cup R)$ of materials by the process given by $\overline{\delta}$. Note that the weights of the operating units given by $\delta[m]$ are disregarded, while the weights of the remaining operating units, *i.e.*, the operating units from $op(\overline{\delta})\backslash op(\delta[m])$, are taken into account. Consequently, $w(\overline{\delta}) \geq g(\omega(\delta[m])) + h^*$ is valid for all $\overline{\delta} \in \omega(\delta[m]) \cap S'$; thus, $g(\delta[m]) + h^*$ is a suitable lower bound. This idea of bounding will be the basis of the refined bounding procedure.

Reduction of the Sizes of the Partial Problems

As the first step in realizing the notion outlined above, the following proposition will be proved.

Lemma 14.1 *Let $\delta[m]$ be an arbitrary regular decision-mapping. If $\delta[m]$ has a feasible solution extension, then its greatest feasible solution extension also exists regarding the inclusion relation of the corresponding* P-*graphs* [1].

Proof Since $S'(\mathbb{S})$ is a finite set, $\delta[m]$ has a finite number of feasible solution extensions denoted by $\delta_1, \delta_2, \ldots, \delta_n$. According to Remark 12.2, the corresponding P-graphs, $\rho^{-1}(\delta_t) = (mat(op(\delta_t)), op(\delta_t))$, $t = 1, 2, \ldots, n$, are pairwise different feasible solutions. Furthermore, their union, denoted by (M', O'), is also a feasible solution from Theorem 3.1. This gives rise to $\rho(M', O') = \delta' \in S'(\mathbb{S})$ again on the basis of Remark 12.2. On the other hand, δ' is obviously a feasible solution extension of $\delta[m]$, *i.e.*, $\delta[m] \leq \delta'$. Thus, (M', O') is the greatest element in the set

$\{\rho^{-1}(\delta_t) : t = 1, 2, \ldots, n\}$ regarding the inclusion relation of P-graphs. Therefore, there exists an integer $1 \leq j \leq n$ such that $\rho^{-1}(\delta_j)$ is the greatest element of the P-graphs, $\rho^{-1}(\delta_t)$, $t = 1, \ldots, n$. But then, δ_j is the greatest feasible solution extension of $\delta[m]$ with respect to the considered ordering. This completes the proof of Lemma 14.1.

Remark 14.2 Let us denote the P-graph belonging to δ_i by (m_i, o_i), $i = 1, 2, \ldots, n$. Moreover, the P-graphs corresponding to $\delta[m]$ and the greatest feasible solution extension among δ_i, $i = 1, 2, \ldots, n$, are denoted by $(\overline{m}, \overline{o})$ and (M', O'), respectively. Then, $(\overline{m}, \overline{o}) \subseteq (m_i, o_i) \subseteq (M', O')$, $i = 1, 2, \ldots, n$. Consequently, each operating unit given by a feasible solution extension of decision-mapping $\delta[m]$ is included in O'.

From practical point of view, Lemma 14.1 does not provide the resolution of the following questions where $\delta[m]$ is an arbitrarily given regular decision-mapping.

Does a feasible solution extension of $\delta[m]$ exist?

If so, how can we obtain the greatest feasible solution extension of $\delta[m]$?

To resolve these questions, an efficient algorithm comprising two major parts will be presented. The first reduces the set of available operating units by eliminating those excluded directly or indirectly by the partial decision system given by $\delta[m]$. The second constructs the required feasible solution extension from the remaining operating units provided that such an extension exists.

Construction of the Greatest Feasible Solution Extension [1]

Reduction

Initialization

- Omit those operating units from O which are directly excluded by the partial decision – system given by $\delta[m]$; stated formally,

$$O_0 = O \backslash \left(\bigcup_{X \in m} (\Delta(X) \backslash \delta(X)) \right).$$

- Moreover, let $M_0 = mat(O_0)$; then, collect those materials, not produced by any operating unit included in O_0, from set $M_0 \backslash R$ into set T_0, *i.e.*,

$$T_0 = \{X : X \in M_0 \backslash R \ \& \ ((\alpha, \beta) \in O_0 \rightarrow X \notin \beta)\}.$$

- Finally, set $r := 0$.

Iteration (r-th iteration)

- *Step 1.1.* Terminate, if $P \nsubseteq M_r$ or $op(\delta[m]) \nsubseteq O_r$; no feasible solution extension of $\delta[m]$ exists. Otherwise, proceed to Step 1.2.
- *Step 1.2.* Proceed to the initialization of the composition phase if $T_r = \emptyset$. Otherwise, proceed to Step 1.3.
- *Step 1.3.* Choose one material from T_r; it is denoted by X. Collect those operating units from O_r, each of which has X as an input material; this set is denoted by O_X, *i.e.*,

$$O_X = \{(\alpha, \beta) : (\alpha, \beta) \in O_r \ \& \ X \in \alpha\}.$$

Let $O_{r+1} = O_r \backslash O_X$. If $O_{r+1} = \emptyset$, terminate; no feasible solution extension of $\delta[m]$ exists. Otherwise, collect those output materials of the operating units of O_X, which are not output materials of any operating unit in O_{r+1} but they are input materials of an operating unit in O_{r+1}, into set $T_{r'}$, *i.e.*,

$$T_{r'} = \{Y : Y \in mat^{out}(O_X) \ \& \ Y \notin mat^{out}(O_{r+1}) \ \& \ Y \in mat^{in}(O_{r+1})\}.$$

Let $M_{r+1} = mat(O_{r+1})$ and $T_{r+1} = (T_r \cap M_{r+1}) \cup T_{r'}$. Set $r := r+1$, and repeat the iteration.

Composition

Initialization

- Let $W_0 = (mat^{in}(op(\delta[m])) \cup P) \backslash (R \cup m)$, $m_0 = m$, $o_0 = op(\delta[m])$, and $\delta^{(0)}[m_0] = \delta[m]$. Furthermore, set $s := 0$.

Iteration (s-th iteration)

- *Step 2.1.* Proceed to Step 2.3 if $W_s = \emptyset$; otherwise, proceed to Step 2.2.
- *Step 2.2.* Choose one material from W_s and denote this material by X. Let $m_{s+1} = m_s \cup \{X\}$. Form set O_X^* of those operating units from O_r, each of which has X as an output material, *i.e.*,

$$O_X^* = \{(\alpha, \beta) : (\alpha, \beta) \in O_r \ \& \in \beta\}.$$

Let

$$o_{s+1} = o_s \cup O_X^*, \quad W_{s+1} = \left(W_s \cup mat^{in}\left(O_X^*\right)\right) \backslash (R \cup m_{s+1}),$$

and

$$\delta^{(s+1)}[m_{s+1}] = \delta^{(s)}[m_s] \cup \left\{\left(X, O_X^*\right)\right\}.$$

Then, set $s := s+1$, and proceed to the succeeding iteration for composition.

- *Step 2.3*. Form the closure, $\bar{\delta}$, of $\delta^{(s)}[m_s]$, *i.e.*, let

$$\bar{\delta}(X) = \{(\alpha, \beta) : (\alpha, \beta) \in op(\delta^{(s)}[m_s]) \ \& \ X \in \beta\},$$

for any $X \in M \backslash R$. Then, terminate; $\bar{\delta}$ is the greatest feasible solution extension of $\delta[m]$.

Prior to the validation of the correctness, the algorithm is illustrated with Example 14.1. For this purpose, let us consider decision-mapping $\delta[m]$ where $m = \{A_1, A_5\}$, $\delta(A_1) = \{u_1\}$, and $\delta(A_5) = \{u_4\}$. Obviously, $\delta[m]$ is a regular decision-mapping. For the unique execution, we shall choose materials from sets T_r and W_s in order of their appearences.

Since $\Delta(A_1) = \{u_1, u_2\}$ and $\Delta(A_5) = \{u_4, u_5, u_6\}$, operating units u_2, u_5, u_6 are eliminated. Therefore, $O_0 = \{u_1, u_3, u_4, u_7, u_8, \ldots, u_{12}\}$, and $M_0 = \{A_1, A_2, \ldots, A_{18}\}$. Forming set T_0, we obtain that $T_0 = \{A_6\}$, and we can proceed to the iteration step in the reduction phase of the procedure.

Obviously, $P \subseteq M_0$ and $op(\delta[m]) \subseteq O_0$. Since T_0 is singleton, material A_6 is chosen from T_0. Then, $O_{A_6} = \{u_3\}$, $O_1 = \{u_1, u_4, u_7, u_8, \ldots, u_{12}\}$, and $T_0' = \emptyset$. Now, $M_1 = M_0 \backslash \{A_6\}$ and $T_1 = \emptyset$.

Continuing by the next iteration step, $P \subseteq M_1$, $op(\delta[m]) \subseteq O_1$, and $T_1 = \emptyset$ are valid, and thus, we are ready to execute the building phase of our procedure.

Determining set W_0, we have $W_0 = \{A_3, A_4, A_8, A_9, A_2\}$, furthermore, $m_0 = \{A_1, A_5\}$, $o_0 = \{u_1, u_4\}$, and $\delta^{(0)}[m_0] = \delta[m]$.

Turning to the iteration part of the building phase, $W_0 \neq \emptyset$, and thus, A_3 need be chosen from W_0. In this case, $m_1 = \{A_1, A_3, A_5\}$, $O_{A_3}^* = \{u_4\}$, $o_1 = \{u_1, u_4\}$, $W_1 = \{A_4, A_8, A_9, A_2\}$, and $\delta^{(1)}[m_1] = \delta^{(0)}[m_0] \cup \{(A_3, \{u_4\})\}$.

Since $W_1 \neq \emptyset$, A_4 is chosen from W_1. Then, $m_2 = \{A_1, A_3, A_4, A_5\}$, $O_{A_4}^* = \{u_4\}$, $o_2 = \{u_1, u_4\}$, $W_2 = \{A_8, A_9, A_2\}$, and the new decision-mapping is $\delta^{(2)}[m_2] = \delta^{(1)}[m_1] \cup \{(A_4, \{u_4\})\}$.

Selecting material A_8 from W_2 in the following iteration step, we obtain that $m_3 = \{A_1, A_3, A_4, A_5, A_8\}$, $O_{A_8}^* = \{u_8\}$, $o_3 = \{u_1, u_4, u_8\}$, $W_3 = \{A_9, A_2, A_{13}, A_{14}\}$, and $\delta^{(3)}[m_3] = \delta^{(2)}[m_2] \cup \{(A_8, \{u_8\})\}$.

Now, choosing A_9 from W_3, we have that $m_4 = \{A_1, A_3, A_4, A_5, A_8, A_9\}$, $O_{A_9}^* = \{u_8, u_9\}$, $o_4 = \{u_1, u_4, u_8, u_9\}$, $W_4 = \{A_2, A_{13}, A_{14}, A_{15}\}$, and the new decision-mapping, $\delta^{(4)}[m_4]$, is equal to $\delta^{(3)}[m_3] \cup \{(A_9, \{u_8, u_9\})\}$.

In the next iteration step, by the choice of material A_2, we get that $m_5 = \{A_1, A_2, \ldots A_5, A_8, A_9\}$, $O^*_{A_2} = \{u_1\}$, $o_5 = \{u_1, u_4, u_8, u_9\}$, $W_5 = \{A_{13}, A_{14}, A_{15}\}$, and $\delta^{(5)}[m_5] = \delta^{(4)}[m_4] \cup \{(A_2, \{u_1\})\}$.

Selecting A_{13} from W_5, we obtain $m_6 = \{A_1, A_2, \ldots, A_5, A_8, A_9, A_{13}\}$, $O^*_{A_{13}} = \{u_{10}, u_{11}\}$, $o_6 = \{u_1, u_4, u_8, u_9, u_{10}, u_{11}\}$, $W_6 = \{A_{14}, A_{15}\}$, and $\delta^{(6)}[m_6] = \delta^{(5)}[m_5] \cup \{(A_{13}, \{u_{10}, u_{11}\})\}$.

Choosing A_{14} from W_6, we get $m_7 = \{A_1, A_2, \ldots, A_5, A_8, A_9, A_{13}, A_{14}\}$, $O^*_{A_{14}} = \{u_{11}, u_{12}\}$, $o_7 = \{u_1, u_4, u_8, u_9, \ldots, u_{12}\}$, $W_7 = \{A_{15}, A_{11}\}$, and $\delta^{(7)}[m_7] = \delta^{(6)}[m_6] \cup \{(A_{14}, \{u_{11}, u_{12}\})\}$.

In the following iteration step, by the choice of material A_{15}, we get $m_8 = \{A_1, A_2, \ldots, A_5, A_8, A_9, A_{13}, A_{14}, A_{15}\}$, $O^*_{A_{15}} = \{u_{12}\}$, $W_8 = \{A_{11}\}$, $o_8 = \{u_1, u_4, u_8, u_9, \ldots, u_{12}\}$, and $\delta^{(8)}[m_8] = \delta^{(7)}[m_7] \cup \{(A_{15}, \{u_{12}\})\}$.

Finally, choosing A_{11} from W_8, we have the following sets and decision-mapping: $m_9 = \{A_1, A_2, \ldots, A_5, A_8, A_9, A_{11}, A_{13}, A_{14}, A_{15}\}$, $O^*_{A_{11}} = \{u_9\}$, $o_9 = \{u_1, u_4, u_8, u_9, \ldots, u_{12}\}$, $W_9 = \emptyset$, $\delta^{(9)}[m_9] = \delta^{(8)}[m_8] \cup \{(A_{11}, \{u_9\})\}$.

Since $W_9 = \emptyset$, the procedure is terminated. P-graph belonging to closure $\overline{\delta}$ of $\delta^{(9)}[m_9]$ is presented in Fig. 14.3.

For illustrating the case when there is no feasible solution extension of a regular decision-mapping, first we use Example 14.1 again. Now, let $\delta'[\{A_2, A_5\}] \in \Omega_{reg}$ be given by $\delta'(A_2) = \{u_2\}$ and $\delta'(A_5) = \{u_4\}$.

Since $\Delta(A_2) = \{u_1, u_2, u_3\}$ and $\Delta(A_5) = \{u_4, u_5, u_6\}$, operating units u_1, u_3, u_5, u_6 are omitted. Thus, $O_0 = \{u_2, u_4, u_7, u_8, \ldots, u_{12}\}$, $M_0 = M$, and $T_0 = \{A_6\}$.

Turning to the iteration part in the reduction phase of the procedure, $P \subseteq M_0$, $op(\delta'[\{A_2, A_5\}]) \subseteq O_0$, and $T_0 = \{A_6\}$. Therefore, we have that $O_{A_6} = \{u_2\}$, furthermore, $O_1 = \{u_4, u_7, u_8, \ldots, u_{12}\}$. Now, $T_{0'} = \emptyset$, $M_1 = \{A_3, A_4, A_5, A_7, A_8, \ldots, A_{18}\}$, and $T_1 = \emptyset$. Then, in the following iteration step, $P \nsubseteq M_1$. Thus, the procedure is terminated; there is no feasible solution extension of the considered decision-mapping, $\delta'[\{A_2, A_5\}]$.

The following example shows such a case when $O_{r+1} = \emptyset$ in the r-th iteration step with $r = 1$.

Example 14.2 Let the reduced structural model of PNS be $\mathbb{S} = (P, R, O)$, where $P = \{A_1\}, R = \{A_3, A_4\}, O = \{u_1, u_2\}$, and the operating units are defined by

$$u_1 = (\{A_2, A_4\}, \{A_1\}), \quad \text{and} \quad u_2 = (\{A_3, A_4\}, \{A_1, A_2\}).$$

Take decision-mapping $\delta[\{A_1\}]$ with $\delta(A_1) = \{u_1\}$. Obviously, $\delta[\{A_1\}]$ is a regular decision-mapping. Then, $O_0 = \{u_1\}$, $M_0 = \{A_1, A_2, A_4\}$, and $T_0 = \{A_2\}$. Since $P \subseteq M_0$, $op(\delta[\{A_1\}]) \subseteq O_0$, and $T_0 \neq \emptyset$, we construct $O_{A_2} = \{u_1\}$. Thus, $O_1 = O_0 \backslash O_{A_2} = \emptyset$, and thus, the procedure is terminated; there is no feasible solution extension of the considered decision-mapping, $\delta[\{A_1\}]$.

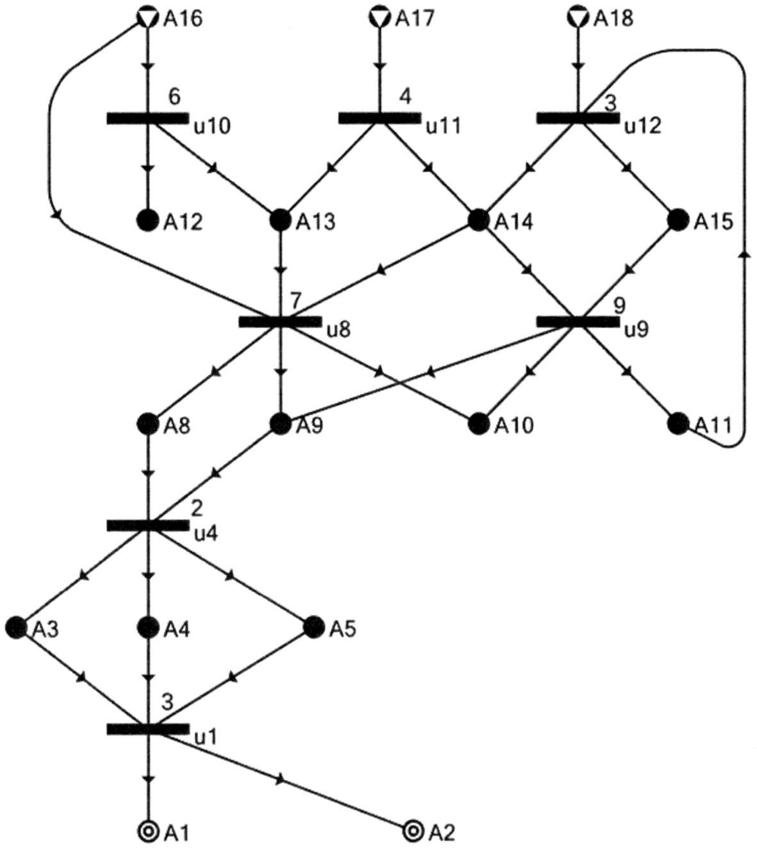

Fig. 14.3 P-graph belonging to closure $\overline{\delta}$ of $\delta^{(9)}[m_9]$

Proof of the Alogrithm

It will be proved that

(i1) the algorithm terminates in a finite number of steps;

(i2) if it is terminated in the reduction phase by $P \not\subseteq M_r$ *or* $op(\delta[m]) \not\subseteq O_r$ or $O_{r+1} = \emptyset$ for some nonnegative integer *r*, then decision-mapping $\delta[m]$ has no feasible solution extension, *i.e.*, $\omega(\delta[m]) \cap S' = \emptyset$, and

(i3) if it is terminated in the composition phase, then $\delta[m]$ has feasible solution extension, *i.e.*, $\omega(\delta[m]) \cap S' \neq \emptyset$, and decision-mapping $\overline{\delta}$ determined by the procedure is the greatest feasible solution extension of $\delta[m]$.

Let us examine the reduction phase of the procedure first; the following properties will be proved by induction on *i*.

(a) $(M_i, O_i) \subseteq (M, O)$,
(b) $\mathbb{S}_i = (P, R, O_i)$ is a structural model of PNS,
(c) for all $\delta' \in \omega(\delta[m])$, $\delta' \in S'$ if and only if the P-graph belonging to δ' is a solution-structure of structural model $\mathbb{S}_i$,
(d) T_i is the set of all materials from $M_i \backslash R$ which are sources in P-graph (M_i, O_i).

At the outset, let $i = 0$; note that $O_0 \neq \varnothing$. Indeed, if $m \neq \varnothing$, then $\delta(X) \subseteq O_0$ for all $X \in m$. Since $\delta[m]$ is a regular decision-mapping, $\delta(X) \neq \varnothing$ if $X \in m$ from Lemma 13.1; therefore, $O_0 \neq \varnothing$. If $m = \varnothing$, then $O_0 = O \neq \varnothing$. Since $O_0 \neq \varnothing$ and $M_0 = mat(O_0)$, we have $M_0 \neq \varnothing$ and $O_0 \subseteq \wp'(M_0) \times \wp'(M_0)$. Since (M, O) is the maximal structure of the reduced structural model of the PNS problem under consideration, equality $M = mat(O)$ is valid. On the other hand, O_0 is obtained by omitting some operating units from O; thus, $O_0 \subseteq O$ and $M_0 \subseteq M$. Hence, $M \cap O = \varnothing$ implies that $M_0 \cap O_0 = \varnothing$; consequently, (M_0, O_0) is a P-graph and $(M_0, O_0) \subseteq (M, O)$, *i.e.*, property (a) is valid.

So far, it has been proved that $O_0 \subseteq O$ and O_0 is a nonempty finite set. Thus, it is evident that $\mathbb{S}_0 = (P, R, O_0)$ is a structural model of PNS according to the definition, *i.e.*, property (b) is satisfied.

For validating property (c), note that $\mathbb{S}_0$ is a structural model of PNS from property (b). Now, let δ' be an arbitrary extension of $\delta[m]$ with rank $|M \backslash R|$; moreover, let $\delta' \in S'$. It will be proved that $\rho^{-1}(\delta')$ is a solution-structure of $\mathbb{S}_0$. First, it will be shown by contradiction that $op(\delta') \subseteq O_0$. To do so, let us suppose that $op(\delta') \nsubseteq O_0$. Then, an operating unit, $u \in op(\delta') \subseteq O$, exists for which $u \notin O_0$; this, in turn, implies that there exists an $X \in m$ with $u \in \Delta(X) \backslash \delta(X)$, *i.e.*, $u \in \Delta(X)$ and $u \notin \delta(X)$. Since $u \in op(\delta')$, there exists a $Y \in M \backslash R$ with $u \in \delta'(Y)$. Moreover, $\delta'(Y) \cap \Delta(X) \subseteq \delta'(X)$ because δ' is a consistent decision-mapping and $X, Y \in M \backslash R$. Since δ' is an extension of $\delta[m]$ and $X \in m$, $\delta'(X) = \delta(X)$; therefore, $\delta'(Y) \cap \Delta(X) \subseteq \delta(X)$. It can be observed that the left-hand side of this inclusion relation contains u while the right-hand side does not, which is a contradiction, thereby indicating that $op(\delta') \subseteq O_0$. Then, $\rho^{-1}(\delta') \subseteq (M_0, O_0)$; moreover, P-graph $\rho^{-1}(\delta')$ satisfies axioms (S1) through (S5) because $\delta' \in S'$. Consequently, $\rho^{-1}(\delta')$ is a solution-structure of $\mathbb{S}_0$.

Conversely, let us suppose that P-graph $(mat(op(\delta')), op(\delta'))$ is a solution-structure of $\mathbb{S}_0$ where $\delta' \in \omega(\delta[m])$. Then, it is obviously a solution-structure of $\mathbb{S}$ also. Thus, on the basis of Remark 12.2, $\delta' \in S'$ which completes the proof of property (c).

Finally, the validity of (d) follows from the definition of T_0.

Now, let i be an arbitrary nonnegative integer; also let properties (a), (b), (c), and (d) be valid for (M_i, O_i) and T_i. Furthermore, it is assumed that the procedure has not yet been terminated, *i.e.*, $P \subseteq M_i$, $op(\delta[m]) \subseteq O_i$, $O_{i+1} \neq \varnothing$, and $T_i \neq \varnothing$ hold.

Let $X \in T_i$ be the selected material; then, by the induction hypothesis, $X \in M_i \backslash R$, and X is a source in P-graph (M_i, O_i). Since $M_i = mat(O_i)$, there exists an operating unit $u = (\alpha, \beta) \in O_i$ with $X \in \alpha$. Thus, $O_X \neq \varnothing$, and consequently, $O_{i+1} \subset O_i$; moreover, from the assumption, $O_{i+1} \neq \varnothing$ is satisfied. It can be seen that $(M_{i+1}, O_{i+1}) \subseteq (M_i, O_i)$; in addition, from the induction hypothesis, $(M_i, O_i) \subseteq (M, O)$. Consequently, $(M_{i+1}, O_{i+1}) \subseteq (M, O)$, thereby validating property (a).

From the assumption, $O_{i+1} \neq \emptyset$, and property (a), $(M_{i+1}, O_{i+1}) \subseteq (M, O)$, thereby giving rise to $M_{i+1} \subseteq M$ and $O_{i+1} \subseteq O$. Then, $\mathbb{S}_{i+1} = (P, R, O_{i+1})$ is a structural model of PNS, and thus, property (b) is valid.

Property (c) will be verified in two parts. Prior to the proof, note that $\mathbb{S}_{i+1}$ is a structural model of PNS according to property (b) and let δ' be an arbitrary element of $\omega(\delta[m])$.

First, let $\delta' \in S'$. It will be shown that $op(\delta') \subseteq O_{i+1}$ is valid. To prove by contradiction, let us suppose that $op(\delta') \nsubseteq O_{i+1}$. From the induction hypothesis, it follows that $\rho^{-1}(\delta')$ is a solution-structure of $\mathbb{S}_i$, and thus, $op(\delta') \subseteq O_i$; therefore, there exists an operating unit, $u \in op(\delta')$, with $u \in O_X$ such that X denotes the selected element of T_i. Now, let us consider the P-graph belonging to δ', *i.e.*,

$$\rho^{-1}(\delta') = (mat(op(\delta')), op(\delta')).$$

This P-graph satisfies $\rho^{-1}(\delta') \subseteq (M_i, O_i)$ and $X \in mat(op(\delta'))$. On the other hand, $X \in T_i$, and by the induction hypothesis, $X \notin R$, and X is a source in (M_i, O_i). Thus, $X \in mat(op(\delta'))$, $X \notin R$, and X is a source in $\rho^{-1}(\delta')$. These and axiom (S2) indicate that $\rho^{-1}(\delta')$ is not a solution-structure of $\mathbb{S}_i$, thereby giving rise to a contradiction. Hence, $op(\delta') \subseteq O_{i+1}$. This inclusion implies that $\rho^{-1}(\delta') \subseteq (M_{i+1}, O_{i+1})$; moreover, $\rho^{-1}(\delta')$ satisfies axioms (S1) through (S5), *i.e.*, $\rho^{-1}(\delta')$ is a solution-structure of $\mathbb{S}_{i+1}$.

Second, conversely, let the P-graph, $(mat(op(\delta')), op(\delta'))$, be a solution-structure of $\mathbb{S}_{i+1}$. Then, it is a solution-structure of $\mathbb{S}$, and thus, $\delta' \in S'$ from Remark 12.2. This completes the proof of the validity of property (c).

With regarding to property (d), it can be observed that $T_{i'}$ contains those materials from $M_{i+1} \backslash R$ which become sources in (M_{i+1}, O_{i+1}) after deleting the elements included in O_X. Furthermore, $T_i \cap M_{i+1}$ contains those materials from $M_{i+1} \backslash R$ which are sources in both (M_i, O_i) and (M_{i+1}, O_{i+1}). Consequently, T_{i+1} contains all the materials from set $M_{i+1} \backslash R$ which are sources in (M_{i+1}, O_{i+1}), *i.e.*, property (d) is valid.

In proceeding to prove statement (i2), let us suppose that the procedure is terminated in the reduction phase by $P \nsubseteq M_r$ or $op(\delta[m]) \nsubseteq O_r$ or $O_{r+1} = \emptyset$ for some nonnegative integer r. It will be shown that there is no feasible solution extension of $\delta[m]$. For verification by contradiction, we assume that $\delta' \in \omega(\delta[m]) \cap S'$.

For the case, $r = 0$, $op(\delta') \subseteq O_0$ as shown previously. Moreover, from Remark 12.2, $\rho^{-1}(\delta') = (mat(op(\delta')), op(\delta'))$ is a solution-structure of $\mathbb{S}$, and thus, $P \subseteq mat(op(\delta'))$ from axiom (S1). Hence, $P \subseteq mat(op(\delta')) \subseteq mat(O_0) = M_0$ thereby leading to $P \subseteq M_0$. On the other hand, δ' is an extension of $\delta[m]$; this implies that $op(\delta[m]) \subseteq op(\delta') \subseteq O_0$. Thus, $P \subseteq M_0$ and $op(\delta[m]) \subseteq O_0$, and subsequently, the procedure must be terminated by $O_1 = \emptyset$. Since $\rho^{-1}(\delta') \in S(\mathbb{S})$, we have $\delta'(Y) \neq \emptyset$ for all $Y \in P$ from axiom (S2). Let $Y_0 \in P$ and $u \in \delta'(Y_0)$ be arbitrary; then, $u \in O_0$ because $op(\delta') \subseteq O_0$. Now, the material selected from T_0 will be denoted by X. According to property (d), $X \notin R$ and X is a source in (M_0, O_0). If $X \notin mat^{in}(u)$, then

$u \notin O_X$, and as a result, $u \in O_1$, thus implying that $O_1 \neq \varnothing$ which is a contradiction. Otherwise, if $X \in mat^{in}(u)$, X is a material point of P-graph $\rho^{-1}(\delta')$. Since $op(\delta') \subseteq O_0$, we have $\rho^{-1}(\delta') \subseteq (M_0, O_0)$. Hence, $X \notin R$ and it is also a source in $\rho^{-1}(\delta')$, thereby contradicting with axiom (S2). Consequently, $O_1 \neq \varnothing$; this is also contradiction; therefore, there is no feasible solution extension of $\delta[m]$.

For the case, $r > 0$, we have the inclusion, $op(\delta') \subseteq O_r$, according to the proof of the case, $i > 0$, and a contradiction arises in the same way as before. This ends the proof of statement (i2).

Regarding the number of the steps of iteration in the reduction phase, note that it is at most $|O| + 1$ since $O \supseteq O_0 \supset O_1 \supset \cdots \supset O_r$.

For proving statement (i3), it is supposed that the reduction phase of the procedure is terminated by $P \subseteq M_r$, $op(\delta[m]) \subseteq O_r$, and $T_r = \varnothing$ for some nonnegative integer r.

Let us now consider the composition phase of the procedure. First, it will be proved by induction on j that the following properties are valid.

(a) $op(\delta^{(j)}[m_j]) = o_j \subseteq O_r$,
(b) $W_j = (mat^{in}(op(\delta^{(j)}[m_j])) \cup P)\backslash(R \cup m_j)$,
(c) $\delta^{(j)}(Y) = \{(\alpha, \beta) : (\alpha, \beta) \in O_r \;\&\; Y \in \beta\}$ for all $Y \in m_j$,
and
(d) $\delta^{(j)}[m_j] \in \Omega_{reg}$.

If $j = 0$, properties (e), (f), and (h) can immediately be obtained from the definitions of o_0, W_0, and $\delta^{(0)}[m_0]$ on the basis of the assumptions, $op(\delta[m]) \subseteq O_r$ and $\delta[m] \in \Omega_{reg}$. Property (g) is obtainable based on the definition of O_0 and inclusions $O \supseteq O_0 \supset O_1 \supset \ldots \supset O_r$.

Now, let $j \geq 0$ be an arbitrary integer, and also let the above properties be valid for j. Moreover, let us assume that the procedure is not terminated, *i.e.*, $W_j \neq \varnothing$. Then, an arbitrary X is selected from W_j, and $\delta^{(j+1)}[m_{j+1}] = \delta^{(j)}[m_j] \cup \{(X, O_X^*)\}$, where $O_X^* = \{(\alpha, \beta) : (\alpha, \beta) \in O_r \;\&X \in \beta\}$.

Now, it will be proved that $O_X^* \neq \varnothing$. The induction hypothesis implies that $op(\delta^{(j)}[m_j]) = o_j \subseteq O_r$, and thus, $mat^{in}(op(\delta^{(j)}[m_j])) \subseteq mat(O_r) = M_r$. Since $P \subseteq M_r$, on the basis of the induction hypothesis on W_j, $X \in W_j$ implies that $X \in M_r\backslash R$. Now, if $O_X^* = \varnothing$, X is a source in (M_r, O_r). Then, property (d) indicates that $X \in T_r$ which is a contradiction; as a consequence, $O_X^* \neq \varnothing$.

The induction hypothesis leads to $op(\delta^{(j)}[m_j]) = o_j \subseteq O_r$. According to the procedure under consideration, $o_{j+1} = o_j \cup O_X^*$ and

$$op\left(\delta^{(j+1)}\left[m_{j+1}\right]\right) = op\left(\delta^{(j)}\left[m_j\right]\right) \cup O_X^*.$$

Moreover, $O_X^* \subseteq O_r$ from the definition of O_X^*, and thus,

$$op\left(\delta^{(j+1)}\left[m_{j+1}\right]\right) = o_j \cup O_X^* = o_{j+1} \subseteq O_r,$$

i.e., property (e) is valid.

From the definition, $W_{j+1} = \left(W_j \cup mat^{in}\left(O_X^*\right)\right) \backslash \left(R \cup m_{j+1}\right)$, and by the induction hypothesis, $W_j = (mat^{in}(op(\delta^{(j)}[m_j])) \cup P) \backslash (R \cup m_j)$. By excluding the elements of set $R \cup m_j$ from the last expression of set W_j and substituting the resultant set into the expression of set W_{j+1},

$$W_{j+1} = \left(mat^{in}\left(op\left(\delta^{(j)}\left[m_j\right]\right)\right) \cup P \cup mat^{in}\left(O_X^*\right)\right) \backslash \left(R \cup m_{j+1}\right).$$

Since

$$op\left(\delta^{(j+1)}\left[m_{j+1}\right]\right) = op\left(\delta^{(j)}\left[m_j\right]\right) \cup O_X^*,$$

we obtain that

$$mat^{in}\left(op\left(\delta^{(j+1)}\left[m_{j+1}\right]\right)\right)$$
$$= mat^{in}\left(op\left(\delta^{(j)}\left[m_j\right]\right) \cup O_X^*\right) = mat^{in}\left(op\left(\delta^{(j)}\left[m_j\right]\right)\right) \cup mat^{in}\left(O_X^*\right).$$

Then,

$$W_{j+1} = \left(mat^{in}\left(op\left(\delta^{(j+1)}\left[m_{j+1}\right]\right)\right) \cup P\right) \backslash \left(R \cup m_{j+1}\right),$$

thereby yielding property (f). From the definitions of $\delta^{(j+1)}\left[m_{j+1}\right]$ and O_X^*, the validity of property (g) is obvious.

To prove property (h), let us observe that on the basis of property (f), material X is chosen in accordance with the input material selection rule. On the other hand, $O_X^* \neq \emptyset$, and thus, it is sufficient to show that $\delta^{(j+1)}[m_{j+1}]$ is a consistent decision-mapping of the reduced structural model $\mathbb{S}$. For this purpose, it will be proved that

$$\text{(i)} \quad \delta^{(j+1)}(U) \cap \Delta(V) \subseteq \delta^{(j+1)}(V) \;\; for \;\; all \;\; U, V \in m_{j+1}.$$

When $U = V$, the above inclusion is obviously valid. If $U \neq V$ and U, $V \in m_j$, this inclusion is the immediate consequence of the induction hypothesis.

At this juncture, let $U \in m_j$ and $V = X$. Then, it has to proved that

$$\delta^{(j+1)}(U) \cap \Delta(X) \subseteq \delta^{(j+1)}(X).$$

Since $U \in m_j$, $\delta^{(j+1)}(U) = \delta^{(j)}(U)$ holds. On the other hand, $\delta^{(j+1)}(X) = O_X^*$, and thus, inclusion (i) is equivalent to the following.

$$\delta^{(j)}(U) \cap \Delta(X) \subseteq O_X^*.$$

Let us suppose that $u = (\alpha, \beta)$ and $u \in \delta^{(j)}(U) \cap \Delta(X)$; then, $X \in \beta$. Furthermore, $u \in o_j \subseteq O_r$ from property (e), and thus, $u \in O_X^*$; consequently, inclusion (i) is valid.

Finally, let $U = X$ and $V \in m_j$; then, inclusion (i) can be given as

$$\delta^{(j+1)}(X) \cap \Delta(V) \subseteq \delta^{(j+1)}(V).$$

Since $\delta^{(j+1)}(X) = O_X^*$ and $\delta^{(j+1)}(V) = \delta^{(j)}(V)$, it must be proved that

$$O_X^* \cap \Delta(V) \subseteq \delta^{(j)}(V).$$

For this purpose, let $u = (\alpha, \beta) \in O_X^* \cap \Delta(V)$. Then, $u \in O_r$ and $V \in \beta$; therefore, $u \in \delta^{(j)}(V)$ by property (g), thereby resulting in the required inclusion, (i). Hence, $\delta^{(j+1)}[m_{j+1}]$ is an extension of $\delta^{(j)}[m_j]$. Since material X is chosen by the input material selection rule and $O_X^* \neq \emptyset$, decision-mapping $\delta^{(j+1)}[m_{j+1}]$ is a regular extension of $\delta^{(j)}[m_j]$. Thus, from the $\delta^{(j)}[m_j] \in \Omega_{reg}$ part of the induction hypothesis, it can be easily seen that $\delta^{(j+1)}[m_{j+1}]$ is a regular decision-mapping, *i.e.*, property (h) is valid.

To proceed to verify statement (i3), it is supposed that the procedure is terminated by $W_s = \emptyset$ for some nonnegative integer s. We shall prove that the resultant decision-mapping, $\overline{\delta}$, is the greatest feasible solution extension of $\delta[m]$.

First, we show that $\overline{\delta}$ is a maximal extension of $\delta[m]$, *i.e.*, $\overline{\delta} \in \omega(\delta[m])$. Let us observe that, the validity of relations given below follows from the proof of property (h)

$$\delta[m] = \delta^{(0)}[m_0] \prec \delta^{(1)}[m_1] \prec \ldots \prec \delta^{(s)}[m_s].$$

The following two cases are distinguished.

Let us suppose that $m_s = M \backslash R$. Then, $\overline{\delta} = \delta^{(s)}[m_s]$ according to the procedure. By the above relations, $\delta^{(s)}[m_s]$ is an extension of $\delta[m]$, furthermore, the rank of $\delta^{(s)}[m_s]$ is $|M \backslash R|$, and thus, it is a maximal extension of $\delta[m]$, *i.e.*, $\overline{\delta} \in \omega(\delta[m])$.

If $m_s \subset M \backslash R$, then we have $\delta^{(s)}[m_s] \in \Omega_{reg} \backslash \Omega_{max}$ from (h). In this case, by $W_s = \emptyset$, property (f), and Lemma 13.4, we obtain that closure $\overline{\delta}$ of $\delta^{(s)}[m_s]$ is a maximal extension of $\delta^{(s)}[m_s]$. From this, by the above relations, we get that $\overline{\delta}$ is a maximal extension of $\delta[m]$, *i.e.*, $\overline{\delta} \in \omega(\delta[m])$.

In what follows, it will be shown that $\overline{\delta} \in S'$. The same cases are distinguished as above.

First, let us assume that $m_s = M \backslash R$. In the light of Theorem 12.1, it is sufficient to prove that

1. $unit(\overline{\delta}) = op(\overline{\delta})$,
2. $\overline{\delta}(X) \neq \emptyset$, for all $X \in mat(op(\overline{\delta})) \backslash R$,
3. $P \subseteq mat(op(\overline{\delta}))$.

Equality (1) can be obtained from property (h) and Lemma 13.3. Since $\overline{\delta} = \delta^{(s)}[m_s]$, property (e) implies $op(\overline{\delta}) = o_s \subseteq O_r$. Thus, $mat(op(\overline{\delta})) = mat(o_s) \subseteq mat(O_r) = M_r \subseteq M$. Therefore, $mat(op(\overline{\delta}))\backslash R \subseteq M\backslash R = m_s$. Now, by (h) and Lemma 13.1, we obtain (2). Finally, by our assumption, $P \subseteq M_r$. Since $P \cap R = \varnothing$, we get that $P \subseteq M_r\backslash R \subseteq M\backslash R = m_s$. On the other hand, by (h) and Lemma 13.1, $m_s \subseteq mat(op(\overline{\delta}))$, and thus, $P \subseteq mat(op(\overline{\delta}))$ resulting in (3).

Now, let us assume that $m_s \subset M\backslash R$. Then, by (f), (h), $W_s = \varnothing$, and Theorem 13.2, we obtain that $\overline{\delta} \in S'$.

It has been proved that decision-mapping $\overline{\delta}$ determined by the procedure is a feasible solution extension of $\delta[m]$, *i.e.*, $\overline{\delta} \in \omega(\delta[m]) \cap S'$.

In what follows, it will be shown that $\overline{\delta}$ is the greatest among the feasible solution extensions of $\delta[m]$. First of all, property (c) implies that $\rho^{-1}(\overline{\delta})$ is a solution-structure of $\mathbb{S}_r$. Moreover, property (g) indicates that $\rho^{-1}(\overline{\delta})$ is the maximal structure of $\mathbb{S}_r$. Then, on the basis of property (c), $\overline{\delta}$ is the greatest feasible solution extension of $\delta[m]$, thus verifying statement (i3).

In reference to the number of iteration steps of the composition phase, $m = m_0 \subset m_1 \subset \cdots \subset m_s \subseteq M_r$, and thus, $s \leq |M_r| \leq |M|$. Since any iteration step and the number of iteration steps are finite for both phases, the procedure terminates in a finite number of steps, *i.e.*, statement (i1) is valid.

This completes the proof of the validity of the procedure for generating the greatest feasible solution extension. Note that by the procedure, a new reduced structural model of PNS can be obtained as described below.

Let $O' = op(\overline{\delta})$, $M' = mat(O')$, and $R' = R \cap M'$; then, $\mathbb{S}' = (P, R', O')$ is a reduced structural model of PNS, and its maximal structure is (M', O'). Furthermore, in the light of Remark 14.2, $(\overline{m}, \overline{o}) \subseteq (m_i, o_i) \subseteq (M', O')$ for $i = 1, 2, \ldots, n$, where $(\overline{m}, \overline{o})$ denotes the P-graph belonging to $\delta[m]$; $\delta_1, \delta_2, \ldots, \delta_n$ are the feasible solution extensions of $\delta[m]$, and $(m_i, o_i) = \rho^{-1}(\delta_i)$, $i = 1, 2, \ldots, n$. For simplicity, the function determined by (M', O') is denoted by $\Delta(X)$ as well, *i.e.*,

$$\Delta(X) = \{(\alpha, \beta) : (\alpha, \beta) \in O' \& X \in \beta\} \;\; for \;\; all \;\; X \in M'.$$

Prior to composing the bounding procedure, a new mapping is introduced. Specifically, the function, $\mu : M' \to O'$, is defined in the following way. Let $X \in M'\backslash R'$ be an arbitrary material and the sequence, $\tilde{o}_0, \tilde{o}_1, \ldots$, be defined by

$$\tilde{o}_0 = \Delta(X) \;\; \text{and} \;\; \tilde{o}_{j+1} = \tilde{o}_j \cup (\cup\{\Delta(Y) : Y \in mat^{in}(\tilde{o}_j)\}).$$

Then, $\tilde{o}_0 \subseteq \tilde{o}_1 \subseteq \ldots$ from the definition of the sequence. Also, it can readily be discerned that if $\tilde{o}_l = \tilde{o}_{l+1}$ for some nonnegative integer l, $\tilde{o}_l = \tilde{o}_j$ for all $j \geq l$. Since $\tilde{o}_j \subseteq O'$ and O' is a finite set, it follows that there exists a nonnegative integer $l < |O'|$ with $\tilde{o}_l = \tilde{o}_{l+1}$.

Now, let us introduce function μ as

$$\mu(X) = \{ \begin{array}{ll} \tilde{o}_l, & X \in M' \backslash R', \\ \varnothing, & else \end{array}$$

where l is the smallest integer with $\widetilde{o}_l = \widetilde{o}_{l+1}$. Then, set $\mu(X)$ contains operating units participating directly or indirectly in the production of X by the process determined by decision-mapping Δ.

Illustration of Bounding for PNS$_w$

We are now in a position to compose the bounding procedure. As mentioned earlier, the present algorithm assigns a value specified as *material bound* to each material, and a lower bound is determined on the basis of this material bound. To facilitate thorough understanding, the bounding procedure is outlined below.

In each iteration step, set M' is divided into two disjoint subsets, $I^{(r)}$ and $J^{(r)}$. Set $I^{(r)}$ contains those materials from M' for which the material bounds have already been determined. Initially, $I^{(0)}$ contains the raw materials and those materials which can be directly produced by the operating units of $op(\delta[m])$; and zero is assigned as the material bounds of the elements of $I^{(0)}$. Set $J^{(r)}$ contains those materials for which the material bounds are yet to be determined but these costs have some upper estimates. An *estimation for the material bound* of a material from $J^{(r)}$ is determined as the minimum of the weights of those operating units of O' that produce this material directly. Either the material bound of a material or its estimate is represented by function $c^{(r)}$ in iteration r. In this iteration step, a material candidate is chosen from $J^{(r)}$ to determine its material bound; this limits the iteration steps to be less than $|M'|$. To realize such a procedure, however, is not simple; some prerequisites need be fulfilled.

First, we constitute the set of the candidates, $L^{(r)}$, from some elements of $J^{(r)}$; specifically, material X from $J^{(r)}$ is a *candidate* if and only if at least one operating unit from $\Delta(X)$ has at least one input material with a known material bound. If $L^{(r)} \neq \varnothing$, then the candidates are assessed in detail. Note that the value given by function $c^{(r)}(X)$ will provide a lower bound for the bound of producing material X.

For each candidate X, we examine all operating units producing X directly; let u_t be one of such operating units. If none of the input materials of u_t has its material bound known, then u_t is assigned with a negative value, $-w(u_t)$, for signifying. Otherwise, if u_t has at least one input material with its material bound known, this and the other input materials of u_t with their material bound known but not involved in a loop containing u_t, are placed into set Q_t. If such a material exists, *i.e.*, $Q_t \neq \varnothing$, then u_t is assigned with the value equal to the sum of $w(u_t)$ and the maximum of the material bound of the materials in Q_t; otherwise, u_t is assigned with the value, $w(u_t)$ if $Q_t = \varnothing$. In particular, if the material bound of each input material of u_t is known, then it is taken into account by variable q_t.

After evaulating all the operating units producing X directly, we identify the minimum of the positive values assigned to the operating units included in set $\Delta(X)$. Since $X \in L^{(r)}$, at least one positive value exists; therefore, a positive value is obtained for candidate X which is denoted by $h(X)$.

If $h(X)$ is derived from operating unit u_{t_0} that has known material bound for each input material and $h(X)$ does not exceed the absolute values of the negative values assigned to some operating units of $\Delta(X)$, then $h(X)$ is assigned to material X as its material cost. In this case, $h(X)$ provides the sharpest possible material bound of X. This means that for any process belonging to a feasible solution extension from $\omega(\delta[m]) \cap S'$, the bound for producing material X is $h(X)$ if we take into account the known material bounds of the input materials to the operating units producing material X under the assumption that u_{t_0} is excluded from any loop of the process under consideration.

If no candidate satisfies the above condition concerning $h(X)$, then those candidates are collected into a separate set L' for which each operating unit producing them directly has at least one input material with its material bound known. In determining L', this property is signified by function $j(X)$ for each candidate X. If $L' \neq \emptyset$, then material X is selected from L', produced by the maximal number of operating units; $h(X)$ is regarded as the material bound of X. Naturally, this material bound is not necessarily the sharpest in contrast to the preceding one; nevertheless, no additional information is available to obtain a sharper than this.

If $L' = \emptyset$, then material X is chosen from $L^{(r)}$ so that function $c^{(r)}$ is maximal. The value of $\min\{h(X), \bar{c}\}$ is assigned to X as its material bound where $\bar{c}$ is the minimum of the weights of those operating units producing X directly; none of the input materials to these operating units has its material bound known.

Finally, if there is no candidate material, *i.e.*, $L^{(r)} = \emptyset$, any of the operating units directly producing material $X \in J^{(r)}$ has no input material with its material bound known. In this case, no additional information available on the cost; hence, an element denoted by X is chosen from $J^{(r)}$, with $c^{(r)}(X)$ is assigned as its material bound.

When the material bound of each material in M' has been established, we take into account those sources of the P-graph belonging to $\delta[m]$ which are not raw materials, and those desired products which are not produced directly by any operating unit of $op(\delta[m])$. Adding the maximum among the material bounds of these materials to $g(\delta[m])$ yields a lower bound of the values of the objective function for any feasible solution extension included in $\omega(\delta[m]) \cap S'$, *i.e.*, a lower bound of

$$w(\delta') \ for \ \delta' \in \omega(\delta[m]) \cap S'.$$

The algorithm outlined above is formalized as the refined bounding procedure.

Refined Bounding Procedure

Initialization

- Define the functions, h, j, and l, on $M'\backslash R'$ by $h(X) = j(X) = 0$ and $l(X) = |\Delta(X)|$ for all $X \in M'\backslash R'$. Form initial set $I^{(0)}$ of the materials with known material bounds as

$$I^{(0)} = mat^{out}(op(\delta[m])) \cup R',$$

and initial set $J^{(0)}$ of materials with unknown material bounds as

$$J^{(0)} = M' \backslash I^{(0)}.$$

Determine the initial function, $c^{(0)}$, on set M' as follows:
For any $Y \in M'$, let

$$c^{(0)}(Y) = \begin{cases} 0, & Y \in I^{(0)}, \\ \min\{w(u) : u \in \Delta(Y)\}, & otherwise. \end{cases}$$

Then, constitute initial set $L^{(0)}$ of the candidates as

$$L^{(0)} = \{Y : Y \in J^{(0)} \ \& \ mat^{in}(\Delta(Y)) \cap I^{(0)} \neq \varnothing\},$$

and determine $\mu(V)$ for any $V \in M'\backslash R'$.
Finally, set $r := 0$.

Iteration (r-th iteration)

- *Test.* If $J^{(r)} = \varnothing$, then $c^{(r)}(Y)$ is the material bound for all $Y \in M'\backslash R'$; then, proceed to Step 7; Otherwise, consider set $L^{(r)}$. If $L^{(r)} = \varnothing$, then choose material X from $J^{(r)}$; set $h(X) = c^{(r)}(X)$ and proceed to Step 6. If $L^{(r)} \neq \varnothing$, then perform Steps 1 through 3 for all $X \in L^{(r)}$, and proceed to Step 4.
- *Step 1.* Consider set $\Delta(X)$ of s elements, $u_1, u_2, \ldots, u_s$.
- *Step 2.* Determine set Q_t for all integer t, $t = 1, 2, \ldots, s$, where

$$Q_t = \{V : V \in mat^{in}(u_t) \cap I^{(r)} \ \& \ u_t \notin \mu(V)\}.$$

Furthermore, let

$$c_t = \begin{cases} -w(u_t), & mat^{in}(u_t) \cap I^{(r)} = \varnothing, \\ w(u_t), & mat^{in}(u_t) \cap I^{(r)} \neq \varnothing, Q_t = \varnothing, \\ (u_t) + \max\{c^{(r)}(V) : V \in Q_t\}, & otherwise, \end{cases}$$

and

$$q_t = \begin{cases} 1, & mat^{in}(u_t) \subseteq I^{(r)} \\ 0, & otherwise \end{cases}$$

- *Step 3*. Compute $\min\{c_t : c_t > 0 \ \& \ 1 \leq t \leq s\}$; moreover, let t_0, $1 \leq t_0 \leq s$, be such an integer for which c_{t_0} is equal to this minimum; redefine $h(X)$ as $h(X) = c_{t_0}$. If $q_{t_0} = 1$ and $c_{t_0} \leq |c_t|$ for all $1 \leq t \leq s$, then proceed to Step 6. Otherwise, redefine the value of function j on X as follows:

$$j(X) = \begin{cases} 0, & c_t < 0 \ for\ some\ 1 \leq t \leq s, \\ 1, & otherwise. \end{cases}$$

- *Step 4*. Form set $L' = \{Y : Y \in L^{(r)} \ \& \ j(Y) = 1\}$; if $L' = \varnothing$, then proceed to Step 5. Otherwise, determine the value of $max\{l(Y) : Y \in L'\}$, and choose material X from L' for which $l(X)$ is equal to the this maximum; then, proceed to Step 6.
- *Step 5*. Choose material X from $L^{(r)}$ for which $c^{(r)}(X)$ is maximal. Determine

$$\bar{c} = min\{-c_t : c_t < 0, 1 \leq t \leq s\},$$

where $c_1, c_2, \ldots, c_s$ are the values assigned to operating units $u_1, u_2, \ldots, u_s$ of $\Delta(X)$ in Steps 1 through 3. Then, redefine $h(X)$ by giving the new value, $\min\{h(X), \bar{c}\}$, to it; proceed to Step 6.
- *Step 6*. Let $I^{(r+1)} = I^{(r)} \cup \{X\}$ and $J^{(r+1)} = J^{(r)} \backslash \{X\}$; then, define function $c^{(r+1)}$ as

$$c^{(r+1)}(Y) = \begin{cases} c^{(r)}(Y), & Y \neq X, \\ h(X), & otherwise, \end{cases}$$

for all $Y \in M'$. In addition, let $O_X = \{(\alpha, \beta) : (\alpha, \beta) \in O' \ \& \ X \in \alpha\}$; then, form set $L^{(r+1)}$ as

$$L^{(r+1)} = \left(L^{(r)} \cup mat^{out}(O_X)\right) \cap J^{(r+1)}.$$

Set r=r+1 and iterate.
- *Step 7. (Bound determination)* Form set F as

Table 14.2 Cost values for Example 14.1

X	1	2	3	4	5	8	9	10	11	12	13	14	15
$c^{(0)}$	0	0	0	0	0	7	7	7	9	6	4	3	3
$c^{(1)}$	0	0	0	0	0	7	7	7	9	6*	4	3	3
$c^{(2)}$	0	0	0	0	0	7	7	7	9	6	4*	3	3
$c^{(3)}$	0	0	0	0	0	7	7	7	9	6	4	3*	3
$c^{(4)}$	0	0	0	0	0	11*	7	7	9	6	4	3	3
$c^{(5)}$	0	0	0	0	0	11	9*	7	9	6	4	3	3
$c^{(6)}$	0	0	0	0	0	11	9	9*	9	6	4	3	3
$c^{(7)}$	0	0	0	0	0	11	9	9	9*	6	4	3	3
$c^{(8)}$	0	0	0	0	0	11	9	9	9	6	4	3	3*

$$F = \left(mat^{in}(op(\delta[m])) \cup P\right) \setminus \left(mat^{out}(op(\delta[m])) \cup R'\right).$$

Determine the maximum of $c^{(r)}(Y)$ for $Y \in F$; it is denoted by h^*.
Terminate; the lower bound determined is $h^* + g(\delta[m])$.

To illustrate our procedure, let us consider Example 14.1 again with the decision mapping, $\delta[\{A_1, A_5\}]$, where $\delta(A_1) = \{u_1\}$ and $\delta(A_5) = \{u_4\}$. As a demonstration of the Greatest Feasible Solution Extension Building Procedure, we determined the greatest feasible solution extension, $\overline{\delta}$, of $\delta[\{A_1, A_5\}]$. The P-graph belonging to $\overline{\delta}$ is given in Fig. 14.3.

Then, $M' = \{A_1, A_2, \ldots, A_5, A_8, \ldots, A_{18}\}$, $O' = \{u_1, u_4, u_8, u_9, \ldots, u_{12}\}$. Set $I^{(0)}$ is equal to $\{A_1, A_2, \ldots, A_5, A_{16}, A_{17}, A_{18}\}$, and its elements have 0 material costs. $J^{(0)} = M' \setminus I^{(0)}$, and for each $X \in J^{(0)}$, we have to take the minimum of the weights of the operating units from $\Delta(X)$ as the estimation for the material cost of the considered material, X. The determined values of functions $c^{(r)}$, $r = 0, 1, \ldots$, are shown in Table 14.2 where we write only the indices of the materials and we mark with star the new values, furthermore, we do not write out the raw materials. The initial set of the candidates is $L^{(0)} = \{A_8, A_9, A_{10}, A_{12}, A_{13}, A_{14}, A_{15}\}$. The P-graph belonging to $\overline{\delta}$ and the initial costs are given in Fig. 14.4.

Since $J^{(0)} \neq \emptyset$ and $L^{(0)} \neq \emptyset$, we begin to execute the first three steps of the iteration part for the elements of $L^{(0)}$. Let us choose the candidates from $L^{(0)}$ corresponding to the smallest index. Then, at the fourth material, A_{12}, we obtain that it is produced by such an operating unit whose all input materials are contained in $I^{(0)}$, *i.e.*, they have material costs.

Then, $h(A_{12}) = 6$, $I^{(1)} = I^{(0)} \cup \{A_{12}\}$, $J^{(1)} = J^{(0)} \setminus \{A_{12}\}$, and the new set of the candidates is $L^{(1)} = \{A_8, A_9, A_{10}, A_{13}, A_{14}, A_{15}\}$. Furthermore, in particular, $c^{(1)}(X) = c^{(0)}(X)$ for all $X \in M'$.

In the following iteration step, at the fourth candidate, A_{13}, we get that the operating units from $\Delta(A_{13})$ have input materials only from $I^{(1)}$, and thus, $I^{(2)} = I^{(1)} \cup \{A_{13}\}$, $J^{(2)} = J^{(1)} \setminus \{A_{13}\}$, $L^{(2)} = \{A_8, A_9, A_{10}, A_{14}, A_{15}\}$. In particular again, $c^{(2)}(X) = c^{(1)}(X)$ for all $X \in M'$.

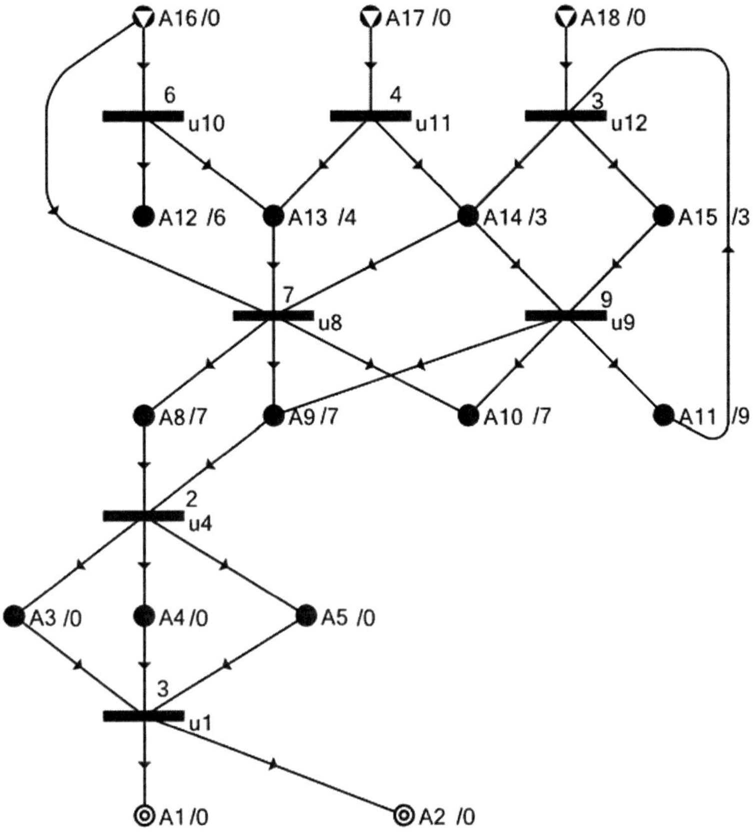

Fig. 14.4 P-graph belonging to $\overline{\delta}$ with the initial costs

Proceeding to the next iteration step, we have to execute the first three steps for all candidates, and we have to apply Step 4. Now, $L' = \{A_8, A_{14}, A_{15}\}$, furthermore, $l(A_8) = 1$, $l(A_{14}) = 2$, and $l(A_{15}) = 1$. Thus, material A_{14} is chosen from L'. Then, $I^{(3)} = I^{(2)} \cup \{A_{14}\}$, $J^{(3)} = J^{(2)} \setminus \{A_{14}\}$, and $L^{(3)} = \{A_8, A_9, A_{10}, A_{11}, A_{15}\}$. Moreover, $c^{(3)}(X) = c^{(2)}(X)$ for all $X \in M'$.

In the next iteration step, for candidate A_8, we have that it is produced by such an operating unit whose all input materials are contained in $I^{(3)}$. Thus, $I^{(4)} = I^{(3)} \cup \{A_8\}$, $J^{(4)} = J^{(3)} \setminus \{A_8\}$, $L^{(4)} = \{A_9, A_{10}, A_{11}, A_{15}\}$, and $h(A_8) = 11$, furthermore, $c^{(4)}(X) = c^{(3)}(X)$ for all $X \neq A_8 \in M'$, while $c^{(4)}(A_8) = 11$.

In the case, $r = 4$, we have that, for the operating units u_8, u_9 producing A_9, each input material of u_8 is contained in $I^{(4)}$ and the value assigned to u_8 is 11, while u_9 is contained in both $\mu(A_{14})$ and $\mu(A_{15})$, and thus, value 9 is assigned to u_9. Similarly, for A_{10}, A_{11}, A_{15}, we obtain the values, 9, 9, 3, respectively, and the value of function j is equal to 1 for all the candidates. Therefore, Step 4 follows. Choosing A_9, $I^{(5)} = I^{(4)} \cup \{A_9\}$, $J^{(5)} = J^{(4)} \setminus \{A_9\}$, $L^{(5)} = \{A_{10}, A_{11}, A_{15}\}$. Thus, $h(A_9) = 9$ and $c^{(5)}(A_9) = 9$, furthermore, $c^{(5)}(X) = c^{(4)}(X)$ for all $X \neq A_9 \in M'$.

Continuing the procedure, we assign to material A_{10} value 9. In this case, $I^{(6)} = I^{(5)} \cup \{A_{10}\}$, $J^{(6)} = J^{(5)} \backslash \{A_{10}\}$, $L^{(6)} = \{A_{11}, A_{15}\}$, $c^{(6)}(A_{10}) = 9$, and $c^{(6)}(X) = c^{(5)}(X)$ for all $X \neq A_{10} \in M'$.

Under $r = 6$, we have to perform the first three steps for both candidates, and we have to apply Step 4. Now, $L' = \{A_{11}, A_{15}\}$, $l(A_{11}) = 1$, and $l(A_{15}) = 1$. Let us choose A_{11}. Then, $I^{(7)} = I^{(6)} \cup \{A_{11}\}$, $J^{(7)} = J^{(6)} \backslash \{A_{11}\}$, $L^{(7)} = \{A_{15}\}$. Since $mat^{in}(u_9) \cap I^{(6)} = \{A_{14}\}$ and $u_9 \in \mu(A_{14})$, we get that $c^{(7)}(A_{11}) = 9$, and therefore, $c^{(7)}(X) = c^{(6)}(X)$ for all $X \in M'$.

In the next iteration step, we have only candidate A_{15}. It is produced by operating unit u_{12} which has input materials from $I^{(7)}$. Therefore, $I^{(8)} = I^{(7)} \cup \{A_{15}\}$, $J^{(8)} = \varnothing$, $L^{(8)} = \varnothing$, $c^{(8)}(A_{15}) = 3$, and in particular, $c^{(8)}(X) = c^{(7)}(X)$ for all $X \in M'$.

Since $J^{(8)} = \varnothing$, we proceed to Step 7. Set F is $\{A_8, A_9\}$. Therefore, $h^* = \max \{c^{(8)}(X) : X \in F\} = 11$, and thus, the determined lower bound is $h^* + g(\delta[m]) = 16$.

It is easy to see that the considered decision-mapping, $\delta[\{A_1, A_5\}]$, has five feasible solution extensions which can be given by the following sets of operating units:

$$\{u_1, u_4, u_8, u_{11}\},$$

$$\{u_1, u_4, u_8, u_{10}, u_{11}\},$$

$$\{u_1, u_4, u_8, u_9, u_{11}, u_{12}\},$$

$$\{u_1, u_4, u_8, u_9, u_{10}, u_{11}, u_{12}\},$$

$$\{u_1, u_4, u_8, u_9, u_{10}, u_{12}\}.$$

The corresponding weights are 16, 22, 28, 34, 30. Therefore, in particular, our lower bound gives the minimum of the corresponding object values.

Prior to proving the validity of the bounding procedure, we recall that $\delta[m]$ is a given regular decision-mapping of $\mathbb{S} = (P, R, O)$ determined in the branching procedure. Its feasible solution extensions are δ_i, $i = 1, 2, \ldots, n$; the greatest one is separately specified as $\overline{\delta}$. For the corresponding P-graphs, we have

$$(\overline{m}, \overline{o}) \subseteq (m_i, o_i) \subseteq (M', O'),$$

where $(\overline{m}, \overline{o})$ denotes the P-graph belonging to $\delta[m]$, $(m_i, o_i) = \rho^{-1}(\delta_i)$, $i = 1, 2, \ldots, n$, and $(M', O') = \rho^{-1}\left(\overline{\delta}\right)$. Moreover, $\mathbb{S}' = (P, R', O')$ with $R' = R \cap M'$ is also a reduced structural model of PNS whose maximal structure is (M', O'); and (m_i, o_i), $i = 1, 2, \ldots, n$, are solution-structures of $\mathbb{S}'$.

Functions $\mu_i : m_i \to o_i$ will be defined for $i = 1, 2, \ldots, n$ similar to function μ. To do so, let $X \in m_i \backslash R'$ be an arbitrary material, and also let the sequence, $o'_0, o'_1, \ldots$, be formed as

$$o'_0 = \delta_i(X) \text{ and } o'_{j+1} = o'_j \cup \left(\cup\left\{\delta_i(Y) : Y \in mat^{in}\left(o'_j\right)\right\}\right).$$

Inclusions $o_{0'} \subseteq o_{1'} \subseteq \ldots$ follow from the definition of this sequence. Naturally, if $o'_l = o'_{l+1}$ for some nonnegative integer l, then $o'_l = o'_j$ for all $j \geq l$. Since $o'_j \subseteq o'_i$, $j = 0, 1, 2, \ldots$, and o_i is a finite set, there exists a nonnegative integer $l < |o_i|$ with $o'_l = o'_{l+1}$. Let l denote the smallest integer with $o'_l = o'_{l+1}$.

At this point, let us define function μ_i as

$$\mu_i(X) = \begin{cases} o'_l, & X \in m_i \backslash R', \\ \varnothing, & otherwise. \end{cases}$$

The significance of function μ_i is as follows: Set $\mu_i(X)$ contains those operating units participating directly or indirectly in the production of X in the process determined by decision-mapping δ_i.

To prove the validity of the procedure, first we ascertain that, for any $1 \leq i \leq n$ and $X \in m_i$,

$$(4) \quad c^{(r)}(X) \leq 0 + \sum_{u \in \mu_i(X) \backslash op(\delta[m])} w(u).$$

The second term in the right-hand side of the above relation varnishes if set $\mu_i(X) \backslash op(\delta[m])$ is empty. This situation is plausible, *e.g.*, when $X \in m_i \cap R'$. Inequality (4) will be proved by induction on r. First, let $r = 0$; then, according to the procedure,

$$I^{(0)} = mat^{out}(op(\delta[m])) \cup R', \quad J^{(0)} = M' \backslash I^{(0)};$$

furthermore,

$$c^{(0)}(Y) = \begin{cases} 0, & Y \in I^{(0)}, \\ \mathrm{m}\{w(u) : u \in \Delta(Y)\}, & Y \in J^{(0)} \end{cases}$$

for any $Y \in M'$. If $X \in I^{(0)}$, then (4) holds obviously. Suppose that $X \in J^{(0)}$; then, $X \notin R'$, and thus, $X \in m_i \backslash R'$, and axiom (S2) implies that $\delta_i(X) \neq \varnothing$. If $\overline{u} = (\alpha, \beta) \in \delta_i(X)$ is an arbitrary operating unit, then $X \in \beta$ and $\overline{u} \in \mu_i(X)$. Furthermore, $\overline{u}$ may not be an element of set $op(\delta[m])$ because if $\overline{u}$ were an element of $op(\delta[m])$, then $X \in mat^{out}(op(\delta[m]))$ which is a contradiction; therefore, $\overline{u} \in \mu_i(X) \backslash op(\delta[m])$. Finally, $\overline{u} \in \Delta(X)$ is obtained because $(m_i, o_i) \subseteq (M', O')$. Hence,

$$c^{(0)}(X) = \mathrm{mim}\{w(u) : u \in \Delta\{X\}\} \leq w(\overline{u}) \leq \sum_{u \in \mu_i(X) \backslash op(\delta[m])} w(u);$$

this validates relation (4) for $r = 0$.

Suppose that r is an arbitrary nonnegative integer, and relation (4) is valid for r. Moreover, let us assume that the procedure is yet to be terminated, *i.e.*, $J^{(r)} \neq \emptyset$, and also let $X \in J^{(r)}$ be the selected material; then, $I^{(r+1)} = I^{(r)} \cup \{X\}$ and $J^{(r+1)} = J^{(r)} \backslash \{X\}$. From the definition of $c^{(r+1)}$, we have $c^{(r+1)}(Y) = c^{(r)}(Y)$ for all $Y \in M' \backslash \{X\}$; consequently, by the induction hypothesis, we obtain

$$c^{(r+1)}(Y) = c^{(r)}(Y) \leq 0 + \sum_{u \in \mu_i(Y) \backslash op(\delta[m])} w(u)$$

for all $Y \in m_i \backslash \{X\}$. It is, therefore, sufficient to prove the validity of relation (4) for the selected material, X; the following cases and subcases are identified.

Case 1. Suppose that $L^{(r)} = \emptyset$; thus, X is an arbitrary element of $J^{(r)}$. According to the procedure, $c^{(r+1)}(X) = c^{(r)}(X)$; then, by the induction hypothesis

$$c^{(r+1)}(X) = c^{(r)}(X) \leq 0 + \sum_{u \in \mu_i(X) \backslash op(\delta[m])} w(u).$$

Case 2. Suppose that $L^{(r)} \neq \emptyset$. In the light of the definition, $Y \in L^{(0)}$ if and only if $mat^{in}(\Delta(Y)) \cap I^{(0)} \neq \emptyset$, for all $Y \in J^{(0)}$; moreover, $L^{(r+1)} = (L^{(r)} \cup mat^{out}(O_X)) \cap J^{(r+1)}$. Hence, it can be proved by induction on r that $Y \in L^{(r)}$ if and only if $mat^{in}(\Delta(Y)) \cap I^{(r)} \neq \emptyset$, for all $Y \in J^{(r)}$. Since set $L^{(r)}$ is not empty and X is one of its elements, X is examined in Steps 1 through 3. We have $X \notin R'$ because $R' \subseteq I^{(0)} \subset I^{(1)} \subset \ldots \subset I^{(r)}$ and $X \in J^{(r)} = M' \backslash I^{(r)}$. Then, $\Delta(X) \neq \emptyset$ is obtained from $X \in m_i \subseteq M'$ and axiom (S2). Let $\Delta(X) = \{u_1, u_2, \ldots, u_s\}$ for some positive integer s. From the property of $L^{(r)}$ mentioned above, $c_t > 0$ is valid for at least one value of t for $1 \leq t \leq s$ where c_1, c_2, …, c_s denote the values determined for operating units u_1, u_2, …, u_s, respectively; as a result, a solution exists for min $\{c_t : c_t > 0 \;\&\; 1 \leq t \leq s\}$. Let t_0 be the smallest integer satisfying $c_{t_0} = \min\{c_t : c_t > 0 \;\&\; 0 \leq t \leq s\}$; it is uniquely determined. The following two subcases will be distinguished.

Subcase 2.1. Suppose that $q_{t_0} = 1$ and $c_{t_0} \leq |c_t|$, $t = 1, 2, \ldots, s$, are valid for material X. Then, from the definition of c_t and $c_{t_0} \leq |c_t|$, for all t, $t = 1, 2, \ldots, s$, we have $c_{t_0} \leq w(u_t)$ if $Q_t = \emptyset$, and

$$c_{t_0} \leq w(u_t) + max\left\{c^{(r)}(V) : V \in Q_t\right\}$$

if $Q_t \neq \emptyset$.

Since $X \in J^{(r)} \cap m_i$, we have $X \notin R'$, and consequently, $\delta_i(X) \neq \emptyset$ according to axiom (S2). Since $\delta_i(X) \subseteq \Delta(X) = \{u_1, u_2, \ldots, u_s\}$, it can be assumed that $(\alpha, \beta) = u_j \in \delta_i(X)$ for some $1 \leq j \leq s$; then, $X \in \beta$, thereby yielding $u_j \in \mu_i(X)$. Moreover, $u_j \notin op(\delta[m])$ because the opposite case implies that $X \in mat^{out}(op(\delta[m])) \subseteq I^{(r)}$ which is a contradiction; therefore, $u_j \in \mu_i(X) \backslash op(\delta[m])$. Two situations are considered below in the light of Q_j.

First, suppose that $Q_j = \varnothing$; then, $c_{t_0} \leq w(u_j)$ and

$$c^{(r+1)}(X) = c_{t_0} \leq w(u_j) \leq \sum_{u \in \mu_i(X) \setminus op(\delta[m])} w(u)$$

thereby yielding relation (4).

Second, assume that $Q_j + \varnothing$; then, $c_{t_0} \leq w(u_j) + max\{c^{(r)}(V) : V \in Q_j\}$. Let V_j denote an element of Q_j, for which $c^{(r)}(V_j) = max\{c^{(r)}(V) : V \in Q_j\}$; then, $c_{t_0} \leq w(u_j) + c^{(r)}(V_j)$. By the induction hypothesis,

$$c^{(r)}(V_j) \leq 0 + \sum_{u \in \mu_i(V_j) \setminus op(\delta[m])} w(u),$$

and thus,

$$c_{t_0} \leq w(u_j) + \sum_{u \in \mu_i(V_j) \setminus op(\delta[m])} w(u).$$

The definition of Q_j implies that $u_j \notin \mu(V_j)$. Since $\mu_i(V_j) \subseteq \mu(V_j)$, we have $u_j \notin \mu_i(V_j)$. Furthermore, $\mu_i(V_j) \subseteq \mu_i(X)$ and $u_j \in \mu_i(X)$; thus, $\mu_i(V_j) \cup \{u_j\} \subseteq \mu_i(X)$. i$(V_j) \cup \{u_j\} \subseteq \mu_i(X)$. Accordingly,

$$c^{(r+1)}(X) = c_{t_0} \leq w(u_j) + \sum_{u \in \mu_i(V_j) \setminus op(\delta[m])} w(u) \leq \sum_{u \in \mu_i(X) \setminus op(\delta[m])} w(u),$$

thus satisfying relation (4).

Subcase 2.2. Suppose that one of the conditions $q_{t_0} = 1$ and $c_{t_0} \leq |c_t|$ $t = 1, 2, \ldots, s$, is not satisfied for any $Y \in L^{(r)}$. Then, set L' is given by $L' = \{V : V \in L^{(r)} \,\&\, j(V) = 1\}$. The following two situations may arise depending on L'.

(a) $L' \neq \varnothing$; here, X is selected from L' with $l(X) = \max\{l(V) : V \in L'\}$; furthermore, $c^{(r+1)}(X) = c_{t_0}$, where c_{t_0} is the value assigned to X in Steps 1 through 3. Since $X \in L'$, we have $c_t > 0$, $t = 1, 2, \ldots, s$. Moreover, we have $c_{t_0} \leq c_t$, $t = 1, 2, \ldots, s$ because $c_{t_0} = \min\{c_t : c_t > 0, 1 \leq t \leq s\}$. Now, the validity of relation (4) can be proved similar to subcase 2.1.

(b) $L' = \varnothing$; here, X is selected from $L^{(r)}$ with $c^{(r)}(X) = max\{c^{(r)}(V) : V \in L^{(r)}\}$. Since $L' = \varnothing$, there exists at least one integer for t, $1 \leq t \leq s$, with $c_t < 0$. Thus, both $\overline{c} = min\{-c_t : c_t \leq 0 \,\&\, 1 \leq t \leq s\}$ and $c^{(r+1)}(X) = \min\{c_{t_0}, \overline{c}\}$ exist. Now, the validity of relation (4) can be proved similar to subcase 2.1.

To complete the proof of the validity, it will be proved that $g(\delta[m]) + h^* \leq w(\delta_i)$ for all $1 \leq i \leq n$, where h^* is the value determined in Step 7 of the procedure. For this purpose, suppose that the procedure is terminated for some nonnegative integer r. Let $1 \leq i \leq n$ be an arbitrarily fixed integer and also let

$$V \in F = \left(mat^{in}(op(\delta[m])) \cup P\right) \setminus (mat^{out}(op(\delta[m])) \cup R')$$

be an arbitrary material; then,

$$c^{(r)}(V) \leq 0 + \sum_{u \in \mu_i(V) \setminus op(\delta[m])} w(u).$$

The definition of g leads to $g(\delta[m]) = \sum_{u \in op(\delta[m])} w(u)$; therefore,

$$g(\delta[m]) + c^{(r)}(V) \leq \sum_{u \in op(\delta[m])} w(u) + \sum_{u \in \mu_i(V) \setminus op(\delta[m])} w(u).$$

Since δ_i is a feasible solution extension of $\delta[m]$, inclusion $op(\delta[m]) \subseteq op(\delta_i)$ results. Moreover, we have inclusion $\mu_i(V) \setminus op(\delta[m]) \subseteq op(\delta_i)$ because $V \in mat^{in}(op(\delta[m])) \cup P$. Thus, both sets $\mu_i(V) \setminus op(\delta[m])$ and $op(\delta[m])$ are included in $op(\delta_i)$; naturally, they are disjoint. As a result,

$$\sum_{u \in op(\delta[m])} w(u) + \sum_{u \in \mu_i(V) \setminus op(\delta[m])} w(u) \leq \sum_{u \in op(\delta_i)} w(u) = w(\delta_i).$$

and hence,

$$g(\delta[m]) + c^{(r)}(V) \leq w(\delta_i)$$

is valid for all $V \in F$; this implies that

$$g(\delta[m]) + \max\{c^{(r)}(V) : V \in F\} \leq w(\delta_i).$$

Since $1 \leq i \leq n$ is an arbitrarily fixed integer, we have

$$g(\delta[m]) + h^* \leq w(\delta_i)$$

for all $i = 1, 2, \ldots, n$; this completes the proof for the validity of the procedure.

Now, let us equip the Accerelated Branch-and-Bound Algorithm with the refined bounding function which is denoted by g^*. This modification results in some changes in the original procedure.

First, we shall show that Step 3 of the Accerelated Branch-and-Bound Algorithm becomes unnecessary, and simultaneously, we need to change the checking part of the bound calculating step. Indeed, if $\delta[m] \in \Omega_{reg} \setminus \Omega_{max}$ and

$$F = \left(mat^{in}(op(\delta[m])) \cup P\right) \setminus (mat^{out}(op(\delta[m])) \cup R) = \emptyset,$$

then, by Theorem 14.1, $\delta^* \in \omega(\delta[m]) \cap S'$, that is, δ^* is a feasible solution extension of $\delta[m]$. Furthermore, by Corollary 14.1, we have that $w(\delta^*) = \min\{w(\delta') : \delta' \in \omega(\delta[m]) \cap S'\}$. Therefore, if the considered $\delta[m]$ is a regular extension of an earlier decision-mapping in the branch-and-bound procedure, *i.e.*, $\omega(\delta[m]) \in \Phi$, then we need not compute its bound; moreover, taking into consideration $w(\delta^*)$ and δ^* at the redefinition of z^* and s, respectively, $\omega(\delta[m])$ does not appear as a live leaf of the actual branch-and-bound tree of the next iteration step, *i.e.*, $\omega(\delta[m])$ is not contained in the redefined L. By this modification, the corresponding set F is not empty for each live leaf of the actual branch-and-bound tree. (In particular, $F = P \backslash R \neq \emptyset$ for the root of the tree.) On the other hand, let us observe that for the set F belonging to $\delta[m]$,

$$F \subseteq \left(mat^{in}(op(\delta[m])) \cup P\right) \backslash (R \cup m)$$

is valid. Thus, $F \neq \emptyset$ implies that the right side of the above inclusion is not empty. But then, the condition of Step 3 of the Accerelated Branch-and-Bound Algorithm does not hold for any live leaf, and thus, this step is unnecassary.

To establish the taking into account of $w(\delta^*)$ and δ^* at the redefinition of z^* and s, respectively, we have also to change Step 5 of the Accerelated Branch-and-Bound Algorithm. Under $\Phi = \{\omega(\delta_i[m_i]) : 1 \leq i \leq k\}$ we investigate set F_i belonging to $\delta_i[m_i]$ for all $1 \leq i \leq k$. If $F_i = \emptyset$, then we determine the corresponding feasible solution extension, δ_i^*, of $\delta[m]$, furthermore, if $w\left(\delta_i^*\right) < z^*$, then we redefine z^* and s by $z^* := w\left(\delta_i^*\right)$ and $s := \left\{\delta_i^*\right\}$. In this case, by the fathoming step, $\omega(\delta_i[m_i])$ will not appear as a live leaf of the actual branch-and-bound tree of the next iteration step. Thus, we need compute the bound for $\omega(\delta_i[m_i])$ when $F_i \neq \emptyset$. On the other hand, if $F_i \neq \emptyset$, then it is obvious that the rank of $\delta_i[m_i]$ is less than $|M \backslash R|$. Therefore, the investigation of the rank of $\delta_i[m_i]$ can be omitted in Step 5 of the Accerelated Branch-and-Bound Algorithm.

As a consequence of the above modifications, we can change the material selection rule in the branching step. If $\omega(\delta[m])$ denotes the selected node, then we have to choose a material according to the input material selection rule, *i.e.*, from set

$$\left(mat^{in}(op(\delta[m])) \cup P\right) \backslash (R \cup m).$$

But set F belonging to $\delta[m]$ is a subset of the above set. Thus, choosing the material with the least index from F, we keep the input material selection rule.

By modifications given above, we obtain the following new procedure for PNS_w.

Accelerated Branch-and-Bound Algorithm with Refined Bounding Function (ABB-RB) [1]

Initialization

- Let $L := \{\omega(\delta[\varnothing])\}$, $z^* := \infty$, $s := \varnothing$ and set $r := 0$. Let $g^*(\delta[\varnothing]) = 0$.

Iteration (r-th iteration)

- *Step 1.* If $L = \varnothing$, then terminate; s contains the optimal solution; z^* provides the optimal value. Otherwise, proceed to Step 2 if $r > 0$ and to Step 3 if $r = 0$.
- *Step 2. (Selection)* Choose an element $\omega(\delta[m])$ from L for which the ratio, $g^*(\delta[m])/|m|$ is minimal. (If there are more candidates with the same minimal value, then choose one of them in a random way.) Let $\omega(\delta[m])$ denote it and proceed to Step 3.
- *Step 3. (Branching)* Choose the material with the least index from set $(mat^{in}(op(\delta[m])) \cup P)\backslash(mat^{out}(op(\delta[m])) \cup R)$, and let X denote it. Form the regular extensions of $\delta[m]$ with respect to X.
  If there is no regular extension of $\delta[m]$ with respect to X, then let $L := L\backslash\{\omega(\delta[m])\}$ and proced to Step 1.
  If there is a regular extension of $\delta[m]$ with respect to X, then let $\delta_i[m_i]$, $i = 1, \ldots, k$, denote all the regular extensions of $\delta[m]$, furthermore, let $\Phi = \{\omega(\delta_i[m_i]) : 1 \leq i \leq k\}$, and proceed to Step 4.
- *Step 4. (Bounding)* For each $1 \leq i \leq k$, perform the following checking.
  Form set $F_i = (mat^{in}(op(\delta_i[m_i])) \cup P)\backslash(mat^{out}(op(\delta_i[m_i])) \cup R)$. If $F_i = \varnothing$, then construct the maximal decision-mapping, δ_i^*, by $\delta_i^*(X) = \{(\alpha, \beta) : (\alpha, \beta) \in op(\delta_i[m_i])\&X \in \beta\}$ for all $X \in M\backslash R$. Furthermore, let $g^*(\delta_i[m_i]) = w(\delta_i^*)$, and if $w(\delta_i^*) < z^*$, then redefine the values of z^* and s by $z^* := w(\delta_i^*)$ and $s := \{\delta_i^*\}$.
  Let $H = \{i : 1 \leq i \leq k \;\&\; F_i \neq \varnothing\}$. If $H = \varnothing$, then proceed to Step 5. In the opposite case, for each $i \in H$, investigate the decision-mapping $\delta_i[m_i]$ in the following way.
  Apply the Greatest Feasible Solution Extension Building Procedure for $\delta_i[m_i]$. If there is no feasible solution extension of $\delta_i[m_i]$, then let $g^*(\delta_i[m_i]) = \infty$. Otherwise, apply the Refined Bounding Procedure for the P-graph of the determined greatest feasible solution extension of $\delta_i[m_i]$ to calculate the bound, $g^*(\delta_i[m_i])$
  After the finishing of these investigations, proceed to Step 5.
- *Step 5. (Fathoming)* Redefine set L in the following way:

$$L := \{\omega(\delta'[m']) : \omega(\delta'[m']) \in (L\backslash\{\omega(\delta[m])\}) \cup \Phi, g^*(\delta'[m']) < z^*\}.$$

Set $r := r + 1$, and proceed to the succeeding iteration.

To illustrate our procedure, let us consider Example 13.2 again. In this case, the bounds provided by the applied different bounding functions can be compared. The maximal structure belonging to this reduced structural model is given in Fig. 13.3.

The procedure is started with $L = \{\omega(\delta[\emptyset])\} = \Omega_{max}$, $s = \emptyset$, $z^* = \infty$, $g^*(\delta[\emptyset]) = 0$. In Step 3, material A_1 is selected. $\Delta(A_1) = \{u_8, u_{11}\}$, and we obtain three regular extensions of $\delta[\emptyset]$. Let us denote them by $\delta_i[\{A_1\}]$, $i = 1, 2, 3$, where $\delta_1(A_1) = \{u_8\}$, $\delta_2(A_1) = \{u_{11}\}$ and $\delta_3(A_1) = \{u_8, u_{11}\}$. For each i, $i = 1, 2, 3$, $F_i \neq \emptyset$, and thus, $H = \{1, 2, 3\}$.

Applying the Greatest Feasible Solution Extension Building Procedure for $\delta_1[\{A_1\}]$, we obtain the greatest feasible solution extension of $\delta_1[\{A_1\}]$. Its P-graph with the corresponding weights is given in Fig. 14.5.

Performing the Refined Bounding Procedure, we get that $c^{(1)}(A_5) = 5$, $c^{(2)}(A_7) = 3$, $c^{(3)}(A_8) = 6$, and $c^{(4)}(A_9) = 5$. By $F_1 = \{A_8, A_9\}$, we obtain the lower bound, $g^*(\delta_1[\{A_1\}]) = 8$.

For $\delta_2[\{A_1\}]$, we obtain the greatest feasible solution extension which consists of the operating units u_1, u_2, ..., u_7, u_9, u_{10}, ..., u_{12}. The obtained material costs are: $c^{(1)}(A_5) = 5$, $c^{(2)}(A_6) = 5$, $c^{(3)}(A_4) = 7$, $c^{(4)}(A_7) = 3$, $c^{(5)}(A_8) = 6$, $c^{(6)}(A_9) = 5$, $c^{(7)}(A_3) = 9$, $c^{(8)}(A_{10}) = 12$, $c^{(9)}(A_2) = 15$, $c^{(10)}(A_{11}) = 7$, and $c^{(11)}(A_{12}) = 9$. Now, $F_2 = \{A_2, A_3, A_{12}\}$, and thus, $g^*(\delta_2[\{A_1\}]) = 22$.

Regarding $\delta_3[\{A_1\}]$, we obtain that the P-graph of the greatest feasible solution extension of $\delta_3[\{A_1\}]$ is equal to the maximal structure, (M, O). Now, $F_3 = \{A_3, A_8, A_9, A_{12}\}$, $c^{(3)}(A_8) = 6$, $c^{(5)}(A_9) = 5$, $c^{(6)}(A_3) = 9$, $c^{(9)}(A_{12}) = 9$, and thus, $g^*(\delta_3[\{A_1\}]) = 18$.

The branch-and-bound tree after the first iteration is shown in Fig. 14.6 where Ω_i denotes the set, $\omega(\delta_i[\{A_1\}])$, $i = 1, 2, 3$.

In the next iteration step, $L = \{\omega(\delta_i[\{A_1\}]) : 1 \leq i \leq 3\}$, and the smallest ratio belongs to $\omega(\delta_1[\{A_1\}])$. In Step 3, material A_8 is chosen. Since $\Delta(A_8) = \{u_4\}$, we obtain only one regular extension of $\delta_1[\{A_1\}]$. Let us denote it by $\delta_4[\{A_1, A_8\}]$ where $\delta_4(A_1) = \{u_8\}$ and $\delta_4(A_8) = \{u_4\}$. Building the greatest feasible solution extension of $\delta_4[\{A_1, A_8\}]$, we obtain the same feasible solution as in the case of $\delta_1[\{A_1\}]$. Executing the refined bounding procedure, we get that $F_4 = \{A_5, A_9\}$, $c^{(1)}(A_5) = 5$, $c^{(3)}(A_9) = 5$, and thus, $g^*(\delta_4[\{A_1, A_8\}]) = 8$. The corresponding branch-and-bound tree is given in Fig. 14.7 where Ω_4 denotes $\omega(\delta_4[\{A_1, A_8\}])$.

Now, $L = \{\omega(\delta_2[\{A_1\}]), \omega(\delta_3[\{A_1\}]), \omega(\delta_4[\{A_1, A_8\}])\}$, and the smallest bound belongs to $\omega(\delta_4[\{A_1, A_8\}])$. Furthermore, $F_4 = \{A_5, A_9\}$, and thus, material A_5 is selected. $\Delta(A_5) = \{u_1, u_3\}$, and in this case, we obtain three regular extensions of $\delta_4[\{A_1, A_8\}]$ which are denoted by $\delta_i\{A_1, A_5, A_8\}]$, $i = 5, 6, 7$, where $\delta_5(A_5) = \{u_1\}$, $\delta_6(A_5) = \{u_3\}$, and $\delta_7(A_5) = \{u_1, u_3\}$. Then, $F_5 = \emptyset$, and determining δ_5^*, $op(\delta_5^*) = \{u_1, u_4, u_8\}$ and $w(\delta_5^*) = 8$. Thus, $g^*(\delta_5[\{A_1, A_5, A_8\}]) = 8$. Since $w(\delta_5^*) < z^*$, we have to redefine the values of z^* and s. The new values: $z^* = 8$ and $s = \{\delta_5^*\}$. For $\delta_6[\{A_1, A_5, A_8\}]$, $F_6 = \emptyset$, and constructing δ_6^*, $op(\delta_6^*) = \{u_3, u_4, u_8\}$ and $w(\delta_6^*) = 10$. Thus, $g^*(\delta_6\{A_1, A_5, A_8\}]) = 10$. Since $w(\delta_6^*) > z^*$, we can continue the procedure with the investigation of $\delta_7[\{A_1, A_5, A_8\}]$. We get that $F_7 = \emptyset$, and taking δ_7^*, $op(\delta_7^*) = \{u_1, u_3, u_4, u_8\}$ and $w(\delta_7^*) = 15$. Then, $g^*(\delta_7[\{A_1,$

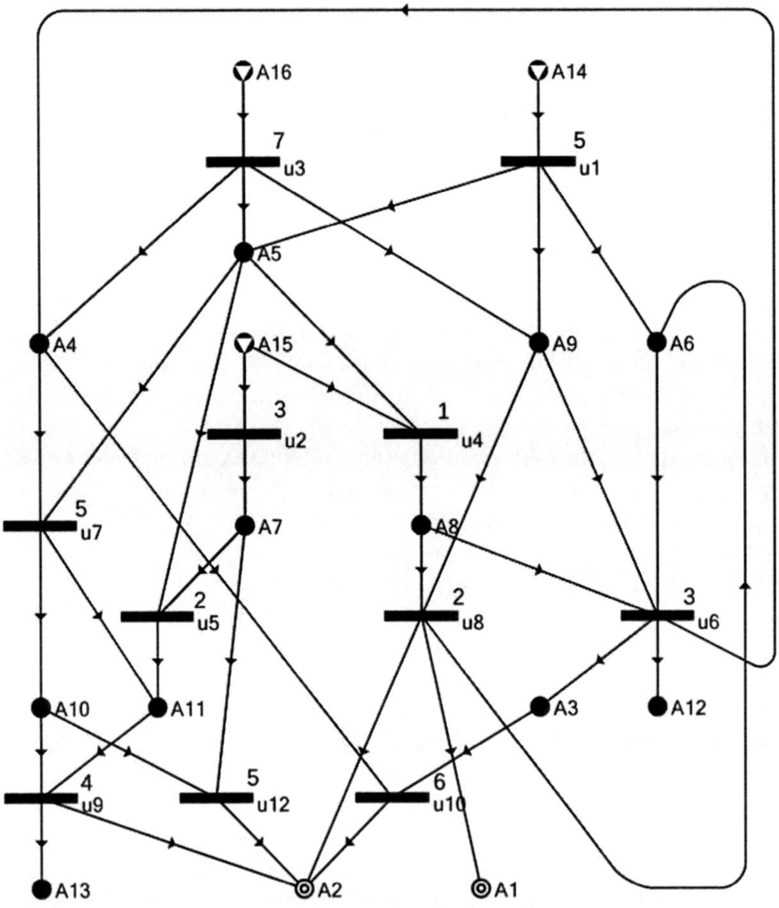

Fig. 14.5 The P-graph of the greatest feasible solution extension of $\delta_1[\{A_1\}]$

Fig. 14.6 Branch-and-bound tree after the first iteration, bounds are indicated at the nodes

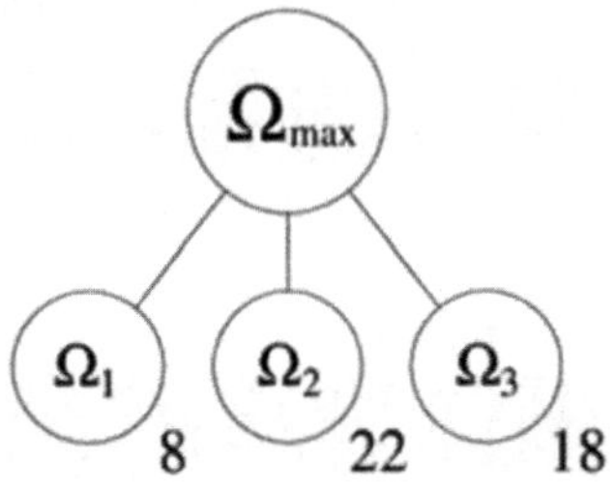

$A_5, A_8\}]) = 15$. Since $w(\delta_7^*) > z^*$, the values of z^* and s do not change. Constructing the set H, we obtain that $H = \varnothing$. Since z^* is not greater than the corresponding bounds, the redefined L is empty in Step 5, and thus, the procedure is terminated. The optimal solution is δ_5^*, and the optimal value is 8.

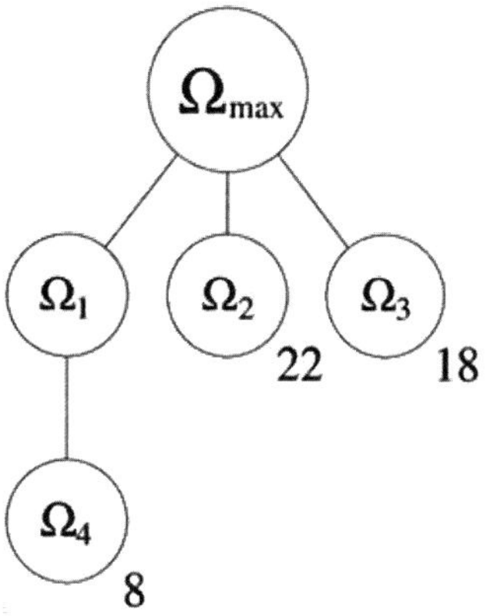

Fig. 14.7 Branch-and-bound tree after the second step, bounds are indicated at the nodes

References

1. Imreh B, Friedler F, Fan LT (1996) An algorithm for improving the bounding procedure in solving process network synthesis by a branch-and-bound method. In: Bonze IM, Csendes T, Horst R, Pardalos PM (eds) Developments in global optimization. Kluwer Academic, Dordrecht, pp 301–348

Chapter 15
A Look Ahead B&B Procedure for PNS_w

Abstract In this chapter, a new algorithm, Look Ahead Branch-and-Bound Algorithm (LABBA), is presented to further enhance the efficiency of the general purpose ABB algorithms for a specific class of PNS problems. The new algorithm improves both the branching, and the bounding aspects of algorithm ABB. Different branching decisions can lead to smaller branching trees, and different branching orders can result in sharper bounding.

Keywords P-graph · Branch-and-bound · Algorithm ABB · Maximal structure · Model transformation

The current chapter utilizes the merging reduction algorithm presented in Chap. 11 to improve algorithm ABB of process network synthesis (PNS) for specific PNS_w problems. This work was originally initiated by Holló [1]. Similar principles can also be applied to general PNS problems with appropriate modifications. The merging idea was based on the observation that the operating units of a PNS problem can be partitioned into equivalence classes in such a way that for each solution-structure, the equivalent operating units simultaneously either appear or do not appear in the solution-structure considered.

When the merging equivalence relation is performed, it can occur that if operating unit u_i is omitted, the maximal structure does not exist for the remaining problem $\mathbb{S}(u_i) = (P, R, O\backslash\{u_i\})$. This means that u_i is contained in every solution-structure. Collecting all of the operating units in set o' which must be in every solution-structure provided that such operating unit there exists, we can start the branch-and-bound (B&B) procedure with such a decision-mapping denoted by $\delta_0[m_0]$ which is defined as follows. Let $m_0 = mat^{out}(o')$ and for every $X \in m_0$, let

$$\delta_0(X) = \{(\alpha, \beta) : (\alpha, \beta) \in o' \& X \in \beta\}.$$

If there is no operating unit in O with this property, then we define $\delta_0[m_0]$ by $m_0 = \varnothing$. Obviously, $\delta_0[m_0]$ is a consistent decision-mapping.

F. Friedler et al., *P-graphs for Process Systems Engineering*,
https://doi.org/10.1007/978-3-030-92216-0_15

Also, the merging equivalence makes possible to fix operating units, as they are equivalent to other fixed units, simultaneously in any iteration step of the procedure. This can affect equivalent operating units, even if they are far from each other in the network, and fix them simultaneously. In the Look Ahead Branch-and-Bound (LABBA) procedure [1], if we fix some operating units in an iteration step, then we fix all of the equivalent operating units to the considered ones simultaneously. Consequently, this new procedure can fix more operating units in an iteration step than the earlier procedures, resulting in the decrease of the size of the B&B tree. Moreover, the bounding functions of the earlier procedures will be improved since we increase the earlier bound with the sum of the weights of the equivalent operating units.

The Look Ahead B&B Algorithm studies the effects of the different possible branching strategies than the one applied in the ABB and ABB-RB (Accerelated Branch-and-Bound Algorithm with Refined Bounding Function of Chap. 14) algorithms. By examining the effects of the different material selections, the size of the B&B tree can be also decreased.

We recall that when the extension of the domain m of a given decision-mapping $\delta[m]$ with one material denoted by Y, then material Y can be selected in different ways. The rule for selecting Y, *i.e.* the material selection rule has a great effect on the size of the branch-and-bound tree, as it determines the actual branching.

The simplest material selection rule, used by the B&B procedure given at the beginning of Chap. 13, is when an element from set $M \backslash (m \cup R)$ is chosen. With that rule, it can occur that Y is an output material for some fixed operating unit, and so the available operating units from $\Delta(Y)$ are uniquely determined. This observation led to another material selection rule, the input material selection rule, which belongs to the ABB procedure presented in Chap. 13. In this case, one can choose a material from set

$$\left(mat^{in}(op(\delta[m]) \cup P\right) \backslash (m \cup R).$$

Let us observe that under such selection strategy, it can occur that we choose such a Y which is output material for some operating units fixed by $\delta[m]$ because of the consistency. In this case in a PNS_w problem, material Y does not require additional operating units to be assigned to it. Other operating units producing Y directly will only be included in the subgraph, if they are assigned to other materials. Consequently, the step of branching on material Y is needless. This problem was solved precisely in the ABB-RB procedure presented in Chap. 14, where the material selection rule is changed so that a material Y is chosen from set

$$\left(mat^{in}(op(\delta[m])) \cup P\right) \backslash \left(mat^{out}(\delta[m]) \cup R\right).$$

To improve this selection, we prefer choosing a material for the current extensions in such a way that the operating units, involved in the process by the material selection, can produce a large number of materials needed currently. As a consequence, the B&B tree will be smaller, and the whole procedure will be more effective.

In the LABBA presented in this section, we are going to establish these three ideas.

First of all, we introduce some particular decision-mappings. Let us assign to every consistent decision-mapping $\delta[m] \in \Omega_M$ a set $O_{\delta[m]} \subseteq O$ of operating units, where $O_{\delta[m]} = op(\delta[m]) \cup (\cup\{C(u) : u \in op(\delta[m])\})$. Here, $C(u)$ denotes the equivalence class containing u according to the notion of Chap. 11. Visually, $O_{\delta[m]}$ contains the operating units fixed directly by $\delta[m]$ or fixed indirectly by the merging equivalence relation.

Now, let $\delta[m] \in \Omega_M$ be a decision-mapping with $|m| < |M \backslash R|$; moreover, let Y be an element of set $(mat^{in}(O_{\delta[m]}) \cup P) \backslash (mat^{out}(O_{\delta[m]}) \cup R)$ provided that this set is not empty. Let $K_1, K_2, \ldots, K_r$ denote the equivalence classes of $\Delta(Y)$ with respect to the restriction of relation $\equiv$ to $\Delta(Y)$. For every nonempty subset J of $\{K_1, K_2, \ldots, K_r\}$, let $K_J = \cup\{K_t : K_t \in J\}$. Then, the consistent decision-mappings of the form, $\delta_t[m \cup \{Y\}] = \delta[m] \cup \{(Y, K_J)\}\ J \subseteq \wp'(\{K_1, K_2, \ldots, K_r\})$, are called *irregular extensions of decision-mapping* $\delta[m]$ *with respect to* Y. Obviously, every irregular extension is an extension as well.

Let us consider the reflexive, transitive closure of the relation of irregular extension on Ω_M. Naturally, the resultant relation is a partial ordering, as it is reflexive, transitive, and antisymmetric. Let us denote this ordering relation by $\sqsubseteq$; furthermore, if $\delta[m] \sqsubseteq \delta'[m']$ and $\delta[m] \neq \delta'[m']$, it is denoted by $\delta[m] \sqsubset \delta'[m']$. Now, set Σ_M is defined by

$$\Sigma_M = \{\delta[m] : \delta[m] \in \Omega_M \ \ \& \ \ \delta_0[m_0] \sqsubseteq \delta[m]\}$$

where $\delta_0[m_0]$ is the decision-mapping defined by the operating units which must be in every solution-structure. The elements of set Σ_M are called *irregular decision-mappings*. In other words, a consistent decision-mapping is irregular if and only if it is equal to $\delta_0[m_0]$ or it can be obtained with a sequence of irregular extension beginning from $\delta_0[m_0]$. The next observation follows immediately from the definitions.

Remark 15.1 *If* $\delta[m] \in \Sigma_M$, *then* $\delta(X) \neq \emptyset$, *for all* $X \in m$.

Let $\delta[m] \in \Sigma_M$ with $(mat^{in}(O_{\delta[m]}) \cup P) \backslash (mat^{out}(O_{\delta[m]}) \cup R) = \emptyset$; also, let decision-mapping δ' be defined as $\delta'(X) = \{(\alpha, \beta) : (\alpha, \beta) \in O_{\delta[m]} \ \& \ X \in \beta\}$, for all $X \in M \backslash R$. Then, δ' is called the *irregular closure* of $\delta[m]$, and it is denoted by $icl(\delta[m])$. The irregular closure is determined uniquely for each regular decision-mapping and it is a maximal consistent decision-mapping. Moreover, if $m = M \backslash R$, then $icl(\delta[m]) = \delta[m]$.

Table 15.1 Operating units of Example 15.1

Unit	Inputs	Outputs	Weight
u_1	A_3, A_4, A_8	A_1, A_2	7
u_2	A_5	A_1, A_2	3
u_3	A_6	A_3	4
u_4	A_6, A_7, A_8	A_4	2
u_5	A_8	A_5	2
u_6	A_9	A_6	3
u_7	A_{10}	A_6, A_7	5
u_8	A_{11}	A_8	2
u_9	A_{12}	A_9, A_{10}	4
u_{10}	A_{13}	A_{10}, A_{11}	3

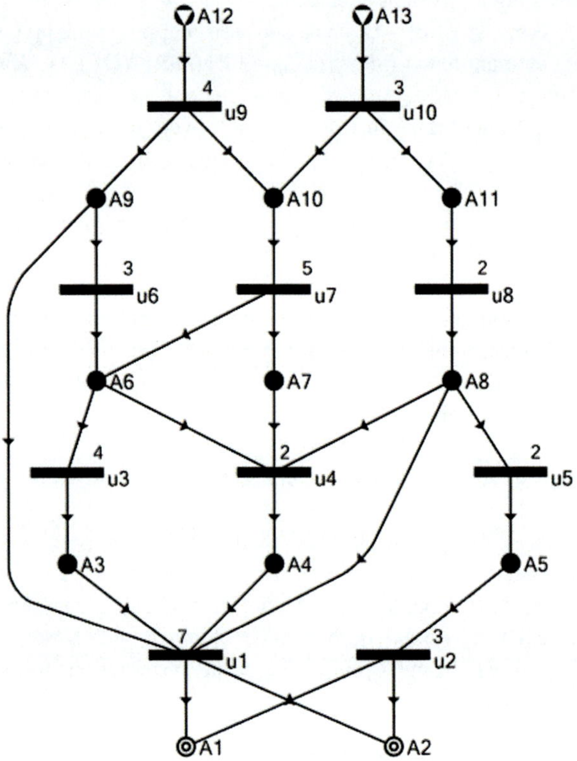

Fig. 15.1 P-graph (M, O) of Example 15.1

The following example is considered for illustration of irregular decision-mappings.

Example 15.1 Let structural model $\mathbb{S} = (P, R, O)$ be given, where $P = \{A_1, A_2\}$; $R = \{A_{12}, A_{13}\}$; $O = \{u_1, u_2, \ldots, u_{10}\}$. The inputs and outputs of the operating units and their weight are listed in Table 15.1 together with their weights. The corresponding P-graph is shown in Fig. 15.1.

Sets O_i, $i = 1, 2, \ldots, 10$ are determined as:

$$O_1 = O_3 = O_4 = O_7 = O_9 = \{u_2, u_5, u_8, u_{10}\},$$

$$O_2 = O_5 = \{u_1, u_3, u_4, u_6, u_7, \ldots u_{10}\}, \quad O_6 = O \backslash \{u_6\}, \quad O_8 = O_{10} = \varnothing.$$

On the basis of sets $O_1, O_2, \ldots, O_{10}$, the equivalence classes for the problem are $V_1 = \{u_6\}$, $V_2 = \{u_1, u_3, u_4, u_7, u_9\}$, $V_3 = \{u_2, u_5\}$, and $V_4 = \{u_8, u_{10}\}$. Since $O_8 = O_{10} = \varnothing$, operating units u_8 and u_{10} must be contained in every solution-structure of $\mathbb{S}$, and then, $m_0 = mat^{out}(\{u_8, u_{10}\}) = \{A_8, A_{10}, A_{11}\}$, furthermore, $\delta_0(A_{10}) = \delta_0(A_{11}) = \{u_{10}\}$ and $\delta_0(A_8) = \{u_8\}$.

As far as the irregular decision-mappings are concerned, the following subset of Σ_M will be important for our procedure. Let

$$S^*(\mathbb{S}) = \left\{icl(\delta[m]) : \delta[m] \in \Sigma_M \& \left(mat^{in}\left(O_{\delta[m]}\right) \cup P\right) \backslash \left(mat^{out}\left(O_{\delta[m]}\right) \cup R\right) = \varnothing\right\}.$$

For set $S^*(\mathbb{S})$, we have the following statements.

Theorem 15.1 $S^*(\mathbb{S}) \subseteq S'(\mathbb{S})$.

Proof Let $\delta \in S^*(\mathbb{S})$ be arbitrary, furthermore, let $o = op(\delta)$ and $m = mat(o)$. Naturally, (m, o) is a P-graph and $(m, o) \subseteq (M, O)$. Now we show that $(m, o) \in S(\mathbb{S})$ and $\rho((m, o)) = \delta$.

Since $\delta \in S^*(\mathbb{S})$, there exists a sequence of irregular decision-mappings, $\delta_0[m_0] \sqsubset \delta_1[m_1] \sqsubset \ldots \sqsubset \delta_k[m_k]$, for which $icl(\delta_k[m_k]) = \delta$, $o = O_{\delta_k[m_k]}$, and $\left(mat^{in}\left(O_{\delta_k[m_k]}\right) \cup P\right) \backslash \left(mat^{out}\left(O_{\delta_k[m_k]}\right) \cup R\right) = \varnothing$. It follows from the last equality that $P \subseteq mat^{out}\left(O_{\delta_k[m_k]}\right) = mat^{out}(o)$.

Now, let $\overline{o} = \{u : u \in o \ \& \ \exists \ path \ [u, Y_n] \ with \ Y_n \in P\}$, and let $(\overline{m}, \overline{o})$ be a P-graph where $\overline{m} = mat(\overline{o})$. First, we prove that $(\overline{m}, \overline{o})$ is a solution-structure of $\mathbb{S}$.

Since $P \subseteq mat^{out}(o)$, by the definition of $\overline{o}$, $P \subseteq \overline{m}$ is also valid, therefore, $(\overline{m}, \overline{o})$ satisfies axiom (S1).

For proving axiom (S2), let $X \in \overline{m}$ be an arbitrary material. If $X \in R$, then there is no (Y, X) arc in $(\overline{m}, \overline{o})$ since $\overline{o} \subseteq O$, and O obviously does not contain any operating unit producing raw material X directly. If $X \notin R$, then $\overline{m} = mat(\overline{o})$ induces that there exists at least one operating unit in $\overline{o}$ producing or consuming X directly. Now, two cases are distinguished. If there exists an operating unit in $\overline{o}$ that produces X directly, then there exists a (Y, X) arc in $(\overline{m}, \overline{o})$, and consequently, $(\overline{m}, \overline{o})$ satisfies axiom (S2). In the other case, where X is not output material for any operating unit in $\overline{o}$, then it is input material for at least one operating unit from $\overline{o}$. Then, by the definition of $\overline{o}$, the same observation also has to be valid for o. This yields that $X \in mat^{in}(op(\delta)) = mat^{in}\left(O_{\delta_k[m_k]}\right)$ and $X \notin mat^{out}(op(\delta)) = mat^{out}\left(O_{\delta k[m_k]}\right)$ which contradicts our assumption. Therefore, this case is impossible, and thus, $(\overline{m}, \overline{o})$ satisfies axiom (S2).

Axiom (S4) is directly validated for $(\overline{m}, \overline{o})$ following from the definition of $\overline{o}$. Finally, since $\overline{m} = mat(\overline{o})$, $(\overline{m}, \overline{o})$ satisfies axiom (S5) as well, and therefore, $(\overline{m}, \overline{o})$ is a solution-structure of $\mathbb{S}$.

The next step is to prove that $o = \overline{o}$. By the definition of $\overline{o}$, we have that $\overline{o} \subseteq o$. On the other hand, we can prove that $O_{\delta_i[m_i]} \subseteq \overline{o}$ by induction on i. If $i = 0$ and $u \in O_{\delta_0[m_0]}$, then there is an operating unit $v \in o$ that is contained in every solution-structure of $\mathbb{S}$, such that $u \equiv v$. Consequently, u is also contained in every solution-structure of $\mathbb{S}$, and in particular, u is contained in solution-structure $(\overline{m}, \overline{o})$, and therefore, $u \in \overline{o}$. Now, for the induction step, let $0 \leq i < k$, and let us suppose that the statement is valid for $O_{\delta_i[m_i]}$. Let $u \in O_{\delta_{i+1}[m_{i+1}]}$. There are two possibilities. If $u \in O_{\delta_i[m_i]}$, then $u \in \overline{o}$ from the induction hypothesis. In the opposite case, there exists an operating unit $v \in o$ such that $u \equiv v$ and v produces at least one input material of some operating unit from $O_{\delta_i[m_i]}$. Then, by the definition of $\overline{o}$, $v \in \overline{o}$. Otherwise, since $u \equiv v$ and $(\overline{m}, \overline{o})$ is a solution-structure, $u \in \overline{o}$ is valid as well. Consequently, $o = \overline{o}$, and thus, following the definitions of m and $\overline{m}$, $(m, o) = (\overline{m}, \overline{o})$.

For the final step of the proof, let us observe that the definition of ρ implies the equality $\rho((m, o)) = \delta$.

Theorem 15.1 means that some of the feasible solutions of problem (13.4) can be obtained as the elements of set $S^*(\mathbb{S})$, however, we can not necessarily obtain every feasible solution of problem (13.4) as an element of set $S^*(\mathbb{S})$. However, a stronger statement can be declared regarding the optimal solutions. The following theorem shows that every optimal solution of problem (13.4) is contained in set $S^*(\mathbb{S})$.

Theorem 15.2 *If (m, o) is an optimal solution-structure of problem* (13.3)*, then $\rho((m, o))$ is contained in $S^*(\mathbb{S})$.*

Proof Let (m, o) be an optimal solution-structure of problem (13.3). Then, Chap. 13 showed that $\rho((m, o)) = \delta$ is an optimal solution of problem (13.4). Therefore, it is sufficient to prove that δ is an element of $S^*(\mathbb{S})$. Let us construct a sequence $\delta_t[m_t]$, $t = 0, 1, \ldots, k$ of irregular decision-mappings such that $\delta_t[m_t] \sqsubset \delta_{t+1}[m_{t+1}]$, $t = 0, 1, \ldots, k-1$, and $icl(\delta_k[m_k]) = \delta$. First, let us suppose that we have a sequence $\delta_t[m_t]$, $t = 0, 1, \ldots, i$, such that $\delta_t[m_t] \sqsubset \delta_{t+1}[m_{t+1}]$, $t = 0, 1, \ldots, i - 1$, and $\delta(X) = \delta_t(X)$ if $X \in m_t$, for all t, $t = 0, 1, \ldots, i$. Obviously, $\delta_0[m_0]$ is suitable for $i = 0$. Now, let us distinguish the following two cases:

Case 1. $\left(mat^{in}\left(O_{\delta_i[m_i]}\right) \cup P\right) \setminus \left(mat^{out}\left(O_{\delta_i[m_i]}\right) \cup R\right) \neq \varnothing$. Let Y be an arbitrarily fixed element of this set. Then, $\Delta(Y) \cap O_{\delta_i[m_i]} = \varnothing$. In the opposite case, there would exist an operating unit $u = (\alpha, \beta)$ in $O_{\delta_i[m_i]}$ such that $Y \in \beta$ which contradicts to the selection of Y. Let us denote $\delta(Y)$ by Q. We prove first that Q has to be a union of some of the equivalence classes of $\Delta(Y)$ with respect to the merging equivalence relation restricted to $\Delta(Y)$. Let $u, v \in \Delta(Y)$, $u \in Q$ and $u \equiv v$. Then, $u \in o$, and since (m, o) is a solution-structure and $u \equiv v$, $v \in o$ is also valid. Then, by the definition of δ, $v \in Q = \delta(Y)$. Consequently, Q is a union of some equivalence classes of $\Delta(Y)$. Now, let us consider the decision-mapping $\delta_{i+1}[m_{i+1}] = \delta_i[m_i] \cup \{(Y, Q)\}$. Obviously, $\delta(X) = \delta_{i+1}(X)$, for all $X \in m_{i+1}$, and hence, $\delta_{i+1}[m_{i+1}]$ is a consistent extension of $\delta_i[m_i]$. On the other hand, Q is a union of some equivalence classes of $\Delta(Y)$, and thus, $\delta_{i+1}[m_{i+1}]$ is an irregular extension of $\delta_i[m_i]$ which is equal to δ on set m_{i+1}. Therefore, a succeeding member of the sequence exists if $\left(mat^{in}\left(O_{\delta_i[m_i]}\right) \cup P\right) \setminus \left(mat^{out}\left(O_{\delta_i[m_i]}\right) \cup R\right) \neq \emptyset$

.

Case 2. $\left(mat^{in}\left(O_{\delta_i[m_i]}\right) \cup P\right) \setminus \left(mat^{out}\left(O_{\delta_i[m_i]}\right) \cup R\right) = \emptyset$. We prove that $icl(\delta_k[m_k]) = \delta$ where $k = i$. To do this, let $o' = op(icl(\delta_k[m_k])$. Then, $o' = O_{\delta_k[m_k]}$. On the other hand, for every operating unit $u \in O_{\delta_k[m_k]}$, there exists a material $X \in m_k$ and an operating unit v (which is not necessarily distinct from u), such that $v \in \delta_k(X)$ and $u \equiv v$. Since $\delta_k(Y) = \delta(Y)$, for all $Y \in m_k$, it follows that $v \in \delta(X)$, and thus, $v \in o$. Now, by $u \equiv v$, we obtain that $u \in o$. Consequently, $o' \subseteq o$.

To prove that $o' = o$, let us suppose that $o' \subset o$. Since $icl(\delta_k[m_k]) \in S^*(\mathbb{S})$, by Theorem 15.1 and the bijective mapping ρ, there exists $(m', o') \in S(\mathbb{S})$ with $\rho((m', o')) = icl(\delta_k[m_k])$. Then, since $o' \subset o$ and all of the weights in the PNS_w problem are positive, the relation

$$\sum_{u \in o'} w(u) = \sum_{u \in op(icl(\delta_k[m_k]))} w(u) < \sum_{u \in o} w(u) = \sum_{u \in op(\delta)} w(u)$$

is a contradiction since δ is an optimal solution of Eq. (13.4). Hence, $o = o'$, and thus, $icl(\delta_k[m_k]) = \delta$ which ends the proof of Theorem 15.2.

By Theorems 15.1 and 15.2 and the fact that ρ is a bijective mapping of $S(\mathbb{S})$ to $S'(\mathbb{S})$, it follows that it is possible to solve the problem (15.1) instead of problem (13.4):

$$\min\left\{\sum_{u \in op(\delta)} w(u) : \delta \in S^*(\mathbb{S})\right\}. \tag{15.1}$$

To solve problem (15.1), now an appropriate B&B procedure is built up. To represent corresponding branching rule of the algorithm, set $\varphi(\delta[m])$ is defined for arbitrary irregular decision-mapping, $\delta[m] \in \Sigma_M$, as follows:

$$\varphi(\delta[m]) = \{\delta' : \delta' \in S^*(\mathbb{S}) \& (\exists\ \bar{\delta}[\bar{m}] \in \Sigma_M)(\delta[m] \sqsubseteq \bar{\delta}[\bar{m}] \&\ icl(\bar{\delta}[\bar{m}]) = \delta')\}.$$

Note that $\varphi(\delta_0[m_0]) = S^*(\mathbb{S})$ and $\varphi(\delta[m])$ visually represents those irregular extensions of $\delta[m]$ which are feasible solutions for problem (15.1); each of them is obtainable by a sequence of irregular extensions and the irregular closure operation. The elements of $\varphi(\delta[m])$ are called *irregular feasible solution extensions* of $\delta[m]$.

For function φ introduced above, if $\delta[m]$ and $\delta'[m']$ are irregular decision-mappings and $\delta(X) \neq \delta'(X)$ for some $X \in m \cap m'$, then

$$\varphi(\delta[m]) \cap \varphi(\delta'[m']) = \emptyset.$$

This observation ensures that a not necessarily nontrivial partition of $\varphi(\delta[m])$ can be defined for any $\delta[m] \in \Sigma_M$ provided that $\delta[m]$ has at least one irregular extension. Now it can be proved that the sets $\varphi(\delta_t[m \cup \{X\}])$, $t = 1, 2, \ldots, k$, that constitute a (not necessarily nontrivial) partition of $\varphi(\delta[m])$ if $\delta_t[m \cup \{X\}] = \delta[m] \cup \{(X, K_{J_t})\}$, $t = 1, 2, \ldots, k$, are indeed the irregular extensions of $\delta[m]$. This partition serves as the branching rule in the presented procedure. Specifically, each leaf of the B&B tree will be given in the form of $\varphi(\delta[m])$; sets $\varphi(\delta_t[m \cup \{X\}])$, $t = 1, 2, \ldots, k$, will be the successors of the leaf under consideration.

The other main component of the B&B procedure is the bounding function. For each leaf of the B&B tree, a lower bound for the objective values of the feasible solutions belonging to this leaf is given by the bounding function. In our procedure, a leaf is given as $\varphi(\delta[m])$, and the irregular feasible solution extensions of $\delta[m]$ belong to this leaf. In a PNS_w problem, the objective function is defined by the sum of weights of the units. Since each operating unit has a positive, the weight of any extension of $\delta[m]$ is not less than the sum of the weights of operating units from $O_{\delta[m]}$. Accordingly, bounding function $g^{\dagger}$ can now be defined as

$$g^{\dagger}(\delta[m]) = \sum_{u \in O_{\delta[m]}} w(u).$$

The following notations, rules, and assumptions are considered for the LABBA procedure.

(a) L denotes the set of live leaves of the B&B tree during the procedure; z^* indicates the current best value of the objective function; while s contains the current best feasible solution.

(b) The node selection rule is given as: Leaf $\varphi(\delta[m])$ is selected from L if

$$g^{\dagger}(\delta[m]) / \mid m \mid$$

is minimal provided that $m \neq \varnothing$. In case there are multiple candidates with the same ratio, choose one randomly.

(c) The effect of material selection is investigated in the following way. Let us observe that the current set, $T = (mat^{in}(O_{\delta[m]}) \cup P) \backslash (mat^{out}(O_{\delta[m]}) \cup R)$ contains those materials which are needed, but they are yet to be produced. Thus, such an X is chosen for which the value $|(mat^{out}(\Delta(X)) \cap T|$ is maximal.

(d) The procedure is completed by a testing step to recognize a feasible solution, *i.e.*,

$$\left(mat^{in}\left(O_{\delta[m]}\right) \cup P\right) \backslash \left(mat^{out}\left(O_{\delta[m]}\right) \cup R\right) = \varnothing.$$

(e) In the branching step, a partial problem is fathomed if there is no irregular extension to $\delta[m]$ with respect to chosen material X.

Look Ahead B&B Algorithm [1]

Initialization

- Compute the merging equivalence relation and determine $\delta_0[m_0]$.
- Let $L := \{\varphi(\delta_0[m_0])\}$, $z^* := \infty$, $s := \emptyset$, and $r := 0$. Compute $g^\dagger(\delta_0[m_0])$.

Iteration (r-th iteration)

- *Step 1. (Termination)* Terminate if $L = \emptyset$: set s contains the optimal solution and z^* provides the optimal value. Otherwise, proceed to Step 2.
- *Step 2. (Selection)* Choose its element if L is a singleton; otherwise choose an element, $\varphi(\delta[m])$, of L for which ratio $g^\dagger(\delta[m])/\mid m\mid$ is minimal. (If there are more than one candidate with the same ratio, then choose one of them randomly.) Denote the selected element of L by $\varphi(\delta[m])$ and proceed to Step 3.
- *Step 3. (Testing)* If $T = (mat^{in}(O_{\delta[m]}) \cup P)\backslash(mat^{out}(O_{\delta[m]}) \cup R) \neq \emptyset$, then proceed to Step 4. Otherwise, form the irregular closure δ' of $\delta[m]$ and redefine the value of z^* and set s by $z^* := w(\delta')$ and $s := \{\delta'\}$, respectively, if $w(\delta') < z^*$. If not, then z^* and s do not change. Let $\Phi := \emptyset$ and proceed to Step 6.
- *Step 4. (Branching)* Choose an X from T for which $|(mat^{out}(\Delta(X)) \cap T|$ is maximal, and form the irregular extensions of $\delta[m]$ with respect to X. Let $L := L\backslash\{\varphi(\delta[m])\}$ if there is no such irregular extension of $\delta[m]$ and proceed to Step 1. Otherwise, denote the irregular extensions of $\delta[m]$ with respect to X by $\delta_i[m_i]$, $i = 1, 2, \ldots, k$. Then, let $\Phi = \{\varphi(\delta_i[m_i]) : 1 \leq i \leq k\}$, and proceed to Step 5.
- *Step 5. (Bounding)* Calculate the value of $g^\dagger(\delta_i[m_i])$, $i = 1, 2, \ldots, k$, and proceed to Step 6.
- *Step 6. (Fathoming)* Redefine set L as

$$L = \{\varphi(\bar{\delta}[\bar{m}]) : \varphi(\bar{\delta}[\bar{m}]) \in (L\backslash\{\varphi(\delta[m])\}) \cup \Phi,\ \ g^\dagger\ (\bar{\delta}[\bar{m}]) < z^*\}.$$

  Set $r := r + 1$ and iterate.

For the demonstration of the procedure, let us consider Example 15.1 again. The procedure is started with $L = \{\varphi(\delta_0[m_0])\}$, $z^* = \infty$, $s = \emptyset$, and $g^\dagger(\delta_0[m_0]) = 5$. Since L is singleton, leaf $\varphi(\delta_0[m_0])$ is selected in Step 2. Then, $T = \{A_1, A_2\}$ and Step 4 follows. Since $\Delta(A_1) = \Delta(A_2)$ any of the elements A_1 and A_2 can be selected. Let us choose material A_1. Then, $\Delta(A_1) = \{u_1, u_2\}$ and the restriction of the merging equivalence relation to $\{u_1, u_2\}$ is the equality relation on this set. Hence, three irregular extensions of $\delta_0[m_0]$ exist; they are

$$\delta_1[m_1] = \delta_0[m_0] \cup \{(A_1, \{u_1\})\},$$
$$\delta_2[m_2] = \delta_0[m_0] \cup \{(A_1, \{u_2\})\},$$
$$\delta_3[m_3] = \delta_0[m_0] \cup \{(A_1, \{u_1, u_2\})\}.$$

Then,

$O_{\delta_1[m_1]} = \{u_1, u_3, u_4, u_7, u_8, u_9, u_{10}\}$, $g^{\dagger}(\delta_1[m_1]) = 27$,
$O_{\delta_2[m_2]} = \{u_2, u_5, u_8, u_{10}\}$, $g^{\dagger}(\delta_2[m_2]) = 10$, and
$O_{\delta_3[m_3]} = \{u_1, u_2, u_3, u_4.u_5, u_7.u_8.u_9, u_{10}\}$, $g^{\dagger}(\delta_3[m_3]) = 32$.

Proceeding to the following iteration,

$$L = \{\varphi(\delta_1[m_1]), \varphi(\delta_2[m_2]), \varphi(\delta_3[m_3])\}$$

and leaf $\varphi(\delta_1[m_1])$ is selected in Step 2. Now, $T = \emptyset$, $z^* = 10$, $s = \{\delta'\}$, where δ' is the irregular closure of $\delta_1[m_1]$, and Step 6 follows. Here, L is changed to the emptyset and the following iteration follows, where the procedure is terminated.

Reference

1. Holló CS (2000) A look ahead branch-and-bound procedure for solving PNS problems. Pure Math Appl 11(2):265–279

Appendix I: Mathematical Basics

Appendix A summarizes the mathematical basics necessary for reading the book and also shows the notations followed in describing the formulas and algorithms. It is assumed that the reader is familiar with the fundamentals of operations research.

The intuitive concept of a *set* is accepted here as a collection of objects, termed *elements* of the set; furthermore, the conventional notations are followed. $A \subseteq B$ signifies that set A is a subset of set B; $\varnothing$ denotes the empty set; and $|A|$ represents the *cardinality* of set A, i.e., the number of its elements. Furthermore, if an element i is an element of set A, it is denoted as $i \in A$.

The basic operations between two sets are *intersection*, *union*, and *difference*. The *intersection* of two sets, denoted by operator $\cap$, consists of those elements that are present in both sets. In other words, *intersection* contains the common elements. The *union* of two sets, denote by operator $\cup$, consists of those elements that are included in at least one of the two sets. This means that the *union* contains all elements in the two sets. The *difference*, denoted by operator $\setminus$, gives the elements that are found in the set on the left, but not in the set on the right. Note that while the *intersection* and *union* are commutative operations, i.e., it does not matter which set is on which side ($A \cap B = B \cap A$ and $A \cup B = B \cup A$), the *difference* is not commutative.

Now, let us illustrate these three operations via a small example, with sets $A = \{O1, O2, O3\}$ and $B = \{O3, O4, O5\}$. The *intersection* contains the common elements, thus $A \cap B = B \cap A = \{O3\}$. The *union* contains all elements in both sets, i.e., $A \cup B = B \cup A = \{O1, O2, O3, O4, O5\}$. The *difference* can be determined both ways. $A \setminus B$ gives the elements in set A that are not in set B, thus $A \setminus B = \{O1, O2\}$. Consequently, $B \setminus A = \{O4, O5\}$.

Another important operation is the *complement*. The *complement* of a set consists of all the elements not included in that set. This operation requires a universal set to be defined that contains all possible elements that can appear in any set. For example, this can be the set of all integer numbers, the set of all nodes of a graph, or, in a P-graph, the set of all materials, or the set of all operating units. Herein, the *complement* of set A is denoted as $\overline{A}$. For instance, let us define the universal set $\boldsymbol{S}$

F. Friedler et al., *P-graphs for Process Systems Engineering*,
https://doi.org/10.1007/978-3-030-92216-0

as the set of units $\{O1, O2, O3, O4, O5, O6\}$. Then, the complement of set $B = \{O3, O4, O5\}$, i.e., $\overline{B} = \{O1, O2, O6\}$.

The operations *union* and *intersection* can be applied to several sets at the same time. The union of a set of sets A_i, $i \in I$, is written as $\cup A_i : i \in I$, and consists of all elements of all the considered sets. Similarly, their intersection is written as $\cap A_i : i \in I$, and it gives the elements that are common in all of the sets.

The sets A and B are said to be disjoint, or mutually exclusive if $A \cap B = \emptyset$.

The *power set* of A is the set of all subsets of A; and it is given by $\wp(A)$. Formally, $\wp(A) = \{X| \ X \subseteq A\}$. For instance, if set A is considered as previously defined, $A = \{O1, O2, O3\}$, then, the power set $\wp(A)$ is equal to

$$\wp(A) = \{\emptyset, \{O1\}, \{O2\}, \{O3\}, \{O1, O2\}, \{O1, O3\}, \{O2, O3\}, \{O1, O2, O3\}\}$$

In particular, the set of all nonempty subsets of A is denoted by $\wp'(A)$.

The *Cartesian product* of two sets contains all possible pairs of elements, where the first element of the pair is from the first set, and the second element is from the second set. Formally, for sets A and B results in set $\{(a, b) : a \in A \ \& \ b \in B\}$ which is denoted by $A \times B$. For instance, the Cartesian product of sets A and B is

$$A \times B = \{(O1, O3), (O1, O4), (O1, O5), (O2, O3), (O2, O4), (O2, O5), (O3, O3), (O3, O4), (O3, O5)\}$$

By a *binary relation* on set A, we mean a subset of $A \times A$. If $\sigma \subseteq A \times A$ is a binary relation, then $(a, b) \in \sigma$ is also expressed as $\check{a}\sigma b$. Binary relation σ on A is said to be:

1. reflexive if $a\sigma a$ for all $a \in A$,
2. *symmetric* if $a\sigma b$ implies $b\sigma a$ for all $a, b \in A$,
3. transitive if $a\sigma b$ and $b\sigma c$ imply $a\sigma c$ for all $a, b, c \in A$,
4. antisymmetric if $a\sigma b$ and $b\sigma a$ imply $a = b$ for all $a, b \in A$,
5. *dichotome* if $a\sigma b$ or $b\sigma a$ is valid for all $a, b \in A$.

On the bases of the above properties, various types of binary relations can be specified. A reflexive, transitive, and antisymmetric binary relation on set A is defined to be a *partial ordering* on A. If σ is a partial ordering on A, then pair (A, σ) is a *partially ordered set*. In general, a partial ordering is denoted by symbol $\leq$; moreover, $a < b$ stands for $a \leq b$ and $a \neq b$.

Let $(A, \leq)$ be an arbitrary, partially ordered set. Element $b \in A$ is a *maximal element* of $(A, \leq)$ if $b \leq a$ implies that $b = a$ for all $a \in A$. In particular, the maximal element, b, is designated as the *greatest element* of $(A, \leq)$ if $a \leq b$ is valid for all $a \in A$. A *minimal element* and the *smallest element* of $(A, \leq)$ are defined correspondingly.

A *linear ordering* on set A, is a dichotome partial ordering. If relation $\leq$ is a linear ordering on A, then pair $(A, \leq)$ is termed a *chain* or a *linearly ordered set*.

Example A.1 Let $A = \{a, b, c, d\}$. Then, it can be readily seen that binary relation

$$\sigma = \{(a,a),(b,b),(c,c),(d,d),(a,b),(c,a),(c,b),(d,b)\}$$

is a partial ordering. Partially ordered set (A, σ) has two minimal elements, *i.e.*, c and d; moreover, it contains one greatest element, *i.e.*, b. If relation σ is restricted to set $B = \{a, b, c\}$, *i.e.*, if we take $\sigma' = \sigma \cap (B \times B)$, then σ' is a linear ordering on B, and the corresponding chain is $c < a < b$.

Another important binary relation is the equivalence relation defined as follows: A binary relation on A is an *equivalence relation* if it is reflexive, symmetric, and transitive. In particular, both *equality relation* $\{(a, a)\colon a \in A\}$ and *complete relation* $\{(a, b)\colon a, b \in A\}$ are equivalence relations on A.

It is known that a strong relationship exists between the partitions of A and the equivalence realtions on A. We recall that set $\{A_i\colon A_i \subseteq A \;\&\; i \in I\}$ is a *partition* of A if $\cup\{A_i\colon i \in I\} = A$ and $A_i \cap A_j = \emptyset$ for all $i \neq j \in I$. Now, let σ be an equivalence relation on A. For any $a \in A$, let $A_a = \{b\colon b \in A \;\&\; a\sigma b\}$; furthermore, let Θ be the set of all $B \subseteq A$ which are of the form, A_a. Obviously, Θ is a partition of A. The subsets contained in Θ are designated as the *equivalence classes*. Conversely, if $\Theta = \{A_i\colon A_i \subseteq A \;\&\; i \in I\}$ is a partition of A, then an equivalence relation with equivalence classes A_i, $i \in I$ can be defined in the following way: for any a, $b \in A$, $a\sigma b$ if and only if $\{a, b\} \subseteq A_i$ for some $i \in I$.

Let set A be given, and also let the set of all partitions of A be denoted by C_A. In addition, let $\Theta_1 = \{A_i\colon i \in I\}$ and $\Theta_2 = \{B_j\colon j \in J\}$ be two partitions of A. Partition Θ_1 is a *refinement* of partition Θ_2 if for any $A_i \in \Theta_1$, there exists $B_j \in \Theta_2$ with $A_i \subseteq B_j$. It can be immediately seen that the refinement relation is reflexive, transitive, and antisymmetric; consequently, it is a partial ordering on C_A. Naturally, the two trivial partitions, *i.e.*, the partition corresponding to the equality relation and that belonging to the complete relation, are the smallest and greatest elements of this partially ordered set of partitions, respectively.

Example A.2 Let $A = \{a, b, c, d\}$; then relations

$$\sigma_1 = \{(a,a),(b,b),(c,c),(d,d),(a,b),(b,a)\}$$

and

$$\sigma_2 = \{(a,a),(b,b),(c,c),(d,d),(a,b),(b,a),(c,d),(d,c)\}.$$

are clearly equivalence relations. The corresponding partitions are $\Theta_1 = \{\{a, b\}, \{c\}, \{d\}\}$, and $\Theta_2 = \{\{a, b\}, \{c, d\}\}$. Moreover, Θ_1 is a refinement of Θ_2.

Let A and B be sets; then, a *mapping* or *function* of A into B is a subset of $A \times B$, φ such that for every $a \in A$, there exists exactly one $b \in B$ with $(a, b) \in \varphi$. As usual, we write $\varphi\colon A \to B$ where A is the *domain* of φ. If $(a, b) \in \varphi$, then element b is termed the *image* of element a under φ this is denoted by $\varphi(a) = b$ or $a\varphi = b$. For subset A' of A, the set of the images of elements in A' under φ is denoted by $\varphi(A')$ or $A'\varphi$.

Mapping $\varphi: A \to B$, is *one-to-one* or *injective* if $a \neq a'$ implies $\varphi(a) \neq \varphi(a')$ for all $a, a' \in A$; φ is *onto* or *surjective* if for any $b \in B$, there exists element $a \in A$ with $b = \varphi(a)$; finally, φ is *bijective* if it is injective and surjective simultaneously. If $\varphi: A \to B$ is injective, then the inverse function, $\varphi^{-1}: \varphi(A) \to A$, of φ is specified by set $\varphi^{-1} = \{(b, a): (b, a) \in B \times A \ \& \ (a, b) \in \varphi\}$. In this case, φ^{-1} is a bijective mapping of $\varphi(A)$ onto A.

Often, mapping $\varphi: A \to B$ is expressed as $\varphi[A]$ to emphasize that its domain is A. From the technical point of view, this notation renders it possible to facilitate the definition of some extensions of φ. Suppose that $X \notin A$ and that φ is to be extended to X. If the image of X under the extended function is Y, which is written as φ^*, then it follows that $\varphi^*[A \cup \{X\}] = \varphi[A] \cup \{(X, Y)\}$. This form representation of functions is used in the definition of decision-mapping.

Directed graph G is signifed by pair $G = (V, E)$ where V is a finite set of *nodes* or *vertices*, and $E \subseteq V \times V$ is the set of *edges* or *arcs*. In drawing a directed graph, each vertex is depicted by a dot; and each edge (u, v), by a line segment directed from vertex u to vertex v with an arrowhead on the line segment.

Let $G = (V, E)$ be a directed graph; if $(u, v) \in E$, *i.e.*, if it is an edge from u to v, then u is the *ancestor* (or *parent*) of v while v is the *successor* (or *child*) of u. The *indegree* of vertex v is the number of its ancestors, the *outdegree* of v is the number of its successors. Node v is a *source* if it has no ancestor, or equivalently, if its indegree is 0; on the other hand, v is a *sink* if it has no successor, or equivalently, its outdegree is 0. A *directed path* comprises a sequence of edges $(v_1, v_2), (v_2, v_3), \ldots, (v_{k-1}, v_k)$ with $v_i \neq v_j$ for all $1 \leq i < j \leq k$; the *length* of the path is the number of edges included in the path; vertex v_1 is the *initial vertex*, and v_k is the *terminal vertex* of the directed path; finally, if the only interest is the two endpoints of the directed path, then the directed path is written as $[v_1, v_k]$. An *undirected path* is a sequence of form $(v_1, v_2), (v_2, v_3), \ldots, (v_{k-1}, v_k)$ where $v_i \neq v_j$ for all $1 \leq i < j \leq k$, and $(v_i, v_{i+1}) \in E$ or $(v_{i+1}, v_i) \in E$ holds for all i, $i = 1, \ldots, k - 1$. Directed graph G is *connected* if there exists an undirected path between any two different vertices of the graph. A maximal connected subgraph of graph G is called a *component* of G.

A sequence of edges, $(v_1, v_2), (v_2, v_3), \ldots, (v_{k-1}, v_k)$, satisfying $v_1 = v_k$ and $v_i \neq v_j$ for all $1 \leq i < j < k$, is a *directed cycle*. Directed graph $G = (V, E)$ is a *directed tree* if it contains no directed cycle and it has exactly one source; furthermore, exactly one directed path exists from the source to any of the vertices distinct from this source. The source of G corresponds to the *root* of the directed tree; the sinks of G, to the *leaves* of G. The *depth* of a directed tree is the maximum among the paths' lengths leading from the source to the leaves. Since exactly one directed path exists from the source of a directed tree to any vertex different from the source, we can disregard the directions of edges by assuming that every edge has a vertical direction, *i.e.*, a direction from the top to the bottom. A directed tree is drawn on the basis of this vertical direction of edges. Let us now consider the following example for illustration.

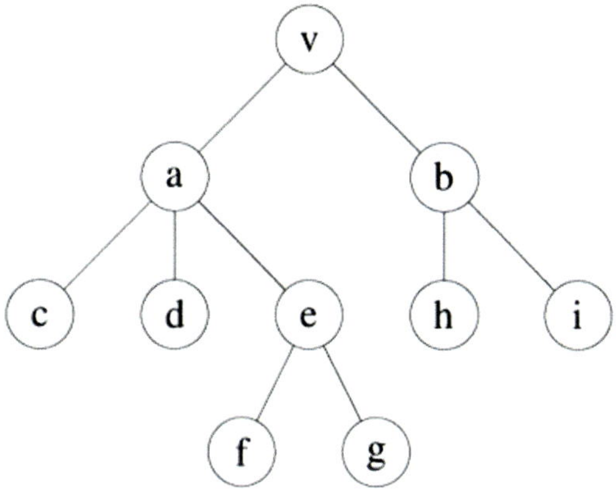

Fig. A.1 Directed tree given in Example A.3

Example A.3 Let $G = (V, E)$ where $V = \{a, b, c, d, e, f, g, h, i, j, v\}$ and E is equal to $\{(v, a), (a, c), (a, d), (a, e), (e, f), (e, g), (v, b), (b, h), (b, i)\}$; the corresponding directed tree is given in Fig. A.1. The source of this tree is v and the leaves are vertices c, d, f, g, h, i; finally, the depth of this tree is 3.

Now, for a given directed tree, let us suppose that for each vertex with at least one successor, the set of successors is linearly ordered. We present a method known as the depth-first search procedure for generating all the vertices of the tree.

Depth-First Search Procedure

Initialization

- Set $X := v$ where v denotes the root of the directed tree. Then, v is regarded as a vertex generated. Set $i = 1$, and proceed to the iteration phase.

  Iteration

- *Step 1.* If X has at least one successor yet to be generated, then choose the first among these successors with respect to their linear ordering; denote it by u. Set $X := u$ and $i := i + 1$. Then, vertex u is considered as a vertex generated. Proceed to the succeeding iteration step. Otherwise, proceed to Step 2 if X has no successor.
- *Step 2.* If $X = v$, then terminate since all the vertices of the directed tree are generated. Otherwise, proceed to Step 3.
- *Step 3.* Take the ancestor of X (it is uniquely determined), and denote it by w. Then let $X := w$, and proceed to Step 1.

For illustration, let us perform the depth-first search procedure for the directed tree in Example A.3 by assuming that the ordering of the successors of any vertex is of the alphabetical ordering. The vertices of the tree are generated according to sequence v, a, c, d, e, f, g, b, h, i. Figure A.2 depicts the graph where each vertex is

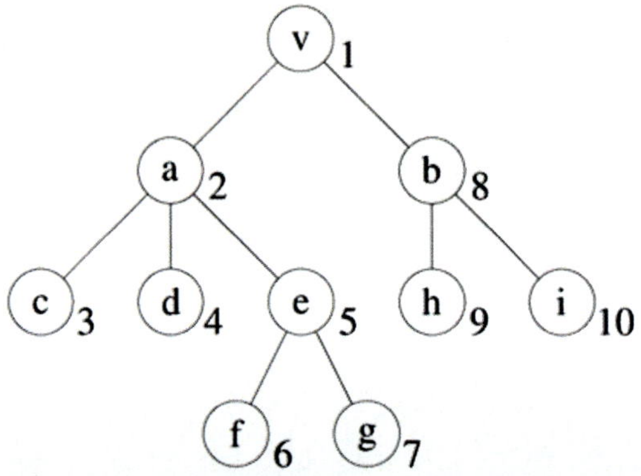

Fig. A.2 Illustration of vertex generation for Example A.3

marked with the number of the iteration step at which the vertex of interest is generated.

We are concerned only with directed graphs throughout this monograph. For shortness, therefore, the adjective, "directed", is omitted from any designation pertaining to graphs, trees, edges, and cycles.

Big O Notation

The so-called „big O" notation is introduced in computer science for describing the asymptotic behavior of algorithms, *e.g.*, for determining an upper boundfor the number of steps when the size of the problem tends to infinity. Formally, $f(N)= O(g(N))$ if there are positive constants c and N_0 such that $f(N) \leq c\, g(N)$ for all $N \geq N_0$. Big O may give an upper bound for the complexity of an algorithm. In Chap. 10, the complexity of algorithm MSG has been determined as polynomial, *i.e.*, it belongs to the class of efficient algoritms.

Appendix II: Step-by-Step Illustration of Algorithm SSG

Appendix II showcases a stepwise presentation of algorithm SSG with more details than what was given in Chap. 5. The figures show the steps of the algorithm for Example 3.2 of Chap. 3. At each step of the algorithm, the depth of recursion, and the current decision mapping are shown. Moreover, the values of every relevant variable are also listed besides the formal algorithm for the first 24 steps.

EXAMPLE 3.2

- Operating units

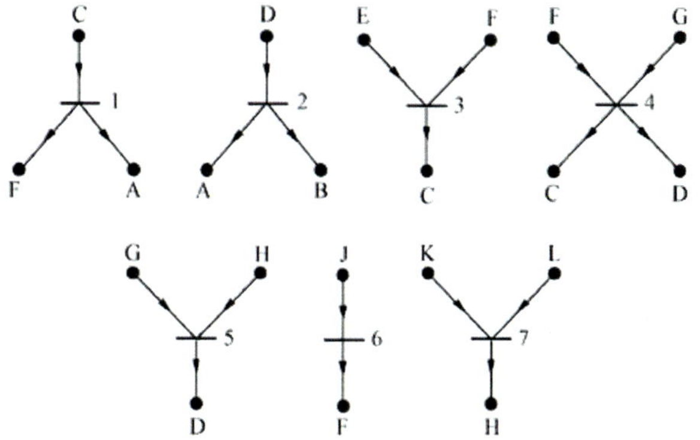

- Available raw materials: E, G, J, K, L
- Product: A

F. Friedler et al., *P-graphs for Process Systems Engineering*,
https://doi.org/10.1007/978-3-030-92216-0

Depth of recursion: 0

{1}*, {2}, {1, 2}

```
⇒ procedure SSG( p, m, δ[m] )  p={A}, m=∅
  begin
  if p = ∅ then begin write δ[m];  p={A}
            return    end
  let x∈p;  p={A}, x=A;
  C:= ℘(Δ(x))\{∅};  Δ(A)={1, 2}
  for all c∈C do  C={ {1}, {2}, {1, 2} }, c={1}
            begin
            if ∀y∈m, c∩δ̄(y) = ∅ & (Δ(x)\c)∩δ(y) = ∅  true
            m=∅, c={1}, Δ(A)={1, 2}
             then
              begin
              δ[m∪{x}]:= δ[m]∪{(x, c)};  m= ∅, x=A, c={1}
⇐             SSG(p∪mat^in (c))\(R∪m∪{x}), m∪{x}, δ[m∪{x}])          δ[∅] = ∅
              p={A}, mat^in ({1}) ={C}, R={E, J, G, K, L}
              m=∅, x=A
              end
            end
  return
  end
```

Depth of recursion: 1

{1}*, {2}, {1, 2}

{3}*, {4}, {3, 4}

```
⇒ procedure SSG( p, m, δ[m] )  p={C}, m={A}
  begin
  if p = ∅ then begin write δ[m];  p={C}
            return    end
  let x∈p;  p={C}, x=C;
  C:= ℘(Δ(x))\{∅};  Δ(C)={3, 4}
  for all c∈C do  C={ {3}, {4}, {3, 4} }, c={3}
            begin
            if ∀y∈m, c∩δ̄(y) = ∅ & (Δ(x)\c)∩δ(y) = ∅  true
            m={A}, c={3}, Δ(C)={3, 4}
             then
              begin
              δ[m∪{x}]:= δ[m]∪{(x, c)};  m={A} , x=C, c={3}
⇐             SSG(p∪mat^in (c))\(R∪m∪{x}), m∪{x}, δ[m∪{x}])
              p={C}, mat^in ({3}) ={E, F}, R={E, J, G, K, L}
              m={A}, x=C
              end
            end
  return
  end
```

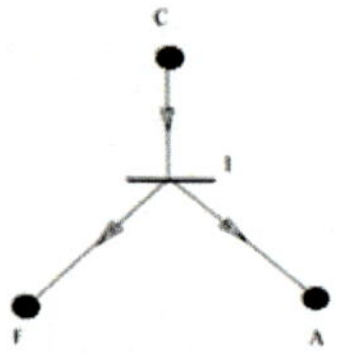

δ[{A}]

Depth of recursion: 2

{1}*, {2}, {1, 2}

{3}*, {4}, {3, 4}

{1}*, {6}, {1, 6}

```
⇒ procedure SSG( p, m, δ[m] )  p={F}, m={A, C}
  begin
  if p = ∅ then begin write δ[m];  p={F}
             return   end
  let x∈p;  p={F}, x=F;
  C:= ℘(Δ(x))\{∅};  Δ(F)={1, 6}
  for all c∈C do  C={ {1}, {6}, {1, 6} }, c={1}
             begin
             if ∀y∈m, c∩δ̄(y) = ∅ & (Δ(x)\c)∩δ(y) = ∅  true
             m={A, C}, c={1}, Δ(F)={1, 6}
              then
               begin
               δ[m∪{x}]:= δ[m]∪{(x, c)};  m={A, C} , x=F, c={1}
⇐              SSG(p∪mat^in (c))\(R∪m∪{x}), m∪{x}, δ[m∪{x}])
               p={F}, mat^in ({1}) ={C}, R={E, J, G, K, L}
               m={A, C}, x=F
               end
             end
  return
  end
```

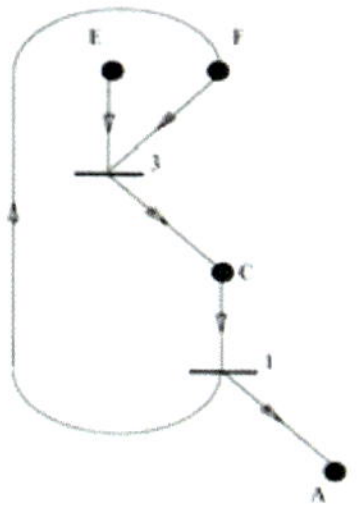

δ[{A, C}]

Depth of recursion: 3

{1}*, {2}, {1, 2}

{3}*, {4}, {3, 4}

{1}*, {6}, {1, 6}

```
⇒ procedure SSG( p, m, δ[m] )  p=∅, m={A, C, F}
  begin
  if p = ∅ then begin write δ[m];  p=∅
⇐            return   end
  let x∈p;
  C:= ℘(Δ(x))\{∅};
  for all c∈C do
             begin
             if ∀y∈m, c∩δ̄(y) = ∅ & (Δ(x)\c)∩δ(y) = ∅
              then
               begin
               δ[m∪{x}]:= δ[m]∪{(x, c)};
               SSG(p∪mat^in (c))\(R∪m∪{x}), m∪{x}, δ[m∪{x}])
               end
             end
  return
  end
```

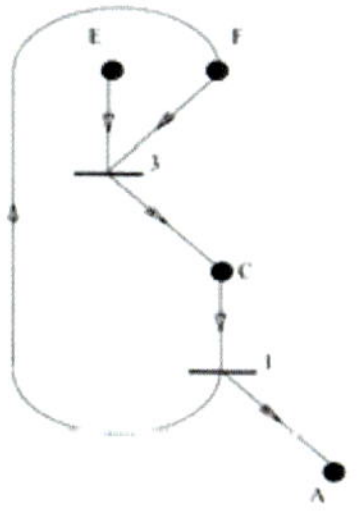

δ[{A, C, F}]

Solution 1

Depth of recursion: 2

{1}*, {2}, {1, 2}
{3}*, {4}, {3, 4}
{1}, **{6}***, {1, 6}

```
procedure SSG( p, m, δ[m] ) p={F}, m={A, C}
begin
if p = ∅ then begin write δ[m]; p={F}
        return    end
let x∈p; p={F}, x=F;
C:= ℘(Δ(x))\{∅}; Δ(F)={1, 6}
for all c∈C do C={ {1}, {6}, {1, 6} }, c={6}
        begin
        if ∀y∈m, c∩δ̄(y) = ∅ & (Δ(x)\c)∩δ(y) = ∅  false
        m={A, C}, c={6}, Δ(F)={1, 6}
         then
          begin
          δ[m∪{x}]:= δ[m]∪{(x, c)};
          SSG(p∪mat^in (c))\(R∪m∪{x}), m∪{x}, δ[m∪{x}])
⇒         end
        end
return
end
```

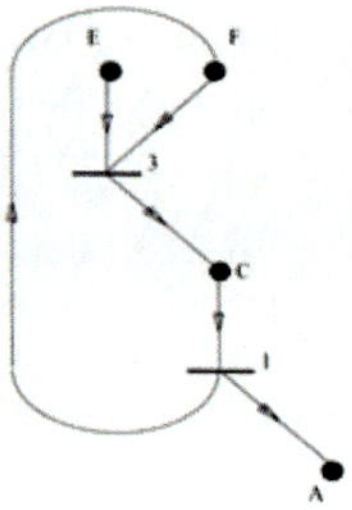

δ[{A, C}]

Depth of recursion: 2

{1}*, {2}, {1, 2}
{3}*, {4}, {3, 4}
{1}, {6}, **{1, 6}***

```
procedure SSG( p, m, δ[m] ) p={F}, m={A, C}
begin
if p = ∅ then begin write δ[m]; p={F}
        return    end
let x∈p; p={F}, x=F;
C:= ℘(Δ(x))\{∅}; Δ(F)={1, 6}
for all c∈C do C={ {1}, {6}, {1, 6} }, c={1, 6}
        begin
        if ∀y∈m, c∩δ̄(y) = ∅ & (Δ(x)\c)∩δ(y) = ∅  true
        m={A, C}, c={1, 6}, Δ(F)={1, 6}
         then
          begin
          δ[m∪{x}]:= δ[m]∪{(x, c)}; m={A, C} , x=F, c={1, 6}
⇐         SSG(p∪mat^in (c))\(R∪m∪{x}), m∪{x}, δ[m∪{x}])
          p={F}, mat^in ({1, 6}) ={C, J}, R={E, J, G, K, L}
          m={A, C}, x=F
          end
        end
return
end
```

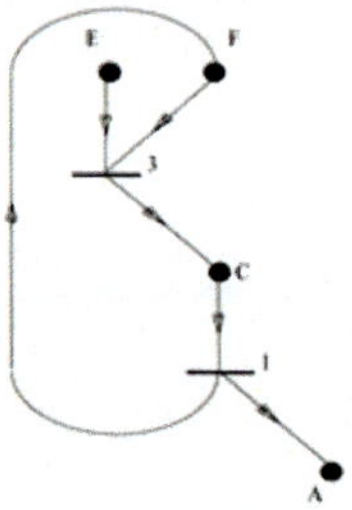

δ[{A, C}]

Depth of recursion: 3

{1}*, {2}, {1, 2}

{3}*, {4}, {3, 4}

{1}, {6}, **{1, 6}***

```
⇒ procedure SSG( p, m, δ[m] ) p=∅, m={A, C, F}
   begin
   if p = ∅ then begin write δ[m]; p=∅
⇐              return   end
   let x∈p;
   C:= ℘(Δ(x))\{∅};
   for all c∈C do
              begin
              if ∀y∈m, c∩δ̄(y) = ∅ & (Δ(x)\c)∩δ(y) = ∅
               then
                begin
                δ[m∪{x}]:= δ[m]∪{(x, c)};
                SSG(p∪mat^in (c))\(R∪m∪{x}), m∪{x}, δ[m∪{x}])
                end
              end
   return
   end
```

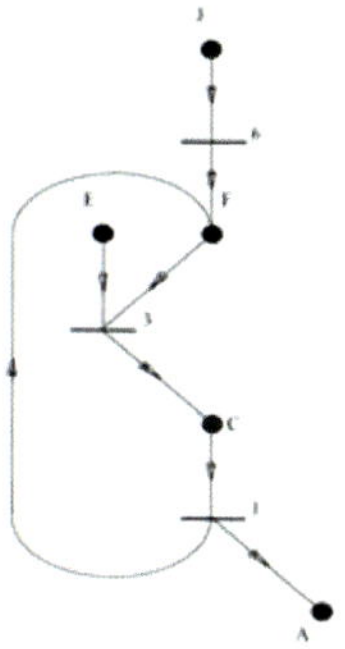

δ[{A, C, F}]

Solution 2

Depth of recursion: 2

{1}*, {2}, {1, 2}

{3}*, {4}, {3, 4}

{1}, {6}, **{1, 6}***

```
   procedure SSG( p, m, δ[m] ) p={F}, m={A, C}
   begin
   if p = ∅ then begin write δ[m]; p={F}
              return   end
   let x∈p; p={F}, x=F;
   C:= ℘(Δ(x))\{∅}; Δ(F)={1, 6}
   for all c∈C do C={ {1}, {6}, {1, 6} }
              begin
              if ∀y∈m, c∩δ̄(y) = ∅ & (Δ(x)\c)∩δ(y) = ∅
               then
                begin
                δ[m∪{x}]:= δ[m]∪{(x, c)};
                SSG(p∪mat^in (c))\(R∪m∪{x}), m∪{x}, δ[m∪{x}])
                end
 ⇒            end
⇐  return
   end
```

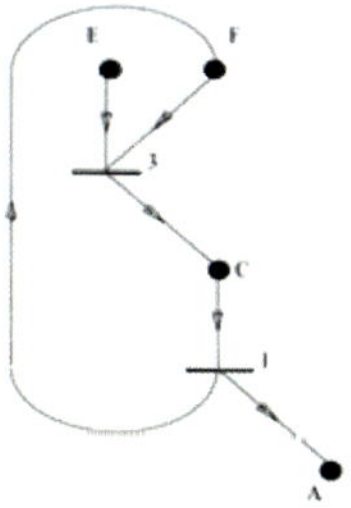

δ[{A, C}]

Depth of recursion: 1

{1}*, {2}, {1, 2}

{3}, {4}*, {3, 4}

```
   procedure SSG( p, m, δ[m] ) p={C}, m={A}
   begin
   if p = ∅ then begin write δ[m]; p={C}
            return    end
   let x∈p; p={C}, x=C;
   C:= ℘(Δ(x))\{∅}; Δ(C)={3, 4}
   for all c∈C do C={ {3}, {4}, {3, 4} }, c={4}
            begin
            if ∀y∈m, c∩δ̄(y) = ∅ & (Δ(x)\c)∩δ(y) = ∅ true
            m={A}, c={4}, Δ(C)={3, 4}
             then
              begin
              δ[m∪{x}]:= δ[m]∪{(x, c)}; m={A} , x=C, c={4}
⇐             SSG(p∪matin (c))\(R∪m∪{x}), m∪{x}, δ[m∪{x}])
              p={C}, matin ({4}) ={F, G}, R={E, J, G, K, L}
              m={A}, x=C
⇒             end
            end
   return
   end
```

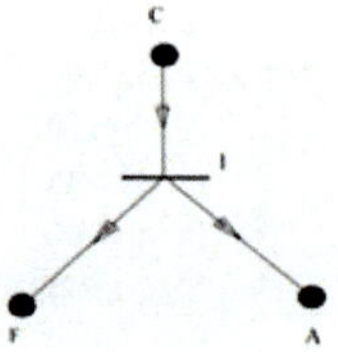

δ[{A}]

Depth of recursion: 2

{1}*, {2}, {1, 2}

{3}, {4}*, {3, 4}

{1}*, {6}, {1, 6}

```
⇒  procedure SSG( p, m, δ[m] ) p={F}, m={A, C}
   begin
   if p = ∅ then begin write δ[m]; p={F}
            return    end
   let x∈p; p={F}, x=F;
   C:= ℘(Δ(x))\{∅}; Δ(F)={1, 6}
   for all c∈C do C={ {1}, {6}, {1, 6} }, c={1}
            begin
            if ∀y∈m, c∩δ̄(y) = ∅ & (Δ(x)\c)∩δ(y) = ∅ true
            m={A, C}, c={1}, Δ(F)={1, 6}
             then
              begin
              δ[m∪{x}]:= δ[m]∪{(x, c)}; m={A, C} , x=F, c={1}
⇐             SSG(p∪matin (c))\(R∪m∪{x}), m∪{x}, δ[m∪{x}])
              p={F}, matin ({1}) ={C}, R={E, J, G, K, L}
              m={A, C}, x=F
              end
            end
   return
   end
```

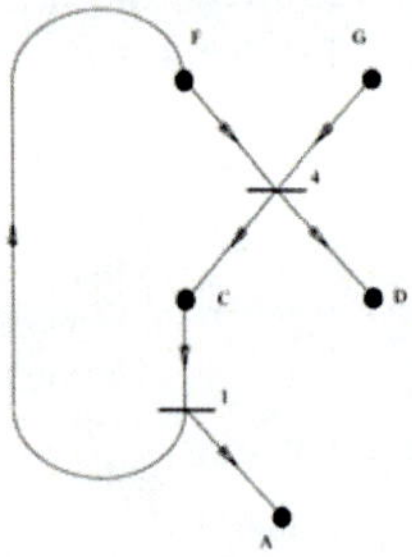

δ[{A, C}]

Depth of recursion: 3

{1}*, {2}, {1, 2}

{3}, {4}*, {3, 4}

{1}*, {6}, {1, 6}

```
⇒ procedure SSG( p,  m, δ[m] )  p=∅, m={A, C, F}
   begin
   if p = ∅ then begin write δ[m];  p=∅
⇐              return    end
   let x∈p;
   C:= ℘(Δ(x))\{∅};
   for all c∈C do
              begin                 _
              if ∀y∈m, c∩δ(y) = ∅ & (Δ(x)\c)∩δ(y) = ∅
               then
                begin
                δ[m∪{x}]:= δ[m]∪{(x, c)};
                SSG(p∪matⁱⁿ (c))\(R∪m∪{x}), m∪{x}, δ[m∪{x}])
                end
              end
   return
   end
```

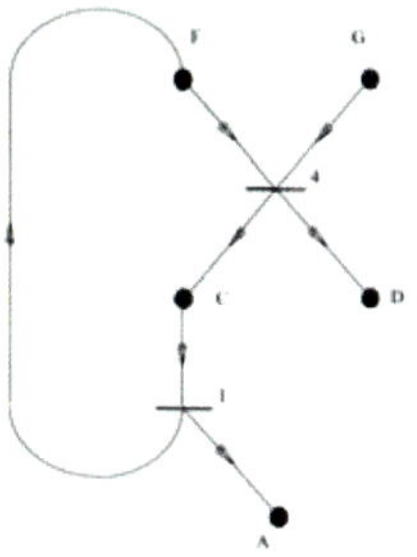

δ[{A, C, F}]

Solution 3

Depth of recursion: 2

{1}*, {2}, {1, 2}

{3}, {4}*, {3, 4}

{1}, {6}*, {1, 6}

```
   procedure SSG( p,  m, δ[m] )  p={F}, m={A, C}
   begin
   if p = ∅ then begin write δ[m];  p={F}
              return    end
   let x∈p;  p={F}, x=F;
   C:= ℘(Δ(x))\{∅};  Δ(F)={1, 6}
   for all c∈C do  C={ {1}, {6}, {1, 6} }, c={6}
              begin                 _
              if ∀y∈m, c∩δ(y) = ∅ & (Δ(x)\c)∩δ(y) = ∅  false
              m={A, C}, c={6}, Δ(F)={1, 6}
               then
                begin
                δ[m∪{x}]:= δ[m]∪{(x, c)};
                SSG(p∪matⁱⁿ (c))\(R∪m∪{x}), m∪{x}, δ[m∪{x}])
 ⇒              end
              end
   return
   end
```

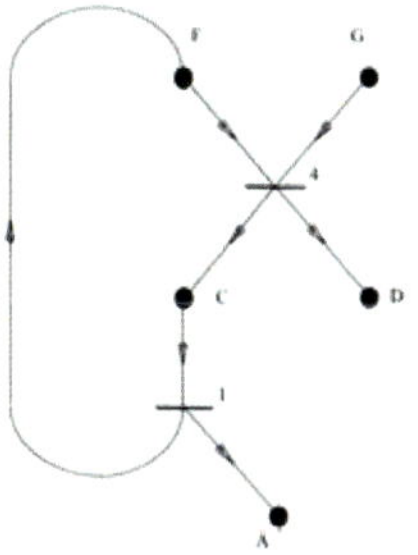

δ[{A, C}]

Depth of recursion: 2

{1}*, {2}, {1, 2}

{3}, {4}*, {3, 4}

{1}, {6}, {1, 6}*

```
   procedure SSG( p, m, δ[m] ) p={F}, m={A, C}
   begin
   if p = ∅ then begin write δ[m]; p={F}
             return   end
   let x∈p; p={F}, x=F;
   C:= ℘(Δ(x))\{∅}; Δ(F)={1, 6}
   for all c∈C do C={ {1}, {6}, {1, 6} }, c={1, 6}
             begin
             if ∀y∈m, c∩δ̄(y) = ∅ & (Δ(x)\c)∩δ(y) = ∅ true
             m={A, C}, c={1, 6}, Δ(F)={1, 6}
              then
               begin
               δ[m∪{x}]:= δ[m]∪{(x, c)}; m={A, C} , x=F, c={1, 6}
⇐              SSG(p∪matⁱⁿ (c))\(R∪m∪{x}), m∪{x}, δ[m∪{x}])
               p={F}, matⁱⁿ ({1, 6}) ={C, J}, R={E, J, G, K, L}
               m={A, C}, x=F
               end
             end
   return
   end
```

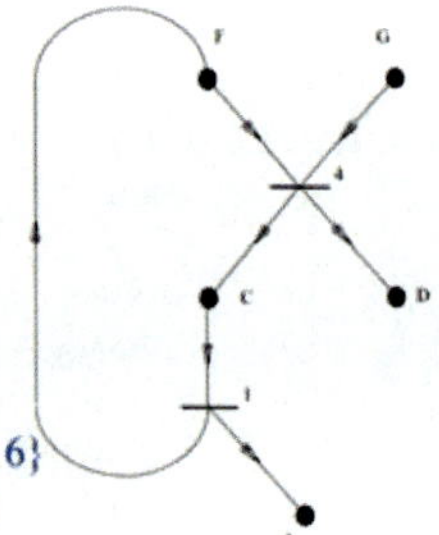

δ[{A, C}]

Depth of recursion: 3

{1}*, {2}, {1, 2}

{3}, {4}*, {3, 4}

{1}, {6}, {1, 6}*

```
⇒  procedure SSG( p, m, δ[m] ) p=∅, m={A, C, F}
   begin
   if p = ∅ then begin write δ[m]; p=∅
⇐            return   end
   let x∈p;
   C:= ℘(Δ(x))\{∅};
   for all c∈C do
             begin
             if ∀y∈m, c∩δ̄(y) = ∅ & (Δ(x)\c)∩δ(y) = ∅
              then
               begin
               δ[m∪{x}]:= δ[m]∪{(x, c)};
               SSG(p∪matⁱⁿ (c))\(R∪m∪{x}), m∪{x}, δ[m∪{x}])
               end
             end
   return
   end
```

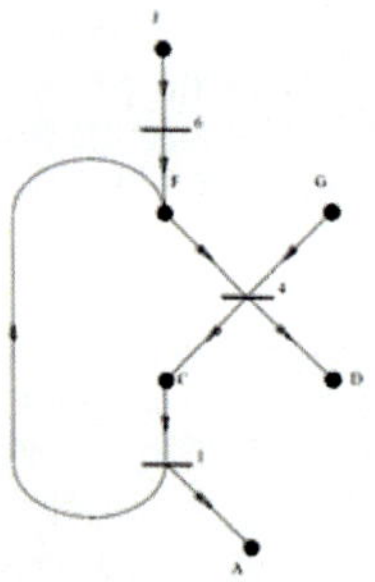

δ[{A, C, F}]

Solution 4

Depth of recursion: 2

{1}*, {2}, {1, 2}

{3}, **{4}***, {3, 4}

{1}, {6}, **{1, 6}***

procedure SSG(p, m, $\delta[m]$) p={F}, m={A, C}
begin
if p = $\emptyset$ **then begin write** $\delta[m]$; p={F}
  return **end**
let $x \in p$; p={F}, x=F;
C:= $\wp(\Delta(x))\backslash\{\emptyset\}$; $\Delta(F)=\{1, 6\}$
for all $c \in C$ **do** C={ {1}, {6}, {1, 6} }
  begin
  if $\forall y \in m,\ c \cap \bar{\delta}(y) = \emptyset\ \&\ (\Delta(x)\backslash c) \cap \delta(y) = \emptyset$
   then
    begin
    $\delta[m \cup \{x\}] := \delta[m] \cup \{(x, c)\}$;
    SSG($(p \cup mat^{in}(c)) \backslash (R \cup m \cup \{x\})$, $m \cup \{x\}$, $\delta[m \cup \{x\}]$)
    end
⇒ **end**
⇐ **return**
end

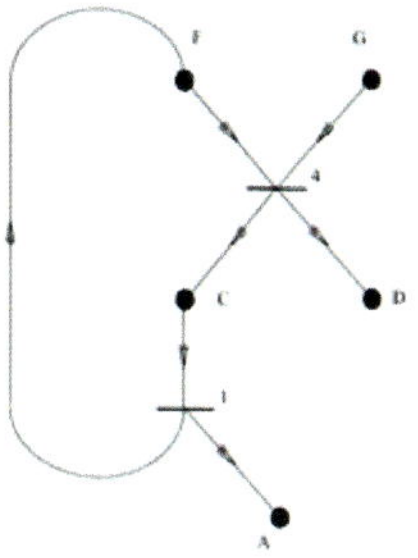

$\delta[\{A, C\}]$

Depth of recursion: 1

{1}*, {2}, {1, 2}

{3}, {4}, **{3, 4}***

procedure SSG(p, m, $\delta[m]$) p={C}, m={A}
begin
if p = $\emptyset$ **then begin write** $\delta[m]$; p={C}
  return **end**
let $x \in p$; p={C}, x=C;
C:= $\wp(\Delta(x))\backslash\{\emptyset\}$; $\Delta(C)=\{3, 4\}$
for all $c \in C$ **do** C={ {3}, {4}, {3, 4} }, c={3, 4}
  begin
  if $\forall y \in m,\ c \cap \bar{\delta}(y) = \emptyset\ \&\ (\Delta(x)\backslash c) \cap \delta(y) = \emptyset$ true
  m={A}, c={3, 4}, $\Delta(C)=\{3, 4\}$
   then
    begin
    $\delta[m \cup \{x\}] := \delta[m] \cup \{(x, c)\}$; m={A} , x=C, c={3, 4}
⇐ SSG($(p \cup mat^{in}(c)) \backslash (R \cup m \cup \{x\})$, $m \cup \{x\}$, $\delta[m \cup \{x\}]$)
    p={C}, $mat^{in}(\{3, 4\})$ ={E, F, G}, R={E, J, G, K, L}
    m={A}, x=C
⇒ **end**
  end
return
end

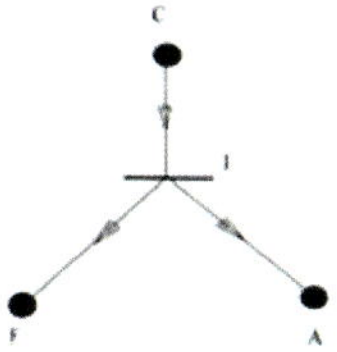

$\delta[\{A\}]$

Depth of recursion: 2

{1}*, {2}, {1, 2}
{3}, {4}, {3, 4}*
{1}*, {6}, {1, 6}

```
⇒ procedure SSG( p, m, δ[m] ) p={F}, m={A, C}
  begin
  if p = ∅ then begin write δ[m]; p={F}
            return   end
  let x∈p; p={F}, x=F;
  C:= ℘(Δ(x))\{∅}; Δ(F)={1, 6}
  for all c∈C do C={ {1}, {6}, {1, 6} }, c={1}
            begin
            if ∀y∈m, c∩δ̄(y) = ∅ & (Δ(x)\c)∩δ(y) = ∅  true
            m={A, C}, c={1}, Δ(F)={1, 6}
             then
              begin
              δ[m∪{x}]:= δ[m]∪{(x, c)}; m={A, C} , x=F, c={1}
⇐             SSG(p∪matⁱⁿ (c))\(R∪m∪{x}), m∪{x}, δ[m∪{x}])
              p={F}, matⁱⁿ ({1}) ={C}, R={E, J, G, K, L}
              m={A, C}, x=F
              end
            end
  return
  end
```

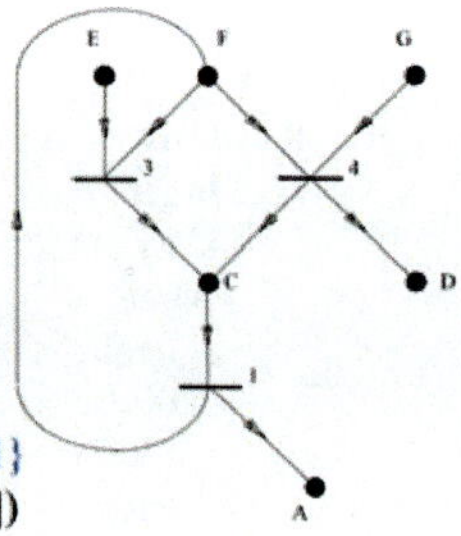

δ[{A, C}]

Depth of recursion: 3

{1}*, {2}, {1, 2}
{3}, {4}, {3, 4}*
{1}*, {6}, {1, 6}

```
⇒ procedure SSG( p, m, δ[m] ) p=∅, m={A, C, F}
  begin
  if p = ∅ then begin write δ[m]; p=∅
⇐           return   end
  let x∈p;
  C:= ℘(Δ(x))\{∅};
  for all c∈C do
            begin
            if ∀y∈m, c∩δ̄(y) = ∅ & (Δ(x)\c)∩δ(y) = ∅
             then
              begin
              δ[m∪{x}]:= δ[m]∪{(x, c)};
              SSG(p∪matⁱⁿ (c))\(R∪m∪{x}), m∪{x}, δ[m∪{x}])
              end
            end
  return
  end
```

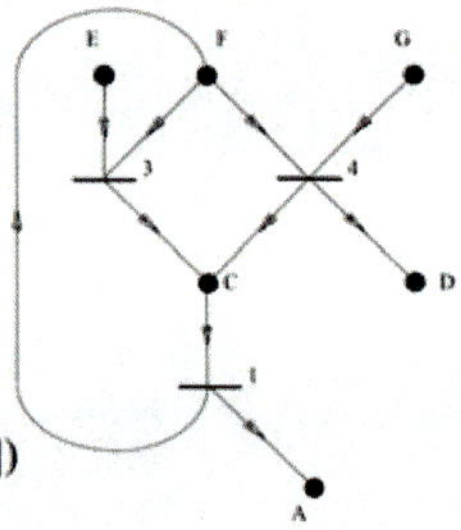

δ[{A, C, F}]

Solution 5

Depth of recursion: 2

{1}*, {2}, {1, 2}

{3}, {4}, **{3, 4}***

{1}, **{6}***, {1, 6}

```
procedure SSG( p, m, δ[m] )  p={F}, m={A, C}
begin
if p = ∅ then begin write δ[m];  p={F}
          return    end
let x∈p;  p={F}, x=F;
C:= ℘(Δ(x))\{∅};  Δ(F)={1, 6}
for all c∈C do  C={ {1}, {6}, {1, 6} }, c={6}
          begin
          if ∀y∈m, c∩δ̄(y) = ∅ & (Δ(x)\c)∩δ(y) = ∅  false
          m={A, C}, c={6}, Δ(F)={1, 6}
           then
            begin
            δ[m∪{x}]:= δ[m]∪{(x, c)};
            SSG(p∪matin (c))\(R∪m∪{x}), m∪{x}, δ[m∪{x}])
⇒           end
          end
return
end
```

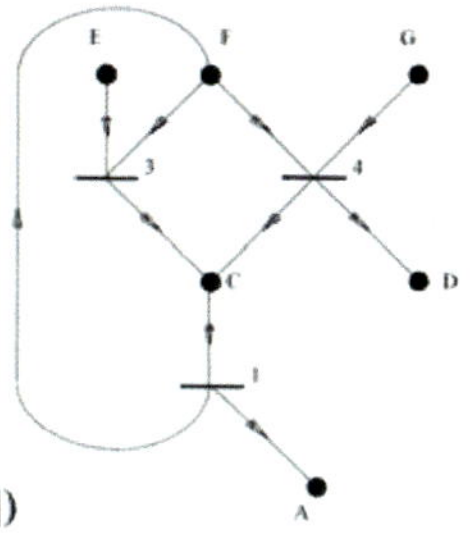

δ[{A, C}]

Depth of recursion: 2

{1}*, {2}, {1, 2}

{3}, {4}, **{3, 4}***

{1}, {6}, **{1, 6}***

```
procedure SSG( p, m, δ[m] )  p={F}, m={A, C}
begin
if p = ∅ then begin write δ[m];  p={F}
          return    end
let x∈p;  p={F}, x=F;
C:= ℘(Δ(x))\{∅};  Δ(F)={1, 6}
for all c∈C do  C={ {1}, {6}, {1, 6} }, c={1, 6}
          begin
          if ∀y∈m, c∩δ̄(y) = ∅ & (Δ(x)\c)∩δ(y) = ∅  true
          m={A, C}, c={1, 6}, Δ(F)={1, 6}
           then
            begin
            δ[m∪{x}]:= δ[m]∪{(x, c)};  m={A, C} , x=F, c={1, 6}
⇐           SSG(p∪matin (c))\(R∪m∪{x}), m∪{x}, δ[m∪{x}])
            p={F}, matin ({1, 6})={C, J}, R={E, J, G, K,  L}
            m={A, C}, x=F
            end
          end
return
end
```

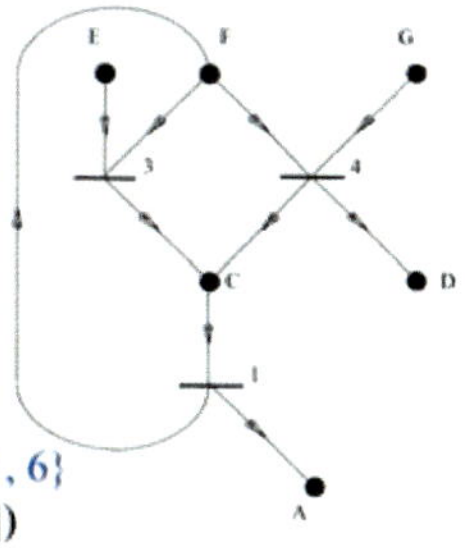

δ[{A, C}]

Depth of recursion: 3

{1}*, {2}, {1, 2}

{3}, {4}, **{3, 4}***

{1}, {6}, **{1, 6}***

```
⇒ procedure SSG( p, m, δ[m] ) p=∅, m={A, C, F}
  begin
  if p = ∅ then begin write δ[m]; p=∅
⇐             return    end
  let x∈p;
  C:= ℘(Δ(x))\{∅};
  for all c∈C do
        begin
        if ∀y∈m, c∩δ̄(y) = ∅ & (Δ(x)\c)∩δ(y) = ∅
         then
          begin
          δ[m∪{x}]:= δ[m]∪{(x, c)};
          SSG(p∪mat^in (c))\(R∪m∪{x}), m∪{x}, δ[m∪{x}])
          end
        end
  return
  end
```

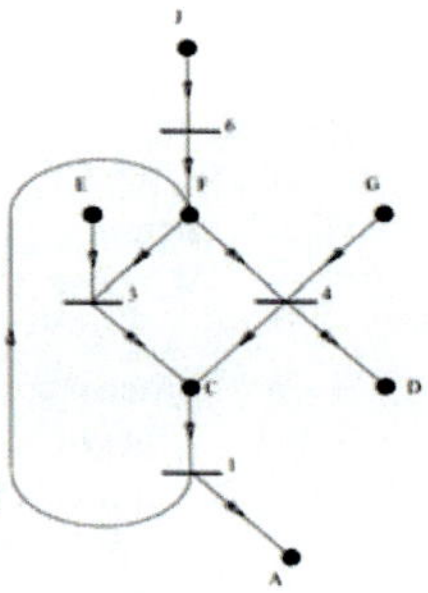

δ[{A, C, F}]

Solution 6

Depth of recursion: 2

{1}*, {2}, {1, 2}

{3}, {4}, **{3, 4}***

{1}, {6}, **{1, 6}***

```
  procedure SSG( p, m, δ[m] ) p={F}, m={A, C}
  begin
  if p = ∅ then begin write δ[m]; p={F}
              return    end
  let x∈p; p={F}, x=F;
  C:= ℘(Δ(x))\{∅}; Δ(F)={1, 6}
  for all c∈C do C={ {1}, {6}, {1, 6} }
        begin
        if ∀y∈m, c∩δ̄(y) = ∅ & (Δ(x)\c)∩δ(y) = ∅
         then
          begin
          δ[m∪{x}]:= δ[m]∪{(x, c)};
          SSG(p∪mat^in (c))\(R∪m∪{x}), m∪{x}, δ[m∪{x}])
⇒         end
        end
⇐ return
  end
```

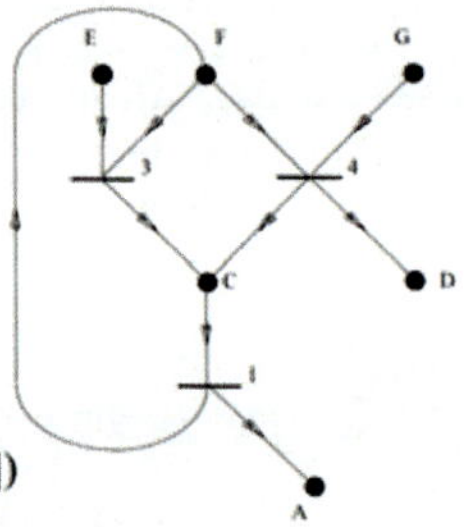

δ[{A, C}]

Depth of recursion: 1

{1}*, {2}, {1, 2}

{3}, {4}, **{3, 4}***

```
    procedure SSG( p, m, δ[m] )  p={C}, m={A}
    begin
    if p = ∅ then begin write δ[m];  p={C}
                return    end
    let x∈p;  p={C}, x=C;
    C:= ℘(Δ(x))\{∅};  Δ(C)={3, 4}
    for all c∈C do  C={ {3}, {4}, {3, 4} }
                begin
                if ∀y∈m, c∩δ̄(y) = ∅ & (Δ(x)\c)∩δ(y) = ∅
                  then
                    begin
                    δ[m∪{x}]:= δ[m]∪{(x, c)};
                    SSG(p∪matⁱⁿ (c))\(R∪m∪{x}), m∪{x}, δ[m∪{x}])
⇒                   end
                end
⇐   return
    end
```

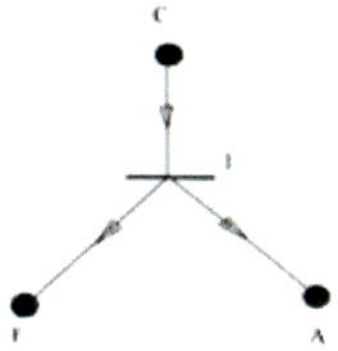

δ[{A}]

Depth of recursion: 0

{1}, **{2}***, {1, 2}

```
    procedure SSG( p, m, δ[m] )  p={A}, m=∅
    begin
    if p = ∅ then begin write δ[m];  p={A}
                return    end
    let x∈p;  p={A}, x=A;
    C:= ℘(Δ(x))\{∅};  Δ(A)={1, 2}
    for all c∈C do  C={ {1}, {2}, {1, 2} }, c={2}
                begin
                if ∀y∈m, c∩δ̄(y) = ∅ & (Δ(x)\c)∩δ(y) = ∅   true
                m=∅, c={2}, Δ(A)={1, 2}
                  then
                    begin
                    δ[m∪{x}]:= δ[m]∪{(x, c)};  m= ∅, x=A, c={2}
⇐                   SSG(p∪matⁱⁿ (c))\(R∪m∪{x}), m∪{x}, δ[m∪{x}])
                    p={A}, matⁱⁿ ({2})={D}, R={E, J, G, K,  L}
                    m=∅, x=A
⇒                   end
                end
    return
    end
```

δ[∅] = ∅

Appendix III: Demonstration Software

Chapters 4, 5, and 6 of the book provided a detailed description of various algorithms for performing Process Network Synthesis (PNS) based on the P-graph framework. One can follow the algorithms presented therein to find potential solutions, the optimal solution, or near-optimal solutions for a given PNS problem. In this regard, various illustrative examples of PNS problems and their algorithmic solutions were also presented in the same chapters. More often than not, however, PNS problems are larger and more complex than those illustrative examples, thus rendering their solutions too complex to be solved manually. Therefore, computer implementations of the algorithms are needed to handle large and complex PNS problems. This appendix introduces two existing software applications for P-graph, one created for demonstration purposes for the basics of the framework, and the other, for showcasing the capabilities in case of real industrial problems. The reader should bear in mind that these software applications are being constantly upgraded and extended; consequently, we will not discuss the fine details about their functionality.

"P-Graph Insight" Software

P-Graph Insight is a free and open-source demonstration software created for realizing the simplest implementations of the original P-graph algorithms. This computer program provides the options for defining a PNS problem and executing the synthesis algorithms, such as MSG, SSG, or ABB, on it. The software is also capable of displaying the steps of the algorithms, thus making it an excellent tool for understanding, learning, or teaching them. However, since the software focuses on the demonstration and not on efficiency, it is not applicable for large-scale, real-world PNS problems. The main concepts embedded in the software are listed below.

F. Friedler et al., *P-graphs for Process Systems Engineering*,
https://doi.org/10.1007/978-3-030-92216-0

1. PNS problems can be defined following the same formalism contained in the book. This means that the set of materials and operating units with their corresponding parameters and connections must be given. Thus, the PNS problem will be specified by sets *P*, *R*, and *O*.
2. After a PNS problem is defined, it is possible to execute the algorithms included. All previous executions for this problem are listed in the software, and they can be inspected further. The software includes the basic P-graph algorithms, such as MSG, SSG, and ABB, and provides some additional options to choose from, such as the number of solutions, or the acceleration techniques applied.
3. The algorithms included in the software are implemented in the purest way possible to exactly represent the formal P-graph algorithms. This means that they are not subject to any optimization or acceleration that would change the essence of these algorithms.
4. For each executed algorithm, the software displays all the generated structures with their parameters. The software also lists the steps the algorithm went through to generate the solutions displayed.

Based on these concepts, the software can explain the details of the P-graph framework's algorithms, thereby making it possible for inspired researchers or engineering students to understand them. We suggest using this program when the main goal is to get familiar with the P-graph framework.

The software is available at: https://github.com/akosorosz/PGraphInsight/releases.

"P-Graph Studio" Software

P-Graph Studio is an advanced software capable of solving real-world PNS problems. This software provides a graphical interface for drawing P-graphs and solving defined PNS problems. In addition, this complex software incorporates the known algorithms of the P-graph framework by focusing on their speed and efficiency. In view of this, most of the implemented algorithms are highly optimized by resorting to several acceleration techniques that are beyond the scope of this book. The main concepts underlying the software are the following.

1. Defining a PNS problem involves drawing the corresponding P-graph and specifying the additional parameters. Each material and operating unit has several parameters that can be given. In addition, the PNS problem has global parameters which can also be changed manually.
2. After the PNS problem is defined, the user can solve it by using one of the algorithms included. The algorithms themselves do not have too many parameters as they always operate according to their currently most efficient implementation based on the parameters of the PNS problem.
3. The software implements the basic algorithms of the P-graph framework in a highly optimized way. This means that instead of being exact representations of these algorithms, they use several acceleration techniques related both to the core

methods and to their software implementations. As a result, the implemented algorithms are capable of handling large-scale problems. Note that these optimized implementations are still proper and valid, i.e., they give exactly the same results as the pure algorithms but with significantly lower computational effort.
4. The results of an algorithm are displayed in the same user interface and the generated solution structures can be seen graphically. In this way, the user can easily identify the properties of any solution.
5. The software also serves as a demonstration platform for some of the new developments for P-graph, such as proposed methods for time-constrained or multi-period PNS problems.

These concepts ensure that P-Graph Studio is an appropriate tool for solving large and complex PNS problems. We recommend this software for solving PNS problems concerning academic or industrial research.

The software can be downloaded at: http://p-graph.org/downloads/.

Epilogue

Various decades of research have pursued the goal of creating tools for the design of processing systems. These tools are envisioned to assist the designers in the conception of process networks, so that the decision-making results in the best choice. In this book we have aimed to provide the underlying methods and proper formalisms of a general framework for Process Network Synthesis (PNS). This resulted in the introduction and overview of the P-graph framework, which constitutes a step forward towards the automatic synthesis of processing systems.

The framework described is a leap forward in the area of process design because it systematically considers the structural properties of the process during its synthesis as opposed to conventional approaches. Such capacity of the framework confers on its strong advantages in solving a PNS problem, for instance, the generation of multiple alternative solutions (or even all possible solutions), or the acceleration of the optimization procedure. Moreover, the P-graph framework has no limitations on the structural complexity of the system considered, or on the system's underlying mathematical model.

The overview of the various applications demonstrates the generality of the framework and its capability to be implemented for solving a plethora of problems in its current state. Needless to say, the research is still open, and the framework has opportunities for further extension. Additional algorithms can be developed and the area of applications can also be extended.

The original axioms and algorithms were developed for product-oriented systems where the main goal is the generation of single product or multiple products. Nonetheless, PNS problems can have different goals and constraints that may require an extension. Independently of the type of extension, the original "inside-out" mathematical modelling philosophy can conveniently be followed. For example, when some of the products are optional and can be ignored if their production would not be beneficial, the formerly used set P of products is to be replaced by two sets, a set for the required and another set for the potential products. Then, only the set of required products appear in axiom (S1).

F. Friedler et al., *P-graphs for Process Systems Engineering*,
https://doi.org/10.1007/978-3-030-92216-0

Another instance would be the case where some of the raw materials can be produced inside the process, while also being purchased if necessary. With an additional axiom, the structural requirement of the circular economy can also be expressed in PNS.

Reversible operations can also be represented for solving a common analysis and synthesis problem of chemical reaction engineering. To do so, reactions can be formulated by bidirectional operating units where the directions of the arcs are decided during the synthesis procedure.

The P-graph framework summarized in the book covers only the structural component of process network synthesis. Naturally, for determining the optimal solution or the n-best solutions of the synthesis problem, all structures generated by P-graph algorithms must be evaluated. The framework has no limitations on the type of evaluation methods, any of them can conveniently be connected to the P-graph algorithms. For example, evaluation can be based on explicit description of the mathematical models of the operating units (linear on nonlinear mathematical models) or implicitly given by simulation program.

In summary, P-graph framework has been developed for designing complex network structures where the nodes of the network are represented by mathematical models. It has wide range of applications and is open for further research. There are areas where the enumeration tools of the framework are essential. It is presumed that this work contributes to solve the increasingly important problems of our globalized World including the design of resilient system.

Index

F. Friedler et al., *P-graphs for Process Systems Engineering*,
https://doi.org/10.1007/978-3-030-92216-0